BASIC APPLIED FLUID POWER: HYDRAULICS

# BASIC APPLIED FLUID POWER: HYDRAULICS

JON OSTER

*President, Capital Engineering Company*
*Minneapolis*

McGraw-Hill Book Company

NEW YORK · ST. LOUIS · SAN FRANCISCO
LONDON · SYDNEY · TORONTO
MEXICO · PANAMA

Library of Congress Catalog Card Number: 68-15738

47895

1 2 3 4 5 6 7 8 9 0 BABA 7 6 5 4 3 2 1 0 6 9

# PREFACE

Rapid expansion of the use of fluid power (hydraulics, pneumatics, and fluidics) in the industrial, mobile, and aerospace fields has created a need for instruction. High schools, industrial schools, junior colleges, and adult education programs have not generally met this need. This workbook, "Basic Applied Fluid Power: Hydraulics," is intended to introduce and teach the basic principles of fluid power.

Fluid power encompasses both hydraulics and pneumatics. This book deals principally with hydraulics. Inasmuch as the characteristics, applications, and uses of liquids (hydraulics) are quite different from those of air (pneumatics), they must be covered by different textbooks.

This book is constructed as a workbook; the text material is designed to introduce the basic principles of certain fundamental physical laws and the practical applications involved. The chapters on fluid power demonstrations and circuits deal with the function and operation of basic fluid power components, as well as their purpose and application.

The demonstrations and circuits in this book have been constructed around, but are not limited to, the Capital Engineering Company's (Minneapolis, Minnesota) training unit. Instructors can assemble their own power plants or acquire them from government surplus or from other manufacturers. Components may be acquired in a similar manner. The author believes that regardless of how the physical facilities are acquired, those used should have *realistic* pressures and *adequate* flows. This permits the use of standard components, which will simplify the expansion of the school's fluid power teaching facilities by future acquisition of hardware from industry, government, and private contributions.

Education, particularly industrial education, necessarily consists of lecture (audio) and text material, demonstration (visual), practice (physical), and explanations (reasons). This is the format we tried to follow, and we sincerely hope that the workbook will serve the instructor and student well.

We gratefully acknowledge the assistance of the student instructors at the Fluid Power Seminar in suggesting requirements for a fluid power course and a training unit. Special thanks go to the directors of the five seminars that were conducted.

We also thank the Educational Committee of the Fluid Power Society for the recommendations made regarding the features of a fluid power training unit.

We also wish to express our gratitude for contributions made by Joseph H. BeVier, B.AE.E (Hydraulic Services), William E. BeVier, Ph.D. ME (Mechanical Technology, Inc.), George E. Carlson (Minnesota Rubber Company), Frederick Porter and Jean Wittke (Capital Engineering).

JON OSTER

# CONTENTS


| | | | |
|---|---|---|---|
| PREFACE | | | *v* |
| SECTION | 1 | INTRODUCTION | *1* |
| | 2 | BASIC SCIENCE | *5* |
| | 3 | MECHANICS | *9* |
| | 4 | ENERGY, WORK, POWER | *17* |
| | 5 | MATHEMATICS | *23* |
| | 6 | FLUID PASSAGES | *29* |
| | 7 | PUMPS | *63* |
| | 8 | ACTUATORS | *85* |
| | 9 | SEALING DEVICES | *95* |
| | 10 | CIRCUIT COMPONENTS | *99* |
| | 11 | FLOW CONTROL COMPONENTS | *107* |
| | 12 | PRESSURE-COMPENSATED FLOW CONTROL | *129* |
| | 13 | HYDRAULIC FLUID | *133* |
| | 14 | FLUID POWER DEMONSTRATIONS | *137* |
| | 15 | CIRCUITS | *163* |
| | 16 | REVIEW QUESTIONS | *203* |
| | 17 | POWER UNIT | *211* |
| | | BASIC COMPONENTS | *211* |
| | | POWER-UNIT CIRCUIT | *213* |
| | | POWER-UNIT INSTALLATION | *215* |
| | | POWER-UNIT OPERATING INSTRUCTIONS | *215* |
| APPENDIXES | A | GLOSSARY | *217* |
| | B | GENERAL FORMULAS AND RULES | *220* |
| | C | GRAPHIC SYMBOLS | *222* |
| | D | TECHNICAL DATA | *228* |

# INTRODUCTION

# 1

Hydraulics, as a known and applied science, has been practiced for many years with water. However, the advent in recent years of oil hydraulics as a power source has not only broadened the application of hydraulics but expanded and refined its science to such a degree that it now occupies a position as a distinct branch of engineering.

The use of hydraulic equipment has grown so rapidly that it has far outstripped existing educational curriculums and facilities. At the time of this writing, most people working in this field—at all levels—are people with appropriate backgrounds who have gained their knowledge and proficiency from experience only. Conscious of the gap between theory and practice, educators and industry are making an effort to close it.

The course to which this text applies is a part of this effort. Together with use of the associated laboratory equipment this text constitutes an introduction to applied hydraulics.

## FLUID POWER: PAST—PRESENT—FUTURE

### The Past

The waterwheel, which provides a form of hydraulic power, was used for many centuries. There is evidence of its use as far back as 100 B.C. Fluid power, as a method of power transmission, stems from experiments made by Pascal in 1650. When he rammed the cork down in a jug completely full of wine, the bottom broke out. From this he went on to develop the law which states that a pressure on a fluid will be equally transmitted in all directions. With Pascal's law in mind it can be seen that a jug with a small opening in the top and a large bottom could have the force increased on the bottom in direct relation to the size of the opening and the bottom.

### From 1850 to Present

In 1850, rams, cylinders, and simple piston pumps were used for industrial presses and shears. Hydraulic power became widely used in England, where a central plant distributed oil under pressure and sold it. For a time it was distributed under franchise as electricity is distributed in this country today.

The development of electricity marked the end of this form of power distribution.

Continual but slow progress was made in fluid power applications for use with auto brakes, jacks, and presses. When World War II began, however, an all-out effort was made to develop fluid power and to utilize its inherent advantages.

A unitized fluid power package was developed (motor, pump controls, and actuator), and this became an efficient manner of power transmission for both industrial and mobile applications. Increased use lowered the cost of components, and today it is, in most applications, the most efficient method for getting a job done.

At the present time industry has a $10 billion sales budget for 1967 and expects a 28 percent increase in 1968. Construction and mobile equipment have converted to fluid power, and agricultural equipment is in the process of being changed. Use of fluid power in industry has been rapidly increasing for a number of years; eventually it will be the principal method of power transmission. The aerospace and aircraft industry is now widely using fluids for both power and controls. Their effectiveness is demonstrated by the fact that a 9-hp motor need have a physical size of only 3 × 3 × 5 in.

The future will see almost complete elimination of lineshafts, chains, and gearboxes on mobile and agricultural equipment. Fluid power transmission to wheels in autos and in the largest tractors will be developed. In the not too distant future pneumatics for controls and hydraulics for power will be the principal method of accomplishing useful work.

## BASIC SYSTEM

The basic hydraulic system (Fig. 1-1) is comprised of a reservoir (oil-supply tank), a pump (power source), a pressure-relief valve (safety valve), a control valve (directional element), and a cylinder or motor (working element). These components are connected, one to another, by piping, hoses, or tubing.

The *reservoir* is a tank made of any suitable material, usually steel, and can be any desired shape, although it is ordinarily either rectangular or cylindrical. Oil is taken from the tank by a pump and, after circulating through the system, the oil is returned to the tank for reuse. In addition to supplying oil, the reservoir also serves to aid cooling and to settle dirt out of the oil.

The *pump* is coupled to and driven by the prime mover of the system, which is most often an electric motor or gasoline engine. The inlet side of the pump is connected to the reservoir; the outlet, or pressure side, is connected to the control valve and thus to the rest of the system. As the pump drive

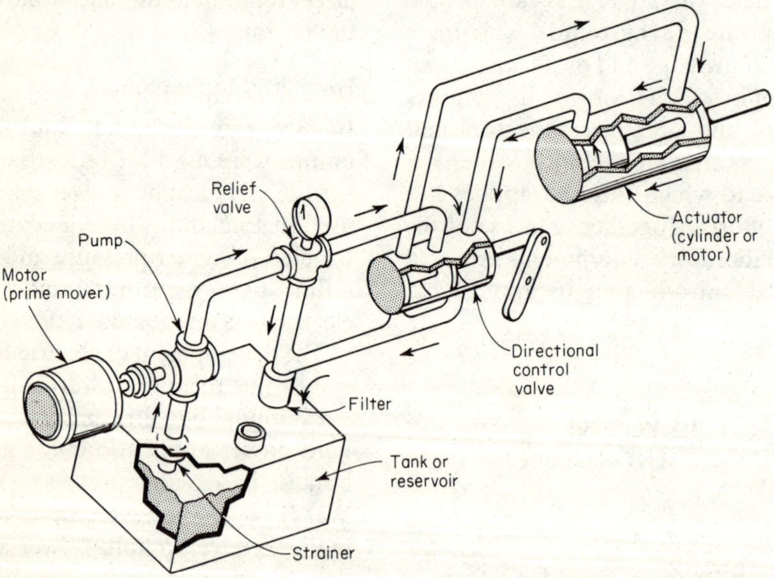


FIG. 1-1 *Complete basic hydraulic circuit.*

 

is rotated, oil is taken into the pump from the reservoir, carried through the pump, and delivered to the system from the outlet port. Since the pumping action creates a less-than-atmospheric pressure at the inlet port, this portion of the pump is operating in a vacuum.

The pumping action delivers oil to the system in volume, under pressure; hence, there is a pressure side. It is this combination of volume and pressure that constitutes the power of the system that performs the work required. Thus the pump is the energy source of the system.

The *pressure-relief valve* is installed near the pump to protect it from overloads and shock. A load at the work end exceeding the design limits, or an overload as a result of a malfunctioning component, will result in excess pressure in the system. The safety, or relief, valve opens and returns the pump flow back to the tank. At the same time it maintains the pressure in the system equal to that preset on the valve.

The *control valve* receives oil from the pump and directs it to the various circuit areas as desired. Since valves are produced in a vast array of configurations (many perform specialized functions), it is difficult to give a general description. However, almost all elementary valves consist of a body having an input port, a discharge port, and one or two cylinder (or control) ports. A movable piece, known as the plunger, or spool, is closely fitted in a matching hole in the body. The relative position of the plunger and the body determines the flow pattern of the valve.

Oil from the pump enters the valve at the intake port. Internal passages carry the oil through the valve in its desired route. Usually, when the valve is set to neutral, oil flows through the valve and out the discharge port, and returns to the reservoir. When oil is desired at a certain point in the system for a specific purpose, such as at a cylinder to lift a load, the plunger is shifted to the appropriate position within the valve body. This action rearranges the internal passages of the valve, and oil flows out of a control port to the cylinder.

Many valves have two control ports. This implies that there are two areas for the oil to serve. The most common example of this is found in the operation of a double-acting cylinder. When the plunger is moved to a second control position, oil flows out of the second port in the same manner as it did from the first. The internal passages are so arranged that when one port is directing oil out to a work area ("open to pressure"), the other port is receiving oil from a work area ("open to discharge") and directing the oil out the discharge port of the valve.

The essential purpose of any hydraulic circuit is to serve the working element of the system. With few exceptions, this is either a cylinder or a motor.

A *cylinder* performs work by moving and applying force in a straight line. Its name is taken from the shape of the piece, which is cylindrical. Its essential parts are a barrel, a piston rod, and a piston. The piston is contained within the barrel and is firmly fixed to the piston rod, which protrudes from one end of the barrel. The other end of the barrel is closed and is attached to the machine; the outboard end of the rod is attached to the load. Oil forced into the barrel either extends the rod from the barrel or retracts it into the barrel and, in so doing, moves the load.

Some cylinders move under oil pressure only in an extending manner. Known as "single-acting" cylinders, they are retracted by the load (or, in many cases, the dead weight) on the end of the piston rod. Since this retracting load acts under the influence of gravity, single-acting cylinders operate essentially in an up-and-down orientation. Examples are the main lift cylinders of loaders used by contractors.

In many cylinders positive oil pressure is applied to the piston in both the extending and retracting directions. Oil entering the cylinder from either end (or, on one side of the piston) discharges from the other end.

Such cylinders are known as "double-acting" cylinders. Since these cylinders are actuated entirely by oil, their orientation (or attitude) does not affect their operation.

Relating the cylinder to the system, oil flows from the control ports of the valve to the cylinder. If it is a single-acting cylinder, only one connection exists between one port of the valve and the closed end of the barrel. The oil returns through the same line by which the cylinder was pressurized. If it is a double-acting cylinder, two control ports of the valve will be connected by separate lines to opposite ends of the cylinder. When oil is being pumped into one end of the cylinder, oil will be returning to the valve from the other end. From the valve, it goes elsewhere in the system, usually to the tank.

It is worth noting that when a cylinder is arranged to reciprocate continuously, it is often called a linear motor. There is also a special kind of cylinder which rotates approximately 300 deg. It is appropriately termed a rotary cylinder.

To all except complete theorists, the hydraulic motor is a device which performs work by rotating its shaft. In appearance it closely resembles a pump, and in operation it is essentially a pump running backwards. In fact, there are some units manufactured that can be either a pump or a motor depending upon how the user wishes to use them.

In a pump, the drive shaft is forcibly rotated and acts upon the oil to produce volume and pressure: Said in another way, work is put into the pump to produce "oil under pressure." A motor is just the opposite. Oil is introduced into the motor to rotate the shaft forcibly; or, oil under pressure is forced into the motor and produces work at the shaft.

It follows that a motor must have an intake and discharge port. Oil coming from the valve enters the motor through the intake port and exits from the discharge port. Depending upon the circuit arrangements, the discharge oil may flow directly to the reservoir or back to the valve and then to some other area, perhaps the reservoir. Whenever possible, the oil should go directly to the tank.

 

# BASIC SCIENCE

# 2

An understanding of the principles of basic physics as related to fluid flow is necessary to any study of fluid technology or power. The prime concepts are the laws governing force, mass, momentum, energy, work, and power. These principles are reviewed in this section of the book. A discussion of the basic relations of fluid flow will follow beginning with Section 5.

## DIMENSIONS

A dimension is a number assigned to a quality of an object; it relates this quality to a defined standard of measure. For example, the weight of a body measured in pounds is a number which relates the body's quality of weight to that of a standard called the pound of force. To be meaningful, therefore, a dimension must describe both quality and magnitude.

The three basic fields of physics to be discussed in this section are *mechanics, heat,* and *electricity*. In mechanics the basic qualities which are given dimensions are *force* and *distance;* in heat, *temperature* and *heat;* in electricity, *force* and *charge*. In all three fields, *time* is important and must be given a dimension also.

### Mechanics

In mechanics the engineering field has adopted the pound-foot system of units. The pound is the dimension of force, and the foot is the dimension of distance. Ounce-inches and ton-miles are included in this system; they are merely fractions or multiples of the base units.

Certain quantities describe the rate at which other quantities change with respect to time. For example, the speed of an object is the rate at which its distance from a certain point changes with respect to time. Thus, if the front wheels of an automobile are on a particular line at some instant and are 10 ft from the line 1 sec later, the automobile has gone 10 ft in 1 sec. We call this motion "speed," saying the average speed of the automobile is 10 ft/sec. We used the word "average" because we did not state that the car was traveling at a constant speed during the second. This raises another important rate of change, that is, *acceleration.*

Acceleration is the rate at which velocity changes with respect to time. If an automobile is standing still at one instant and 1 sec later is moving at a speed of 5 ft/sec, the average acceleration of the automobile is 5 ft/sec/sec. Again, "average" denotes that we did not state that the acceleration was constant. We define these relations by the following equations:

Average velocity

$$= \frac{\text{final distance} - \text{initial distance}}{\text{final time} - \text{initial time}}$$

or
$$v_a = \frac{d_2 - d_1}{t_2 - t_1}$$

Average acceleration

$$= \frac{\text{final velocity} - \text{initial velocity}}{\text{final time} - \text{initial time}}$$

or
$$a_a = \frac{v_2 - v_1}{t_2 - t_1}$$

*Example:* An auto starts from rest and travels 500 ft in 10 sec. The average velocity is as follows:

$$v_a = \frac{d_2 - d_1}{t_2 - t_1}$$

or
$$\frac{500\text{ ft} - 0}{10\text{ sec} - 0} = 50\text{ ft/sec}$$

The average acceleration is:

$$A_a = \frac{v_2 - v_1}{t_2 - t_1}$$

or
$$\frac{100\text{ ft/sec} - 0}{10\text{ sec} - 0} = 10\text{ ft/sec/sec}$$

*Note:* Since average velocity $v_a$ is the mean between the final and initial speeds, for computation purposes we can assume that the average is half the final.

### Heat

In the field of heat, the dimensions of importance are *temperature* and *heat.* Temperature measures the amount of heat contained in a body. In engineering in the United States, the Rankine scale is used for absolute measurements and the Fahrenheit scale for relative measurements. The difference between these two scales is 460°, so that temperature in degrees Rankine equals 460 plus the temperature in degrees Fahrenheit.

The Rankine temperature is called the absolute scale because at 0°R an object has no heat to radiate. This is the absolute minimum temperature which can exist.

Heat is measured in British thermal units (Btu). One Btu is the amount of heat necessary to raise 1 lb of water 1°R. In equation form:

$$\text{Btu} = (T_2 - T_1)\text{ per 1 lb of water}$$

*Example:* One lb of water is increased in temperature from 495°R to 500°R. The number of Btu necessary for this increase is

$$\text{Btu} = T_2 - T_1 = 500°\text{R} - 495°\text{R}$$
$$= 5°\text{R} = 5\text{ Btu}$$

### Electricity

The electrical measures of importance are *force* and *charge.* Force is measured in *volts* and charge is measured in *coulombs* (coul). Probably a more familiar term is *amperes* (amp). The amperage of a current is the rate at which the charge is flowing and 1 amp is the equivalent of 1 coul/sec. This is analogous to velocity in mechanics.

All three of these systems of measurement (mechanics, heat, and electricity) are analogous to a certain extent. Each has a term which measures position on some relative scale and each has a quantity which tends to force the position to change. Table 2-1 summarizes the above discussions.

TABLE 2-1 MEASUREMENT SYSTEMS IN MECHANICS, HEAT, AND ELECTRICITY

| | FORCING QUANTITY | | POSITION QUANTITY | | RATE OF CHANGE QUANTITY | |
|---|---|---|---|---|---|---|
| | *Name* | *Units* | *Name* | *Units* | *Name* | *Units* |
| Mechanics | Force | Pounds | Distance | Feet | Speed | ft/sec |
| Heat | Temperature | °R | Heat | Btu | Heat exchange | Btu/sec |
| Electricity | Voltage | Volts | Charge | Coulombs | Current | Amperes |

 

## PROBLEMS

**1.** A stone weighs 3½ tons. How many pounds does it weigh? ______

**2.** It is 90 ft from home plate to first base. How many inches is this? ______

**3.** An airplane has recently flown to 90,000 ft. How many miles is this? ______

**4.** An automobile on the Salt Flats traveled between two poles which were 4 miles apart. It passed the first pole at 9:37:12 and passed the second pole at 9:37:48. What was the speed of the automobile in miles per second? ______ In miles per hour? ______ In feet per second? ______

**5.** If a baseball player runs to first base in 4.5 sec, what is his average speed in feet per second? ______ In miles per hour? ______

**6.** If the same ball player is running at 14 mph when he goes past first base, what is his average acceleration in miles per hour per second? ______ In miles per hour per hour? ______ In feet per second per second? ______

**7.** There are 3 lb of water in a jar. The water is initially at 493°R. The jar is placed over a burner. Later it is removed from the burner and found to be at 543°R. How many Btu were added to the water? ______

**8.** One cu ft of water weighs 62.4 lb. How many pounds of water can be held in a pan which is 2 ft long by 1½ ft wide by 1 ft deep? ______

**9.** How many Btu are required to raise the temperature of the same full pan of water from 43 to 57°F? ______

**10.** A round lake which is 6.28 miles around has an average depth of 6 ft. On June 1, this lake had an average temperature of 37°F. On July 31, it had an average temperature of 59°F. How many Btu were added to the lake during this time? ______

**11.** What was the average rate at which heat was added to the water in Btu per day? ______ In Btu per hour? ______ In Btu per minute? ______

**12.** If, during the same time period, 1 million Btu were added to the water, what would its final temperature be? ______

**13.** A copper plate is charged with 530 coul. A wire is placed between the plate and ground. Two sec later the wire is disconnected, and the charge on the plate is measured to be 225 coul. What was the average current in the wire for those 2 sec? ______

**14.** One plate is charged with 1000 coul and another is charged with 200 coul. A wire is placed between the two plates, and current flows until the charges on the two plates are equal. What are the new charges on the plates? ______

**15.** The time it took for the phenomena of Prob. 14 to occur was 28 sec. What was the average current in the wire during this time? ______

# MECHANICS

# 3

Newton's principles of force and momentum will be reviewed in this section. The concepts of gravity, weight, and mass will also be reviewed. All concepts discussed will be illustrated with examples, and several problems will be given at the conclusion of the section.

## NEWTON'S LAWS OF MOTION

Sir Isaac Newton, an Englishman of the seventeenth century, formulated the basic laws of mechanics. Newtonian physics is that branch of physics based on Newton's three laws of mechanics. These laws form the basis of the entire field of mechanics and will be discussed at some length here. They are:

1. Every body continues in its state of rest, or in a state of uniform motion in a straight line, unless it is compelled to change that state by forces impressed upon it.
2. The rate of change of momentum produced by an unbalanced force is proportional to the unbalanced force, and is in the direction in which the force acts.
3. To every action there is always opposed an equal reaction.

The understanding of all Newton's laws requires the introduction of two concepts basic to mechanics—the concepts of *mass* and *momentum.* Mass is the property a body has that is dependent upon its size and density. The weight a body has is due in part to its mass and in part to the attraction of gravity. The weight and mass of a body are related by the equation:

$$W = mg$$

where $W$ = weight
$m$ = mass
$g$ = gravitational constant

The weight of a body is a force which the body exerts vertically upon any object upon which it rests. The gravitational constant is a number which measures the earth's gravitational attraction to the body. Mass, then, can be thought of as a property which relates a body's weight and gravity.

The second concept of importance is momentum. The momentum of a body is the product of the body's mass and its velocity. Up to this time we have discussed only speed. Velocity is different from speed in that it has two properties—*magnitude* and *direction.* The magnitude of the velocity is speed; but to completely describe a velocity we must give both its speed and its direction. Because it is the product of mass and velocity, the momentum, therefore, also has these two properties—magnitude and direction. Its magnitude is the product of the mass and the speed. Its direction is that of the velocity. The fact that it has a direction is indicated in equations by placing an arrow over the quantity. Thus:

$$\vec{M} = \overrightarrow{mv}$$

states that the momentum $\vec{M}$ is equal to the mass $m$ times the velocity $\vec{v}$.

### First Law

Newton's first law is often paraphrased as follows: "A body at rest tends to remain at rest and a body in motion tends to remain in motion." Applying the concept of momentum to this law, we see that $\vec{M}$ remains constant unless some force or forces act on the body to change its $\vec{M}$.

That a body at rest tends to remain at rest can be readily appreciated. Such an object's velocity and momentum are zero. Any object that is set in place will move from that position only when a force is applied against the object.

The fact that a body in motion tends to remain in motion is more difficult to comprehend. Experience indicates that all free-moving bodies eventually stop moving without any apparent forces acting on them. Friction is the unseen force that slows and stops motion. A ball rolling on the ground and a boat coasting in water provide examples of friction between a body and its environment.

When a body drops from a ledge, the weight of the body is the force pulling it downward. A body free-falling through air ultimately reaches its terminal velocity—where air resistance exactly opposes the weight force of the body. At lower altitudes a man falling, for example, will reach a maximum speed of approximately 140 mph. Other objects will attain other maximum speeds depending on their size and weight.

Perhaps the clearest example of bodies continuing in motion is that of satellites in orbit around the earth. Once in orbit, they will move at a constant velocity virtually indefinitely because there is no environmental resistance to their motion.

### Second Law

Newton's second law is that most often quoted. It is often paraphrased by the statement "Force equals mass times acceleration." Let us consider momentum ($\vec{M} = \overrightarrow{mv}$) in light of the second law. How can the momentum be changed? It will change if the mass changes (and this occurs in, for example, rockets, in which part of the mass is the rocket fuel which is burned up), and it will also change if the velocity changes. Usually, the mass of an object remains constant, and therefore a change in momentum means a change in velocity.

We have already seen that the acceleration of a body is the time rate of change of its velocity. Therefore, the second law says that if the mass is constant, the product of the mass and its acceleration is equal to the force. It further says that both the force and the acceleration have directions. The equation expressing the second law is:

$$\vec{F} = \overrightarrow{ma}$$

A free-body diagram will help us understand and use the second law. We represent the body in question by a rectangle or a replica of the actual body. We then represent the forces acting on the body by arrows pointed in the direction of their actions on the body, and represent velocities with arrows in the direction of the velocity. We arbitrarily assign one direction a positive value. This allows us to add up the forces acting on the body and determine its acceleration. Thus, for the example of a ball rolling on the ground, we draw the free-body diagram shown in Fig. 3-1.

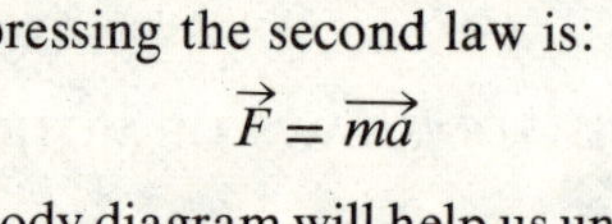


FIG. 3-1 *Horizontal free-body diagram.*

 

$F$ represents the frictional force and $v$ is velocity of the ball. Since $\vec{F} = m\vec{a}$, which means that $\vec{F}$ and $\vec{a}$ are in the same direction, we expect that the rate of change of the velocity will be in a direction opposite to the velocity. We now recall the relation between acceleration and velocity:

$$\vec{a} = \frac{\vec{v_2} - \vec{v_1}}{t_2 - t_1}$$

When the acceleration direction is opposite to the velocity direction, the acceleration is negative. For this to happen $v_1$ must be larger than $v_2$. In this case, the velocity grows continually smaller, and eventually the ball comes to rest.

For another example, let us consider a body falling through air (Fig. 3-2). In this diagram $\overrightarrow{W}$ represents the weight of the body and $\overleftarrow{F}$ represents the air resistance to the motion of the body. The newtonian equation for the body's motion is this:

$$\overrightarrow{W} - \overleftarrow{F} = \overleftarrow{ma}$$

There are two forces, so we must add their effects on the body. Since $\overrightarrow{W}$ and $\overleftarrow{F}$ act in opposite directions, we will arbitrarily make the downward direction positive, calling $\overrightarrow{W}$ a positive force and $\overleftarrow{F}$ a negative force. Adding them is the same as subtracting $\overleftarrow{F}$ from $\overrightarrow{W}$ and gives the result a downward direction. If $\vec{a} = 0$, then we see from the equation that $\overrightarrow{W} = \overleftarrow{F}$. Also, if $\vec{a}$ is zero, $\vec{v_1} = \vec{v_2}$. Thus, the velocity is constant. This

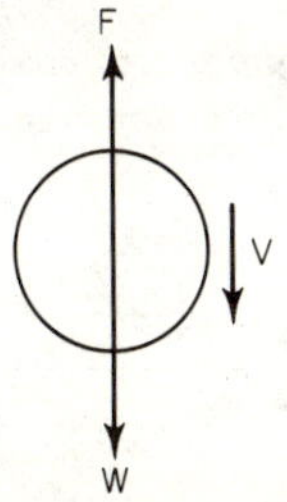


FIG. 3-2 *Vertical free-body diagram.*

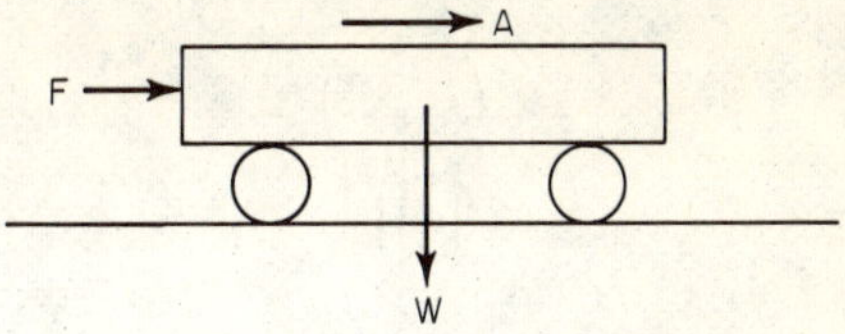


FIG. 3-3 *Forces acting on an object.*

constant velocity is called "terminal" velocity.

As another example of Newton's second law, suppose that we push a wheeled cart weighing 10 lb with a force of 10 lb. Disregarding frictional forces opposing the motion, we wish to know how fast the cart will accelerate (see Fig. 3-3). We know that

$$\vec{F} = m\vec{a}$$

From our definition of mass

$$m = \frac{w}{g}$$

Therefore

$$\vec{F} = \frac{w}{g}\vec{a}$$

The value of $g$ is roughly 32 ft/sec/sec. Substituting values of $g$, $F$, and $W$ in the equation to determine $\vec{a}$, we get the following solution:

$$\vec{10} = \frac{10}{32}\vec{a} \qquad \vec{a} = 32 \text{ ft/sec/sec}$$

### Third Law

Newton's third law can be paraphrased "Every action has an equal and opposite reaction"; put in this form, the law describes equilibrium. Equilibrium describes a condition in which the summation of all forces acting on a body, or within a system of forces, equals zero. This applies to bodies in motion as well as to bodies at rest.

Constant velocity in a straight line constitutes equilibrium in a body in motion. Figures 3-4 and 3-5 are examples. Figure 3-4 depicts the terminal velocity of a body free-falling in air and Fig. 3-5 shows an automobile driving at a steady speed on a level

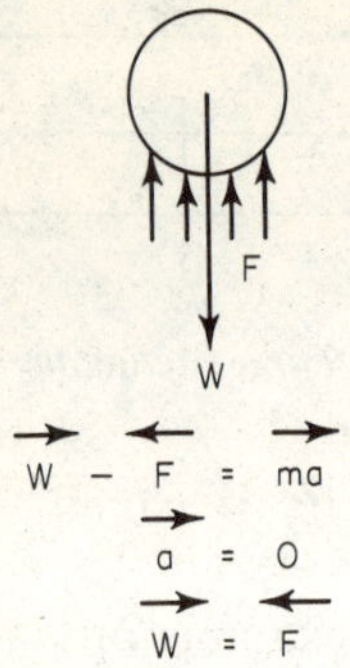


FIG. 3-4 *Forces acting on a free-falling object.*

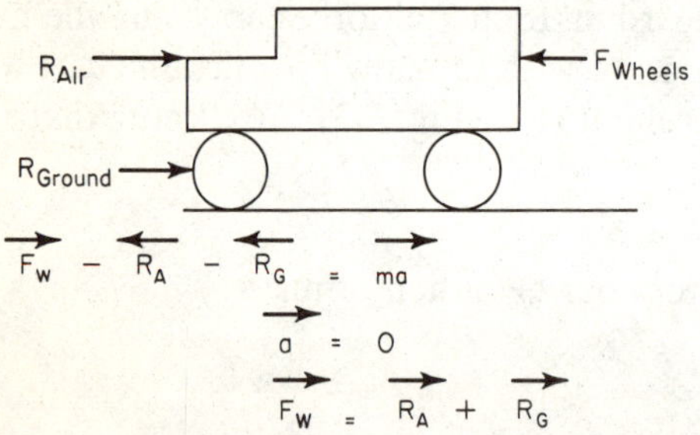


FIG. 3-5 *Constant-speed automobile. R = resistance.*

highway. In each case the forces acting on the body—those moving it in the direction it is going and those that are in resistance to this movement—are exactly equal. In the free-falling body, the weight force pulling the body down is exactly balanced by the air resistance pushing up. In the car with constant velocity, the forward force applied by the wheels is exactly balanced by the ground resistance and the air resistance opposing the motion.

As we have seen, constant velocity means that there is no acceleration; acceleration would mean an unbalanced force acting. This is another way of explaining equilibrium as an absence of unbalanced forces.

It should be pointed out that when an unbalanced force is applied to a body, the force meets with an equal and opposite reaction. For example, an accelerating force applied by wheels to the automobile is reacted to by a force in the pavement which is directed against the wheels.

A bridge spanning a river is an example of a body in static equilibrium. The weight force of the bridge acting downward is exactly resisted by the supports at each end of the bridge. If this equality of forces did not exist, the bridge would move (Fig. 3-6).

Note that in the example of the bridge, we are unable to say how much $R_L$ and $R_R$ are individually. We know only their total. In order to determine each force separately we must introduce a different type of force, called a *torque,* or a *moment.* Torque, or moment, is the product of a force and the distance of that force from a given point about which it acts. When a bolt is tightened with a wrench, the torque applied is the force on the wrench times the distance from the force to the bolt, or

$$T = Fd$$

All of Newton's laws apply to torques as well as to forces. However, now let us consider *angular velocity,* or how fast a body is turning, and *angular acceleration,* or how fast the angular velocity is changing. The position units normally used to measure turns are radians (rad). One rad is an arc equal to 57.296 deg and $2\pi$ rad equal one complete 360-deg revolution. (That is, $2 \times 3.1416 \times 57.296 = 360$ deg.)

Directions of turns are just as important as direction of motion; we shall describe directions as *clockwise* or *counterclockwise.*

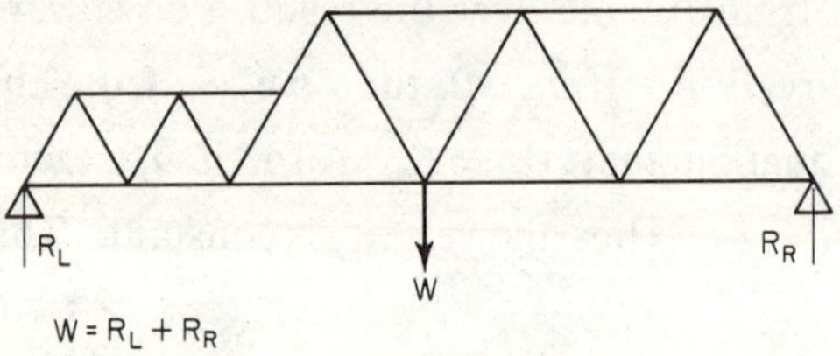


FIG. 3-6 *Weight and reaction relationship.* $R_L$ = *reaction left;* $R_R$ = *reaction right.*

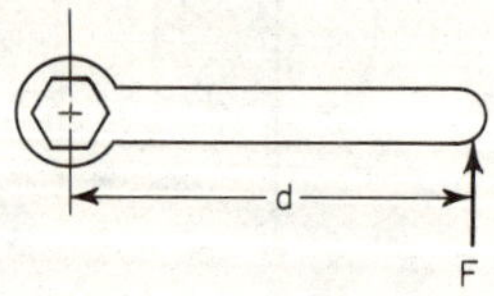


FIG. 3-7 *Torque developed with force 90 deg.*

 

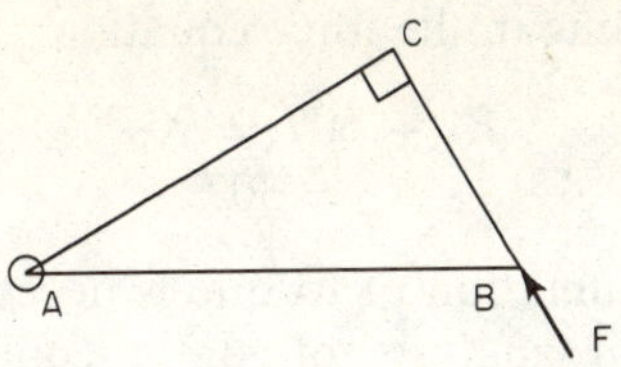


FIG. 3-8 *Torque developed with force other than 90 deg.*

Finally, the distance between the point of application and the force is always measured along a line perpendicular to the force. To illustrate this, suppose we have a line *AB* and a force F being applied at point *B* (Fig. 3-8). The torque which the force F exerts about the point *A* is equal to F*d*; *d* is the distance between *A* and *C*. The line *AC* must be perpendicular to the line *CB* along which the force F lies.

When we use torques, we must alter Newton's equation for the second law. Instead of $\vec{F} = \overrightarrow{ma}$, we have

$$\vec{T} = I\vec{\alpha}$$

Here $\vec{T}$ is the torque and the arrow means that it has a direction either clockwise (cw) or counterclockwise (ccw); $\vec{\alpha}$ is the angular acceleration and is in the same direction as the torque; *I* represents the inertia of the body to which the torque is applied. The inertia of a body is similar to the mass of a body in that it is a property which relates the torque to the angular acceleration. Usually it is more difficult to determine than the mass.

Another concept which is necessary for the understanding of torques is that of *center of gravity.* The center of gravity is the point at which all the mass of the body could be concentrated and give the same effect in equilibrium calculations. Therefore, if we know where the center of gravity of a body is located, for the purpose of calculation we can assume that all the body weight is concentrated at that point.

Consider the wrench-bolt example (Fig. 3-9). Let us assume that the wrench has no weight. Then let us hang a weight on the end of the wrench. The torque is simply $Wl_1$ (clockwise). Now, if the wrench has weight, this weight must be applying a torque to the bolt in the same manner as the weight we hung on the wrench. But what is *l* in this case? It is the distance from the bolt center to the center of gravity of the wrench. Thus, the torque applied to the bolt by just "hanging" the wrench on the bolt is *W* (weight of the wrench) times $l_2$ (distance from the bolt to the center of gravity of the wrench). Once again, however, we must remember that *l* is measured along a line perpendicular to the direction in which the weight acts. Since weight always acts vertically, *l* must be the horizontal distance from the bolt to the center of gravity.

Equilibrium is just as applicable to torque as it is to force. Thus, if a body does not turn, it means that the torques on the body are balanced. This leads us to the basic rules of static equilibrium. In order for a body to be in static equilibrium, the following must be true:

1. The sum of the forces in all directions must be zero.

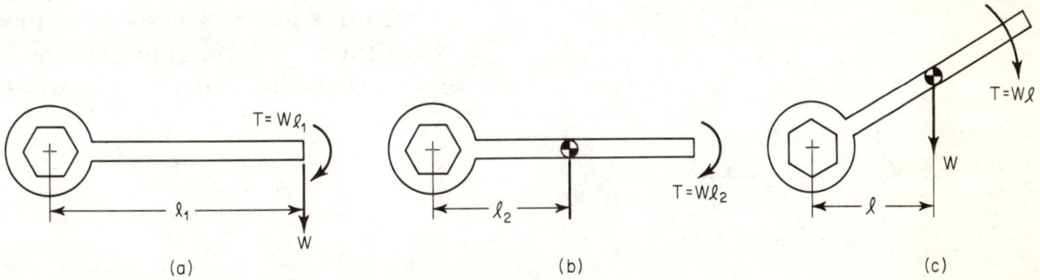


FIG. 3-9 *Torque development.*

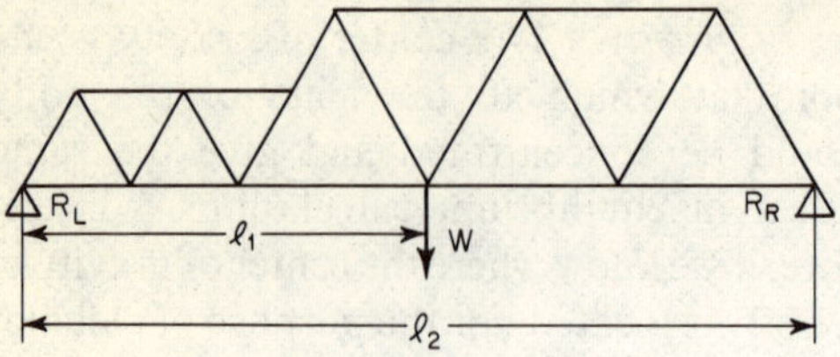


FIG. 3-10 *Taking moments about point $R_L$.*

2. The sum of the torques in all directions must be zero.

We shall now return to the bridge example. See Fig. 3-10. The bridge is in static equilibrium (neither accelerating up or down nor turning), and, therefore,

$$R_L + R_R = W \qquad (1)$$

$$R_L(0) + Wl_1 \text{ (cw)} - R_Rl_2 \text{ (ccw)} = 0 \qquad (2)$$

$$R_Rl_2 = Wl_1 \qquad \therefore R_R = W\frac{l_1}{l_2}$$

To find $R_L$, we substitute this value of $R_R$ in Eq. (1), and

$$R_L + W\frac{l_1}{l_2} = W$$

$$R_L = W - W\frac{l_1}{l_2} = W\left(1 - \frac{l_1}{l_2}\right)$$

For clarity (see Fig. 3-11), we translate this into numbers:

Summation of forces:

$$R_L + R_R = W = 90T$$

Summation of moments about $R_L$:

$$R_L(0) + 90T \times 50 \text{ ft} - 75 \text{ ft } R_R = 0$$

$$R_R = \frac{90T \times 50 \text{ ft}}{75 \text{ ft}} = 60T$$

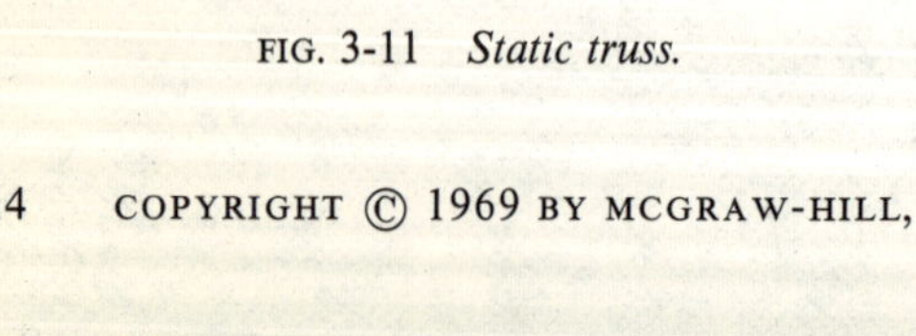


FIG. 3-11 *Static truss.*

Substituting in the force equation:

$$R_L + 60T = 90T$$
$$R_L = 30T$$

Determination of torque is necessary for successful analysis of static equilibrium. Torque is a common force application in many mechanisms and deserves further discussion.

Regardless of what mechanism is involved, torque is always the turning effort applied to a point by a force operating through a lever arm (distance) about the point. Common mechanisms utilizing torque are pulleys, winches, brake drums, and gear trains. In each case the mechanism is a circular piece mounted on a shaft at its center with provision for applying a load on the periphery of the circle.

The lever arm is the radius of the circle. The load always applies tangentially to the circle, since a tangent is always perpendicular to the radius of a circle at the point of tangency. The total torque applied to the center of the circle is the product of the tangential force and the radius. Since torque is a combination of force and distance, its dimension is always in inch-pounds, foot-pounds, gram-centimeters, etc.

*Example:* In a V-belt drive, a belt operates on a 12-in.-dia. pulley and pulls with a 50-lb force. The torque is 300 in.-lb, or 25 ft-lb (see Fig. 3-12). Thus we have

$$T = Fr$$
$$T = 50 \text{ lb} \times 6 \text{ in.}$$
$$T = 300 \text{ in.-lb} = 25 \text{ ft-lb}$$

*Example:* Torque can also be applied to the shaft of a mechanism to move a load on the periphery. If the drum diameter of a winch is 2 ft, how many pounds will the

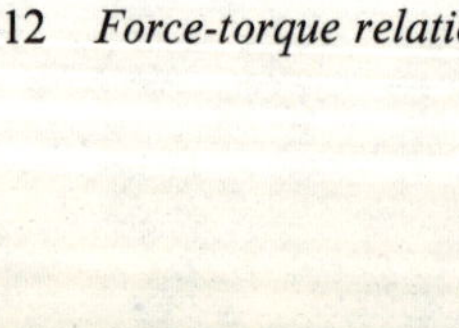


FIG. 3-12 *Force-torque relationship.*

 

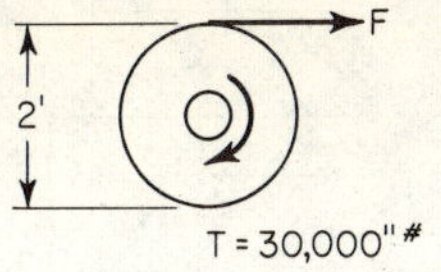


FIG. 3-13 *Force-torque relationship.*

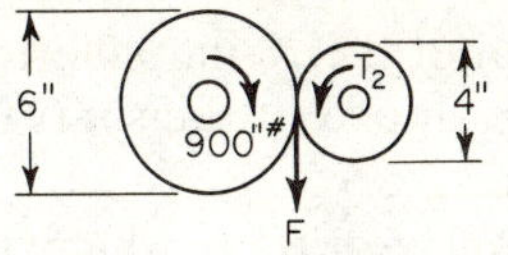


FIG. 3-14 *Torque-transmission relationship.*

winch move if a torque of 30,000 in.-lb is applied to the shaft (Fig. 3-13)? We have

$$T = Fr$$
$$30{,}000 \text{ in.-lb} = 12F$$
$$F = \frac{30{,}000}{12} = 2500 \text{ lb}$$

*Example:* A gear train represents a mechanism which transfers torque from one shaft to another through application of force at their mutual perimeters. A 6-in.-dia. gear is meshed with a 4-in. gear. If 900 in.-lb of torque is applied to the shaft of the 6-in. gear, what torque will the second shaft deliver? (See Fig. 3-14.)

| Gear 1 | Gear 2 |
|---|---|
| $T = Fr$ | $T = Fr$ |
| $900 \text{ in.-lb} = 3 \text{ in. } F$ | $T = 300 \text{ lb} \times 2 \text{ in.}$ |
| $F = 300 \text{ lb}$ | $T = 600 \text{ in.-lb}$ |

### Relationships among Acceleration, Velocity, and Distance

We have determined that distance, velocity, and acceleration are interrelated by the time rate of change. For example, velocity is the time rate of change of distance, and acceleration is the time rate of change of velocity. From this interrelationship we can formulate three postulates which are basic to the force equation and to subsequent problems in mechanics as well.

If acceleration is constant:

1. Final velocity equals acceleration $a$ multiplied by time $t$:

$$v_f = at$$

From this we can also state that the *average* velocity is the mean of the initial and the final velocities:

$$v_a = \frac{v_f + v_i}{2}$$

If the body starts from rest ($v_i$ equals 0), the average velocity is one-half the final velocity. Therefore,

$$v_a = \frac{v_f}{2}$$

2. Distance traversed equals one-half of the acceleration multiplied by the square of the time, or

$$d = \tfrac{1}{2}at^2$$

3. Average velocity equals the square root of twice acceleration times distance, or

$$v_a = \sqrt{2ad}$$

### PROBLEMS

**1.** An automobile is traveling on a flat road at 60 mph. The wind resistance against the car is 565 lb and the road resistance is 50 lb. How much forward force must the wheels exert to keep the car moving at that constant velocity? The weight of the automobile is 3200 lb. _______

**2.** What will the speed change be for the car in Prob. 1 if the wheels exert a forward force of (*a*) 500 lb? _______ ; (*b*) 700 lb? _______

**3.** If the engine is disengaged, what will be the deceleration of the car, disregarding wind resistance? _______

**4.** (*a*) How long will it take the car to coast to a stop? _______

(*b*) How far will the car coast? _______

**5.** Assuming the same amounts of wind and road resistance as used in Prob. 1, determine how much force must be exerted by the wheels to accelerate the car from 30 mph to 60 mph in 10 sec _______

**6.** If the force calculated in Prob. 5 were continued, what would be the terminal velocity of the car assuming that the ground resistance remains constant and the wind resistance is proportionate to velocity? ______

**7.** If air resistance increases at the rate of 0.001 lb/sq in. per foot per second, what will the terminal velocity be of a free-falling 100-lb weight with area of 100 sq in.? ______

**8.** How much pull must be applied to the rim of a 10-in. pulley to develop a 500 in.-lb torque on the shaft? ______

**9.** If the automobile in Prob. 1 has 30-in.-dia. wheels, what torque must the axles apply?

______

**10.** A belt runs around a pair of pulleys, one 20 in. in diameter and one 30 in. in diameter. If 1000 in.-lb of torque is applied to the shaft of the 20-in. pulley, what torque will be developed at the shaft of the 30-in. pulley? ______

**11.** In a four-gear train, what will the torque be on the last shaft if the torque on the first shaft is (*a*) 100 ft-lb? ______; (*b*) 200 ft-lb? ______ (The gear diameters, in order, are 4, 5, 6, and 8 in.)

**12.** Calculate the support loads on a crane carrying 90 tons. The crane weighs 25 tons (Fig. 3-15). $R_L$ ______; $R_R$ ______

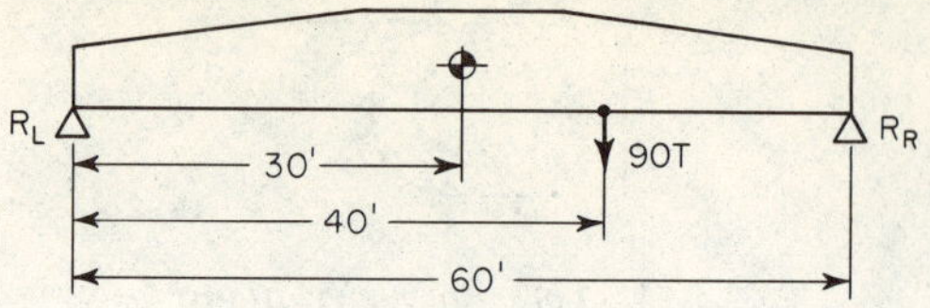

FIG. 3-15 *Simple beam span.*

 

# ENERGY, WORK, POWER

# 4

## ENERGY

As we have seen, force is the essence of all productive effort. *Energy, work,* and *power* are the logical end products of force. Energy is sometimes used synonymously with work, but it is more accurate to say that it is the generative source of work. Power is the measure of work per unit time. Each of these concepts will be described individually.

In the engineering world, there are three broad fields of endeavor: mechanical engineering, thermal engineering, and electrical engineering. In each of these fields, work is accomplished by the particular forces pertinent to that field.

Energy is the capacity a body has for doing work. A body possesses one of two kinds of energy, depending on its position, or condition. If a body is held in such a position that upon its release it can perform work, it possesses energy of position, that is, *potential energy*. Potential energy may also be described as stored, or passive, energy. If a body is in motion, thereby possessing momentum, it possesses *kinetic energy*.

Both forms of energy can be illustrated by the example of a weight supported at some elevation by a rope. While the weight is suspended by the rope, it possesses potential energy; if the rope is removed and the weight drops, it gains kinetic energy. As it is falling, its potential energy is converted to kinetic energy.

We use the same units of measure in calculating energy that we do in calculating work. The potential energy $E_p$ of a body suspended by a rope is the product of the body's weight $w$ and the height to which it is raised $h$. The equation is

$$E_p = wh$$

The kinetic energy of a body dropping from an elevated position is the product of the force of the falling body and the distance it falls. The equation is

$$E_k = \frac{1}{2}\frac{w}{g}v^2$$

where $E_k$ = kinetic energy
$g$ = gravity
$v$ = velocity

The pull of gravity is generally 32 ft/sec/sec, but it varies slightly at different parts of the earth depending on location and elevation. In outer space, in a case in which the pull of gravity from two planets on a body was equal, $g$ would be zero.

Velocity is measured in feet per second, inches per minute, etc.

Energy and work have the same units of measure because they are comprised of the same elements—force and distance. This leads to the conclusion that potential energy,

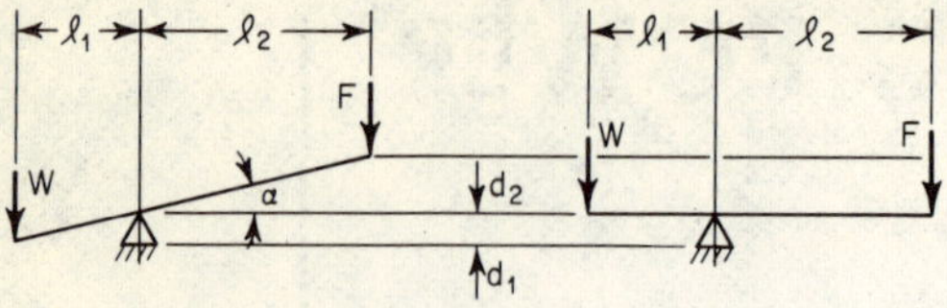


FIG. 4-1 *Work-force relationship.*

kinetic energy, and work are interchangeable, or more correctly, reversible.

Examples of potential energy are a wound-up clock spring, water behind a dam, pressure in a fluid power system, and heat stored in fuel. Examples of kinetic energy are a rotating flywheel, water from the dam flowing over a waterwheel, and liquid or gases flowing in a fluid power system.

## WORK

Work is defined as a force acting vertically through a distance. Conversely, when a body is displaced against resistance, work is done upon it. Force $F$ has units of measurement such as pounds, grams, and tons. Distance $d$ is measured in units such as inches and feet. Thus, work is described in foot-pounds, ton-miles, and gram-centimeters. The equation is:

$$\text{Work} = dF$$

Consider a lever (Fig. 4-1). $l_1$ and $l_2$ can be assumed to remain constant through the two positions shown, since the angle $a$ is small. A load $W$ is raised a distance $d_1$ by a force $F$ acting through distance $d_2$. We determine that the force required is:

$$W\frac{l_1}{l_2} = F$$

The work expended is:

$$Wk = Fd_2 \qquad \text{or} \qquad Wk = W\frac{l_1}{l_2}d_2$$

Note that the potential energy of $W$ has been increased by $Wd_1$, and it is the same as the work done with force $F$ through distance $d_2$.

Just as the linear-force equation has a rotational counterpart, so does the work equation. Rotation is caused by the force couple $Fr$ that is created by the force $F$ applied at the periphery which acts through the radius $r$ of the circle (see Fig. 4-2*a*). The work done by the couple is $Fr\theta$, where $\theta$ is the angle passed over. The angle $\theta$ is measured in radians. Since there are $2\pi$ rad in a circle (Fig. 4-2*b*), $\theta$ can be written as $2\pi n$, where $n$ is the number of turns or fraction thereof. The equation for work done on or by a rotating body is:

$$Wk = Fr2\pi n \qquad \text{(Fig. 4-2}c\text{)}$$

*Example 1:* What work is done by a 100-lb force pulling a 6-in. wheel through an arc of 120 deg?

*Solution:*

$$Wk = 100 \times 3 \times 2\pi \times \frac{120}{360} = 200\pi \text{ in.-lb}$$

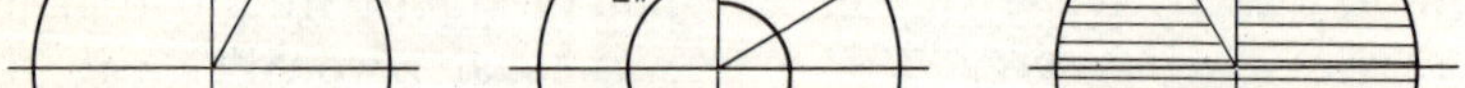


FIG. 4-2 *Rotary work-force relationship.*

 

*Example 2:* How much work is done if the wheel of Example 1 rotates eight turns?

*Solution:*

$$Wk = 100 \times 3 \times 2\pi \times 8 = 4800\pi \text{ in.-lb}$$

*Example 3:* If a torque of 1000 in.-lb is applied to the shaft of a 12-in. pulley and turns it through eight turns, how much work is done?

*Solution:*

$$Wk = 1000 \times 2\pi \times 8 = 16{,}000\pi \text{ in.-lb}$$

Since torque equals $Fr$, the force and radius of the pulley are included in the 1000 in.-lb.

## POWER

Power is the rate at which work is done, that is, the amount of work per unit time. The units of measure are foot-pounds per minute, gram-centimeters per second, etc. Engineers have adopted the horsepower as a convenient power rating. One horsepower equals 33,000 ft-lb/min or 550 ft-lb/sec.

*Example:* How many horsepower are required to apply 1200 in.-lb of torque to a shaft turning at 900 rpm?

*Solution:*

$$\text{Power} = 1200 \times 2\pi \times 900 = 2{,}160{,}000\pi \text{ in.-lb/min}$$

$$\text{hp} = \frac{2{,}160{,}000\pi}{12 \times 33{,}000} = 5.45\pi = 17.11$$

Almost all power trains combine electrical or heat energy with mechanical energy. Therefore, the concepts of energy, work, and power are present in all engineering fields and are interrelated and reversible.

### Electrical Energy

As in the mechanical system, electrical energy and work are essentially synonymous. The unit of measure is one joule (1 joule = 1 volt × 1 coul).

Power is the measure of work per unit time; the unit of electrical power is one watt.

$$1 \text{ watt} = 1 \text{ joule/sec}$$
$$1 \text{ joule/sec} = 1 \text{ volt} \times 1 \text{ coul/sec}$$
$$1 \text{ coul/sec} = 1 \text{ amp}$$

Therefore

$$1 \text{ watt} = 1 \text{ volt} \times 1 \text{ amp}$$

Because a watt is a rather small quantity, electrical power is more often measured in kilowatts (kw).

$$1 \text{ kw} = 1000 \text{ watts}$$

Of direct use to us is the relation between electrical and mechanical power:

$$1 \text{ hp} = 745.5 \text{ watts} = 0.7455 \text{ kw}$$
(commonly rounded to 0.746)
$$1 \text{ kw} = 1.341 \text{ hp}$$

### Heat

Heat energy is measured in Btu's; power is measured in Btu's per second. The relation between mechanical and heat power is:

$$1 \text{ hp} = 0.7067 \text{ Btu/sec}$$
$$1 \text{ Btu/sec} = 1.415 \text{ hp}$$

The relation between electrical and heat power is:

$$1 \text{ Btu/sec} = 1.055 \text{ kw}$$
$$1 \text{ kw} = 0.943 \text{ Btu/sec}$$

## FLUID MECHANICS

Fluids are a class of substances which yield to any force tending to change their shapes without appreciably altering their volume; thus they are capable of flowing. Fluids include liquids and gases. Liquids are only slightly compressible, whereas gases (which include air) are highly compressible.

Liquids differ from gases in their lack of compressibility and their material rather than ethereal character. Liquids differ from solids in their capability of flowing in their normal state. In a gas the molecules tend to repel themselves, but readily move closer together under pressure. In a liquid they will compress slightly, but will slip by each other under pressure. Molecules hold their position in a solid, but under pressure will very slightly slip by each other and move closer together.

Changes in temperature cause all materials to change character. At extremely low temperatures, almost all materials become solid. At elevated temperatures, most liquids become gaseous. As temperatures continue to increase, solids liquefy, and at superelevated temperatures most solids gasify.

### Pascal's Law

Pascal's law states that the pressure at any point within a static fluid acts equally in all directions. Postulates to this law are: (1) A pressure applied at any point in a confined liquid immediately translates to all parts of the liquid; and (2) pressure, as well as liquid in conduit, flows from a position of high toward low. These principles are basic to all hydraulic theory.

The hydraulic press is a direct example of this law. The essentials of a hydraulic press are a small pressure (or actuator) cylinder and a large force (or work) cylinder connected by a fluid passage which transmits fluid under pressure from the small (pressure) to the large (force) cylinder (Fig. 4-3).

Let us assume that $A_L$ equals $100A_s$. If we apply a force of $F_s$ on the piston of the small cylinder, it will create a pressure $p_s$. This pressure will be the same in the large cylinder $p_L$, and when it is applied to the large piston it will create a force $F_L$, which will be 100 times greater than $F_s$.

From this we see that force acting over a piston area exposed to fluid creates pressure in the fluid. Conversely, pressure acting on an area creates force. Therefore,

where $F$ = total force, lb
$p$ = unit pressure, psi
$A$ = area, sq in.

then $F = pA$

*Example:* A 200-ton press has a 24-in.-dia. cylinder. What diameter plunger must be used to develop this tonnage if the force on the plunger will be 500 lb?

*Solution:*

Large cylinder

$$p = \frac{F}{A}$$

$$F = 200 \times 2000 = 400{,}000 \text{ lb}$$

$$A = \frac{\pi}{4}(24)^2 = 452 \text{ sq in.}$$

$$p = \frac{400{,}000}{452} = 885 \text{ psi}$$

Plunger

$$A = \frac{F}{p}$$

$$F = 500 \text{ lb}$$

$$p = 885 \text{ psi}$$

$$A = \frac{500}{885} = 0.565 \text{ sq in.}$$

$$d = \sqrt{\frac{4A}{\pi}} = \sqrt{\frac{4 \times .565}{\pi}} = 0.848 \text{ in.}$$

*Elevation Head*

Consider a column of liquid (Fig. 4-4) of unit weight $w$, height $h$, and area $A$. The

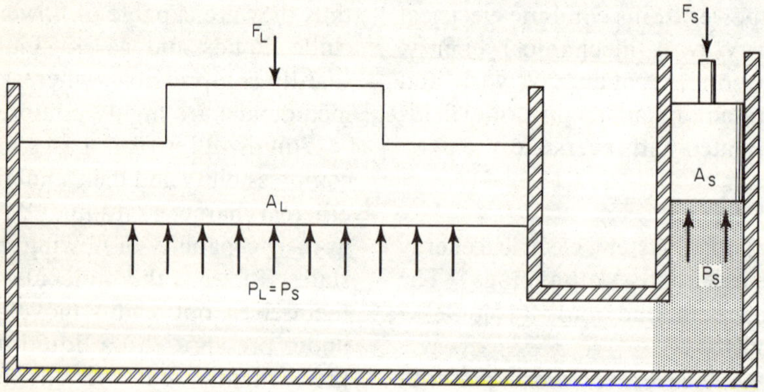


FIG. 4-3 *Pascal's law illustrated.*

 

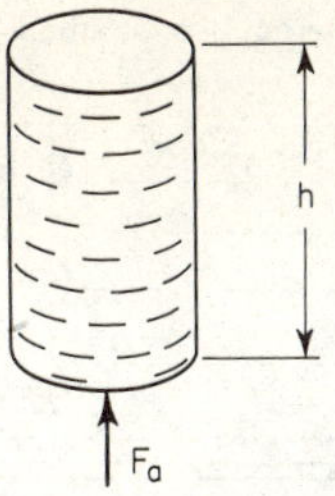


FIG. 4-4 *Elevation head.*

total weight of the liquid will be $whA$, which will be the force $F_a$ on the bottom.

Thus, $F_a = whA$

Also, $F_a = pA$ (Pascal's law)

Therefore, $p = wh$ or $h = \frac{p}{w}$

Figure 4-4 shows what is known as the *elevation head* and depicts the pressure exerted by a liquid solely as a result of its elevation. This is a state of potential energy.

If a pressure $p_0$ is exerted on top of the liquid, the equation becomes $p_a = p_0 + wh$.

*Example:* A town wishes to provide water pressure of 60 psi to their citizens. Assuming the town is on flat ground, how high should they build the standpipe? (Water weighs 62.4 lb/cu ft.)

*Solution:* We know that $h = p/w$; therefore $h = (60 \times 144)/62.4 = 138.3$ ft.

#### Continuity Equation

The quantity of fluid flowing in a pipe is the product of the velocity of the fluid and the area of the pipe. Therefore,

$$Q = VA$$

where $Q$ = quantity in gallons, cubic inches, etc., in a specific time
$V$ = velocity, ft/min or ft/sec, etc.
$A$ = area of pipe, sq in.

If the quantity of fluid flowing in a pipe remains constant, the velocity of the fluid must vary inversely as the area of the pipe (assuming no voids in the pipe). If the area is doubled, the velocity will reduce by one-half; if the area is reduced by one-half, the velocity will double.

Across any two points (1 and 2) in the pipe, we can write the equation $V_1A_1 = V_2A_2$. Since volume of flow $Q$ is constant, we can write:

$$V_1A_1 = V_2A_2 = Q$$

*Example:* If 10 gpm are flowing through a ½-in.-dia. pipe, what is the velocity?

*Solution:* We know that $Q = VA$.

$$10 \times 231 = V\frac{(\pi)(\frac{1}{2})^2}{4}$$

$$V = 11{,}770 \text{ in./min} = 16.33 \text{ ft/sec}$$

*Example:* What will the velocity be if the pipe is enlarged to ¾-in. dia.?

*Solution:*

$$10 \times 231 = V\frac{(\pi)(\frac{3}{4})^2}{4}$$

$$V = 7.26 \text{ ft/sec}$$

#### Bernoulli's Equation

As we learned in mechanics, all solids have weight and mass, and can gain potential energy and obtain kinetic energy. A mechanical system, comprised of masses and forces in various states of rest and motion, will have a total energy level equal to the summation of energies of the many components.

Fluids, like solids, have mass and weight and potential and kinetic energy. It follows that the total energy of a fluid system will be the summation of various energy levels within the fluid. This statement can be applied at any point in a system, and the total energy will be the summation of energies to the point being considered. Further, the total energy of a system remains constant throughout that system. Thus, the total energy extant at any two points in the system will be equal. The forms of energy may change—potential may become kinetic and pressure may become velocity—but the total remains constant. Work done on the system adds energy, and work done by the system subtracts energy. This concept is compatible with the law of conservation of energy and is expressed mathematically by Bernoulli's equation.

The forms of energy are potential and kinetic. In hydraulics, potential energy is either due to pressure applied to the fluid (called "pressure head") or to the height to which the fluid is raised (called "elevation head"). Kinetic energy is the energy of the flowing fluid mass (called "velocity head"). Work done on or by the system is included as a form of kinetic energy. Bernoulli's equation states:

Pressure head + elevation head + velocity head = constant

Pressure head = $p$ (Pascal's law)

Elevation head = $hw$ (Pascal's law)

$$\text{Kinetic energy} = \frac{1}{2}\frac{w}{g}v^2$$

where $p$, $hw$, and $\frac{1}{2}(w/g)v$ are measured in pounds per square inch and $w$ is unit weight in pounds per cubic inch. Inserting these values in the above equation

$$p + hw + \frac{1}{2}\frac{w}{g}v^2 = K$$

If we divide by $w$, the equation becomes dimensional in inches:

$$\frac{p}{w} + h + \frac{v^2}{2g} = K$$

This equation is the foundation upon which all systems analyses, particularly investigation of power distribution, are based. The pure equation applies to a theoretical fluid which is frictionless and flows without loss through an entire system. Similarly, all work done on or by the theoretical system is included at 100-percent efficiency.

In an actual system, usable energy is continually being dissipated by friction in the pipes, restrictions in orifices and valves, and by internal losses (inefficiencies) in actuators. This lost energy is not available to perform useful work. The losses are usually incorporated into the equation as percentages of velocity head. The summation of these losses determines the overall system efficiency.

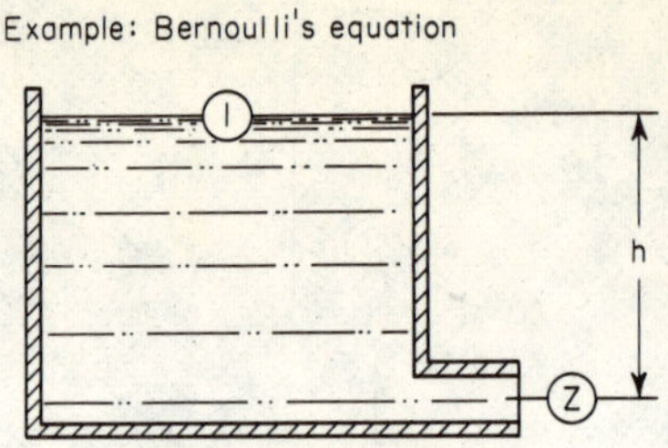


FIG. 4-5 *Bernoulli's law.*

*Example:* Given a tank of height $h$ (Fig. 4-5), open to atmosphere, determine the velocity of fluid discharged into the atmosphere from a nozzle in the bottom of the tank.

*Solution:* Writing Bernoulli's equation across points 1 and 2, we have

$$\frac{P_a}{w} + h + 0 = \frac{P_a}{w} + 0 + \frac{v^2}{2g}$$

$$h = \frac{v^2}{2g} \qquad \text{or} \qquad v = \sqrt{2gh}$$

**PROBLEMS**

**1.** A shop press has a 4-in.-dia. ram. How many tons of force will it develop if actuated by a ¼-in. plunger under a 100-lb force? ______

**2.** Assume the plunger in Prob. 1 is actuated by a lever 1½ ft long. The fulcrum of the lever is 4 in. from one end. (*a*) How much force must be applied to the opposite end of the lever to develop 5,000 psi? ______
(*b*) What tonnage will the press develop at this pressure? ______

**3.** What water pressure is exerted against the base of a dam holding water 550 ft high? ______

 

# MATHEMATICS

# 5

**Review of Common Fractions**

**1.** Calculate the total length of a bolt in Fig. 5-1. _______

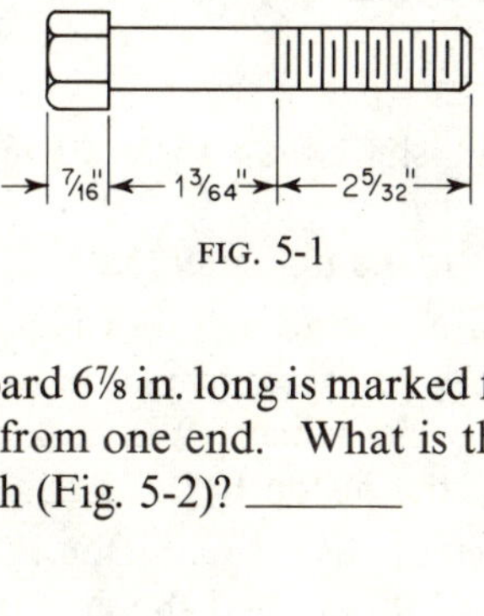


FIG. 5-1

**2.** A board 6⅞ in. long is marked for cutting 2 15/16 in. from one end. What is the remaining length (Fig. 5-2)? _______

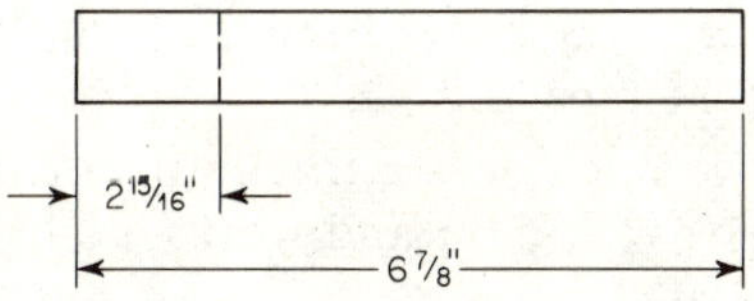


FIG. 5-2

**3.** Several pieces of wire, 9 in., 17⅝ in., 7¼ in., 13 in., and 18 in., were cut from a 10-ft spool of wire.

(*a*) What was the total length of wire used? _______

(*b*) How much remains? _______

**4.** Determine distance *x* in Fig. 5-3. _______

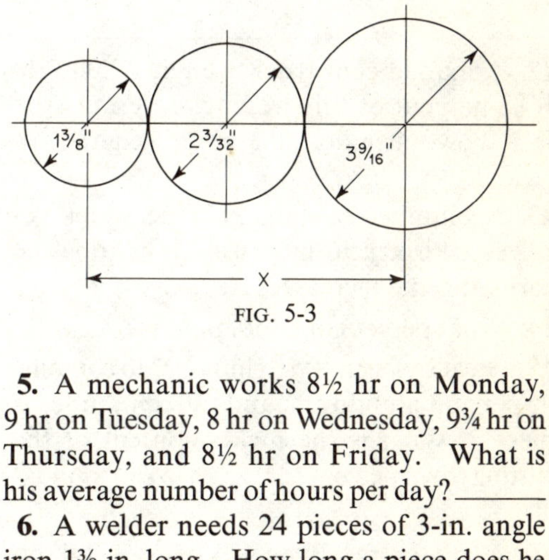


FIG. 5-3

**5.** A mechanic works 8½ hr on Monday, 9 hr on Tuesday, 8 hr on Wednesday, 9¾ hr on Thursday, and 8½ hr on Friday. What is his average number of hours per day? _______

**6.** A welder needs 24 pieces of 3-in. angle iron 1⅜ in. long. How long a piece does he need, making no allowance for waste? _______

**7.** Six pieces, each 3⅝ in. long, were cut from a 1-in. brass rod 3 ft long. Allowing ⅛ in. of waste per cut, how much rod was left? _______

**8.** A mechanic finds that he needs three pieces of hose, 7¾ in., 9⅜ in., and 18½ in., to assemble a hydraulic unit. How much hose must he order to assemble 14 units?

_______

**9.** A board 58⅛ in. long is divided into five equal lengths. What is the length of one division? _______

**10.** How many cubic feet are there in a tank which holds 105 gal? [A cubic foot contains approximately 7½ gal (7.48).] ______

**Review of Percentage**

**11.** Rewrite the following items as a common fraction, decimal fraction, and percent:

(*a*) Three-tenths

______

(*b*) Five-eighths

______

(*c*) Ninety-five thousandths

______

(*d*) Two hundred seventeen thousandths

______

(*e*) Five-ninths

______

**12.** A loaded coal truck weighs 32,000 lb. If 82 percent of this weight represents the load, how much does the truck weigh?

______

**13.** A plumber earning $125 per week receives a 12-percent increase. What does he earn after the increase? ______

**14.** What percent of a foot is ⅜ in.? ______

**15.** An alloy casting weighing 207 lb contains 45 percent aluminum, and the remainder is nickel. What is the nickel content of the casting (*a*) in percent? ______; (*b*) in pounds?

______

**16.** A motor receiving 9 hp loses 1.4 hp through inefficiency. What percent of the hp is lost? ______

**17.** 1½-in. galvanized pipe is listed at 11 cents/ft with a 3-percent discount allowed for cash. Find the cash cost for 350 feet of pipe. ______

**18.** A man sells a car for $240 and loses 20 percent of his original investment. What did the car cost him? ______

**19.** An electrical transmitting line has a 4-percent voltage loss. If the generator output is 140 volts, what is the voltage at the end of the line? ______

**Review of Decimal Fractions**

**20.** Rewrite the following items as decimals:

(*a*) One-hundred-seven thousandths

______

(*b*) Sixty-four ten-thousandths

______

(*c*) Fifty-six and nine-hundredths

______

(*d*) Sixty-five hundred-thousandths

______

(*e*) Six-tenths ______

(*f*) one hundred hundred-thousandths

______

**21.** Convert the following common fractions to decimals.

(*a*) $\frac{3}{20}$ ______

(*b*) ⅜ ______

(*c*) $\frac{3}{32}$ ______

(*d*) ⅚ ______

**22.** A small shaft was to be 0.685 in. in diameter but it was made 0.0018 in. oversize. What is the diameter? ______

**23.** A certain engine cylinder is $3\frac{3}{16}$ in. in diameter. A properly fit piston must have a 0.0045-in. clearance. What should the diameter of the piston be? ______

**24.** The actual inside diameter of a one-inch pipe is 1.04 in., and the actual outside diameter is 1.315 in. Find the wall thickness of the pipe. ______

**25.** How much does a truck driver earn if he receives $2.75/hr and works a 6-hr day for 5½ days? ______

**26.** A man wishes to divide a $25 reward among seven boys. How much does each boy receive? ______

**27.** A radio tower built in six equal sections is 289 ft high. What is the height to the top of the fourth section? ______

**28.** Calculate the overall length of the belt in Fig. 5-4. ______

 

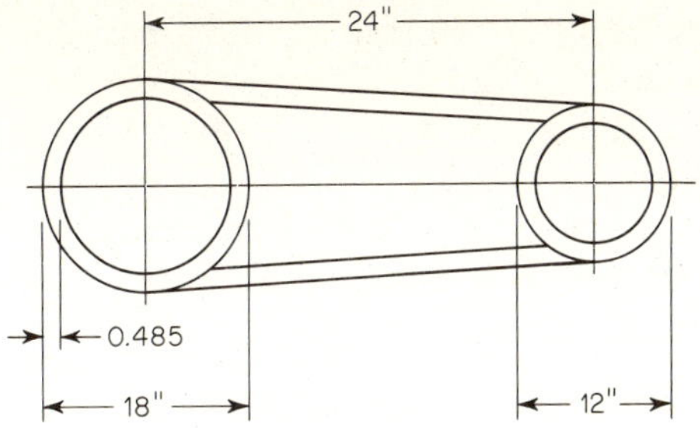


FIG. 5-4

**29.** A garage used 215 gal of gasoline a year at a cost of \$.1575/gal. What did the gasoline cost: (*a*) per year? ______ (*b*) per month? ______

**Review of Ratio and Proportion**

**30.** Express the following as simple ratios:
(*a*) 3 in. to 12 in. ______
(*b*) 12 in. to 2 yd ______
(*c*) 50 cents to 2 dollars ______
(*d*) 3 ft to 3 yd ______

**31.** The smaller of two pulleys turns at 300 rpm and the larger at 60 rpm. What is the ratio of their speeds? ______

**32.** Find the missing term:
(*a*) $2:3 = 6:$ ______
(*b*) $7:8 = 21:$ ______
(*c*) $4:5 = 16:$ ______
(*d*) $9:$ ______ $= 81:18$

**33.** On a road map ¼ in. represents 5 miles. Two cities are located 3⅛ in. apart. Find the actual distance between them. ______

**34.** If an automobile runs 72 miles on 4 gal of gasoline, how far will it go on 18 gal? ______

**35.** If a 12-lb can of grease costs \$3.72, what would a 5-lb can cost? ______

**36.** The small pulley in Fig. 5-5 is 11 in. in diameter and the larger is 20 in. in diameter.

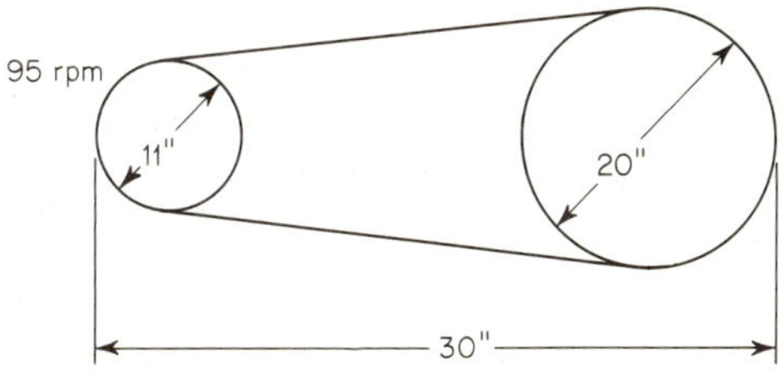


FIG. 5-5

How many rpm does the large pulley turn if the smaller revolves at 95 rpm? ______

**37.** A 21-in. pulley revolving at 65 rpm turns a small pulley at 210 rpm. What is the diameter of the small pulley (Fig. 5-6)? ______

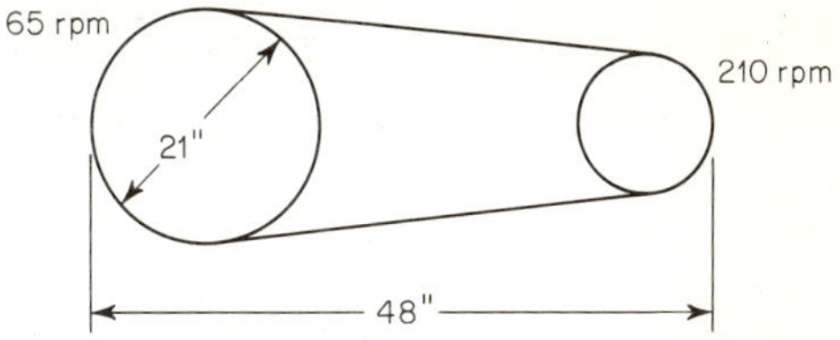


FIG. 5-6

**38.** Pulley A is 11 in. in diameter, B is 8 in., C is 14 in., and D is 10 in. Pulley A turns at 240 rpm. How fast does D turn? (See Fig. 5-7.) ______

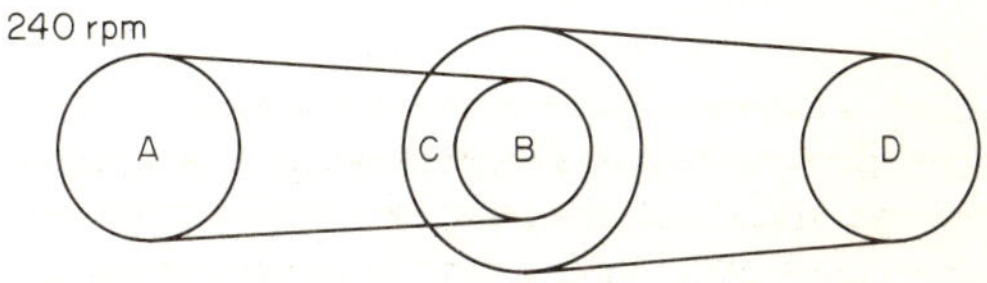


FIG. 5-7

**39.** A step pulley (Fig. 5-8) with positions of 13 in., 10 in., and 5 in. in diameter revolves at 135 rpm. It drives another step pulley of 3 in., 6 in., and 11 in. in diameter. What speeds can be obtained in lines A? ______; line B? ______; line C? ______

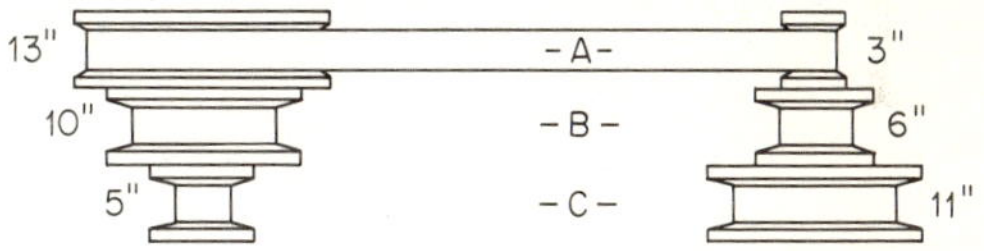


FIG. 5-8

**Review of Perimeter, Circumference, and Area**

(Use $\pi = 3.14$)

**40.** Find the perimeter of a shop 25 ft × 75 ft (Fig. 5-9). ______

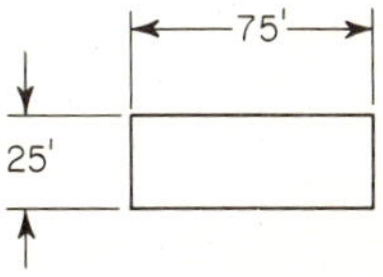


FIG. 5-9

**41.** Two rows of tape are placed around a cardboard carton 8 in. × 10 in. How many inches of tape are required (Fig. 5-10)? ______

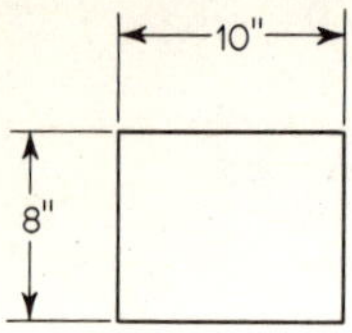

FIG. 5-10

**42.** A hydraulic tube is to be placed around a room 21 ft 6 in. × 11 ft 6 in. How many feet of tubing should be ordered (Fig. 5-11)? ______

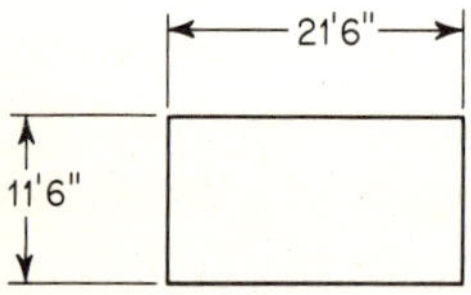

FIG. 5-11

**43.** Find the area of a sheet of steel 8 in. wide × 11 in. long. ______

**44.** Calculate the area in square inches (Fig. 5-12). ______

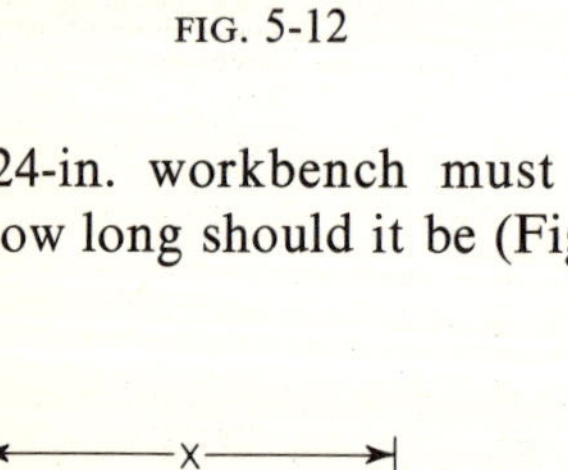

FIG. 5-12

**45.** If a 24-in. workbench must contain 21 sq ft, how long should it be (Fig. 5-13)? ______

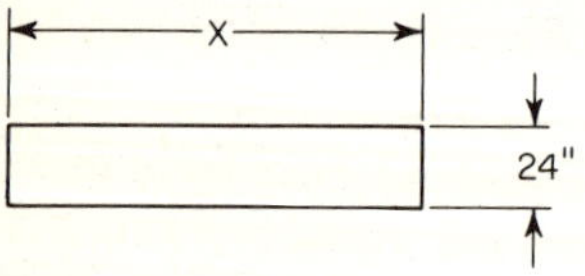

FIG. 5-13

**46.** Find the distance around a pipe 2 in. in diameter. ______

**47.** The radius of an automobile tire is 12 in. What is its circumference? ______

**48.** Find the length of a belt connecting two 4-in. pulleys which are 2 ft center to center (Fig. 5-14). ______

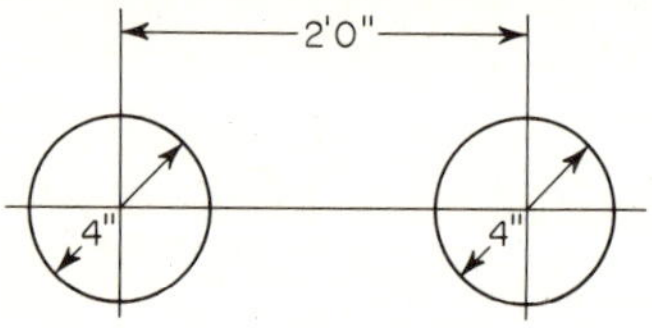

FIG. 5-14

**49.** Find the area of a circle with a 4-in. radius. ______

**50.** Calculate the area of a piston 6 in. in diameter. ______

**51.** How many square inches of metal are needed for the bottom of a pail that is 12 in. in diameter? ______

**52.** (*a*) Calculate the open area of a cross section of a pipe whose inside diameter is 11 in. and whose outside diameter is 12 in. (Fig. 5-15). ______

(*b*) How many square inches of metal are exposed in the pipe? ______

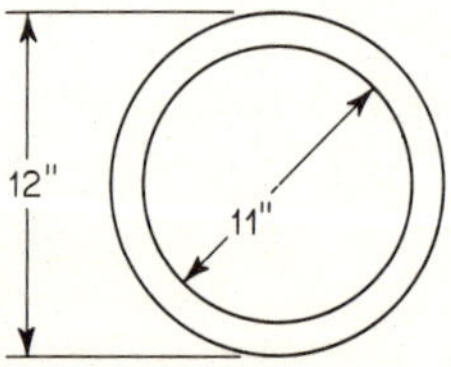

FIG. 5-15

**53.** A circular post has a cross-sectional area of 38.465 sq in. What is its diameter? ______

### Review of Volumes of Prisms and Cylinders

(Use $\pi = 3.14$; 1 gal = 231 cu in.)

**54.** Find the volume of a rectangular pan whose base is 12 in. × 9 in. and whose altitude is 3 in. (Fig. 5-16). ______

 

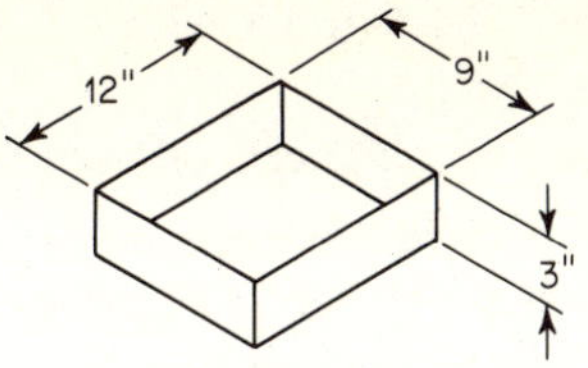


FIG. 5-16

**55.** Find the volume of a storage box whose base is 9 ft 6 in. × 12 ft and whose height is 8 ft 3 in. _______

**56.** An accumulator has an inside diameter of 4 in. It contains ½ gal of fluid. What is the height of the fluid? _______

**57.** Find the capacity in gallons of a rectangular reservoir 9 in. × 11 in. × 1 ft. _______

**58.** Calculate the weight of a bar of gold 3 in. × 5 in. × 12 in. (Gold weighs 0.6975 lb/cu in.)

**59.** Find the volume of an 8-in.-dia. tank which is 10 in. high. _______

**60.** How many cubic feet are in a water reservoir which is 24 ft in diameter and 70 ft tall? _______

**61.** Find the height of a 2-gal can if the diameter is 8 in. _______

**62.** Calculate the weight of a 1-in. steel rod which is 12 ft long. (Steel weighs 0.284 lb/cu in.) _______

### Problems Involving Hydraulic Cylinders

(Use $\pi = 3.14$; 1 gal = 231 cu in.)

**63.** Calculate the areas of the following pistons used in hydraulic cylinders. $A = \pi r^2$ or $\pi D^2/4$.

(*a*) 1-in. dia. _______
(*b*) 2-in. dia. _______
(*c*) 5-in. dia. _______
(*d*) 10-in. dia. _______

**64.** Find the volume of a 6-in.-dia. hydraulic cylinder in which the rod moves out 22 in.; ignore volume occupied by the piston. Find answers only for cylinder while rod is extending (Fig. 5-17). _______ cu in.; _______ gal.

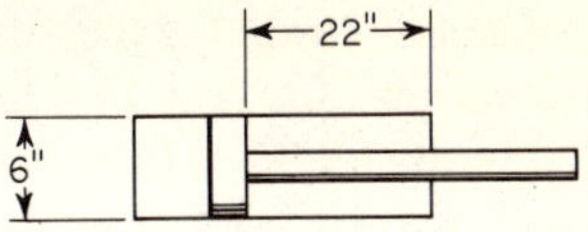


FIG. 5-17

**65.** Compute the displacement of a 1½-in.-dia. cylinder with the rod moving out 9 in. (Fig. 5-18). _______ cu in.; _______ gal.

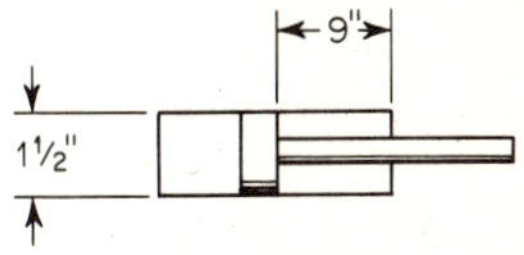


FIG. 5-18

**66.** A rod travels a distance of 4 in. in 2 sec. What is its speed in inches per second?

**67.** A rod completes a 24-ft stroke in 4 min. What is its speed of travel in feet per minute? _______

**68.** If a 4-in.-dia. piston completes a 3-in. stroke, what is the cylinder's volume in gallons? Consider the rod moving out. _______

**69.** (*a*) What is the displacement of a 7-in.-dia. cylinder with a 2-in.-dia. rod and an 18-in. stroke while extending (Fig. 5-19)? _______ cu in.; _______ gal.

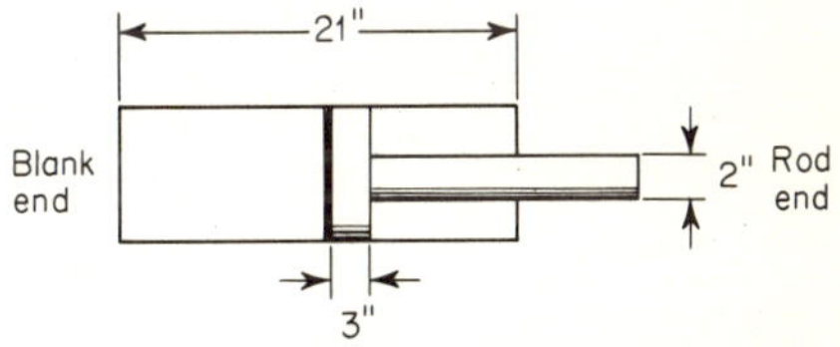


FIG. 5-19

(*b*) What is total displacement of this cylinder for its retracting stroke? _______ cu in.; _______ gal.

**70.** If a 5-in.-dia. piston with a 2-in.-dia. rod completes a 20-in. stroke in 30 sec while extending, what is the gpm delivered to the cylinder? _______

**71.** What gpm are required for a cylinder with a 1-in.-dia. piston to complete a 12-in. extending stroke in 2 sec? _______

**72.** A 4-in.-dia. cylinder receives 4 gpm. What is the speed of travel while rod is extending? ______

**73.** Find the areas of the following ring sections (the end-view hydraulic piston and rod). The net area of the piston = area of piston minus area of rod, or area $A$ − area $B$ (Fig. 5-20).

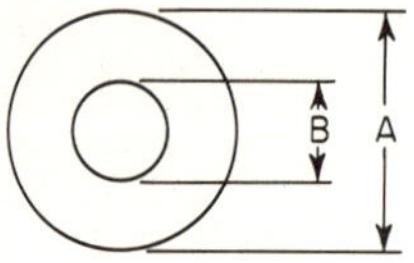


FIG. 5-20

(*a*) $A$ = 4-in.-dia. piston and 1-in.-dia. rod ______

(*b*) $A$ = 3-in.-dia. piston and 1½-in.-dia. rod ______

(*c*) $A$ = 9-in.-dia. piston and 2-in.-dia. rod ______

**74.** Calculate the extending volume or displacement for a 6-in.-dia. cylinder with an 18-in. stroke and 1-in. rod (Fig. 5-21).

(*a*) Extending: ______ cu in.; ______ gal.

(*b*) Retracting: ______ cu in.; ______ gal.

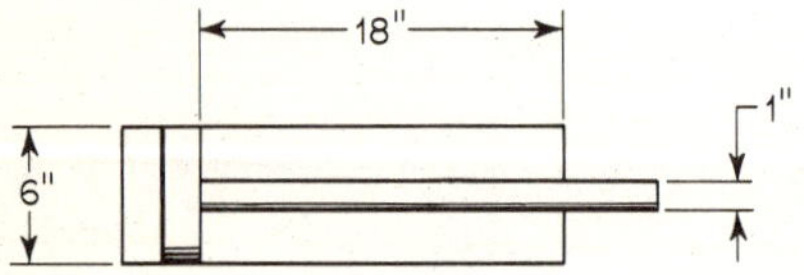


FIG. 5-21

**75.** A 6-in.-dia. cylinder with a 1-in. rod received 8 gpm.

(*a*) What is the speed of travel extended in feet per minute? ______

(*b*) What is the speed of travel retracted in feet per minute? ______

(*c*) What is the difference in the speed of travel between the two strokes? ______

**76.** On the same machine is another 6-in.-dia. cylinder receiving 8 gpm, but the cylinder has a 3-in. rod.

(*a*) What is the speed of travel extending in feet per minute? ______

(*b*) What is the speed of travel retracting in feet per minute? ______

(*c*) What is the difference in the speed of travel between the two strokes? ______

**77.** A 1-in.-dia. cylinder with a 12-in. stroke has a speed of 60 ft/min. How much time will be required to complete its extending stroke? ______

**78.** A cylinder stroke is 2 ft long. How long will it take to complete its extending stroke if the speed is 10 ft/min? ______

**79.** Calculate the load for a 6-in.-dia. cylinder while extending and operating at the following pressures:

(*a*) 100 psi ______
(*b*) 500 psi ______
(*c*) 1000 psi ______

**80.** Calculate the return load for a 6-in.-dia. cylinder with a 2-in.-dia. rod at the following pressures:

(*a*) 100 psi ______
(*b*) 500 psi ______
(*c*) 1000 psi ______

 

# FLUID PASSAGES

# 6

## IRON PIPE

Piping is the oldest known means of conveying fluids. Its use goes back to the clay piping utilized by the early Grecian and Roman civilizations for carrying water. Usage of iron pipe in hydraulic systems grew out of its extensive use in water systems. Iron pipe can be used for conveying all types of hydraulic fluids, from the Skydrol and nonflammable phosphate ester hydraulic oils to the standard petroleum-based hydraulic oils, without corrosion. The cost of materials for pipe and pipe fittings is less than that for rubber-base or hydraulic steel tubing and fittings; piping thus provides a very economical method of plumbing the hydraulic system. The limitations in the use of iron pipe are the result of its inherent bulkiness and lack of flexibility.

### Pipe Application

Iron piping is used in hydraulic systems because of its economy and because it is available in relatively large sizes. It is generally best suited for long, straight, permanent lines with a minimum of bends. It is usually connected by threading into a tapped port onto another pipe fitting. Its heavy wall thickness and the fact that it is not annealed prohibit bending. Fittings must be used wherever a turn or bend must be made.

### Pipe Sizes

Pipe sizes are specified by inside diameter (I.D.) dimensions. Note (see Table 6-1) that the actual I.D. for a pipe of very small diameter, such as ½ in. or ¾ in., is not equal to its designated diameter for any wall thickness or weight of pipe. In any given pipe size, the outside threads of the pipe remain the same regardless of any varying wall thickness. For various wall thicknesses in any size pipe, the outside diameter (O.D.) of that size remains constant, but the I.D. varies (see Table 6-1). When specifying iron pipe in a system, two dimensions are necessary: the nominal I.D. and the schedule (wall thickness). Table 6-1 shows the I.D.'s and burst pressures of pipe used in hydraulic applications. The pressure used in the system, the anticipated surge pressures, and the required flows determine the pipe size to use.

It is important to note that while iron pipe has one set of pressure ratings (shown in Table 6-1), the ratings of pipe fittings are dependent upon the method by which they were manufactured. For example, pipe fittings found in hardware stores and used in domestic water systems (30 to 40 psi) are generally cast-iron fittings. Cast-iron fittings for all sizes have rated working pressures of 150 psi in steam applications or 300 psi when used in conjunction with water or hydraulic oil. With a 4 to 1 safety factor, these fittings

TABLE 6-1 INSIDE DIAMETERS AND BURST PRESSURES OF PIPE USED IN HYDRAULIC APPLICATIONS*

| NOMINAL PIPE SIZE, IN. | OUTSIDE DIAMETER OF PIPE, IN. | NUMBER OF THREADS PER INCH | LENGTH OF EFFECTIVE THREADS, IN. | SCHEDULE 40 (STANDARD) | | SCHEDULE 80 (EXTRA HEAVY) | | SCHEDULE 160 | | DOUBLE EXTRA HEAVY | |
|---|---|---|---|---|---|---|---|---|---|---|---|
| | | | | *Pipe I.D., in.* | *Burst press., psi* | *Pipe I.D., in.* | *Burst press., psi* | *Pipe I.D., in.* | *Burst press., psi* | *Pipe I.D., in.* | *Burst press., psi* |
| ⅛ | 0.405 | 27 | 0.26 | | | | | | | | |
| ¼ | 0.540 | 18 | 0.40 | 0.364 | 16,000 | 0.302 | 22,000 | | | | |
| ⅜ | 0.675 | 18 | 0.41 | 0.493 | 13,500 | 0.423 | 19,000 | | | | |
| ½ | 0.840 | 14 | 0.53 | 0.622 | 13,200 | 0.546 | 17,500 | 0.466 | 21,000 | 0.252 | 35,000 |
| ¾ | 1.050 | 14 | 0.55 | 0.824 | 11,000 | 0.742 | 15,000 | 0.614 | 21,000 | 0.434 | 30,000 |
| 1 | 1.315 | 11½ | 0.68 | 1.049 | 10,000 | 0.957 | 13,600 | 0.815 | 19,000 | 0.599 | 27,000 |
| 1¼ | 1.660 | 11½ | 0.71 | 1.380 | 8,400 | 1.278 | 11,500 | 1.160 | 15,000 | 0.896 | 23,000 |
| 1½ | 1.900 | 11½ | 0.72 | 1.610 | 7,600 | 1.500 | 10,500 | 1.338 | 14,800 | 1.100 | 21,000 |
| 2 | 2.375 | 11½ | 0.76 | 2.067 | 6,500 | 1.939 | 9,100 | 1.689 | 14,500 | 1.503 | 19,000 |
| 2½ | 2.875 | 8 | 1.14 | 2.469 | 7,000 | 2.323 | 9,600 | 2.125 | 13,000 | 1.771 | 18,000 |
| 3 | 3.500 | 8 | 1.20 | 3.068 | 6,100 | 2.900 | 8,500 | 2.624 | 12,500 | | |

* Working pressures for various schedule pipes are obtained by dividing burst pressure by the safety factor.

*Source:* Hydraulics and Pneumatics.

 

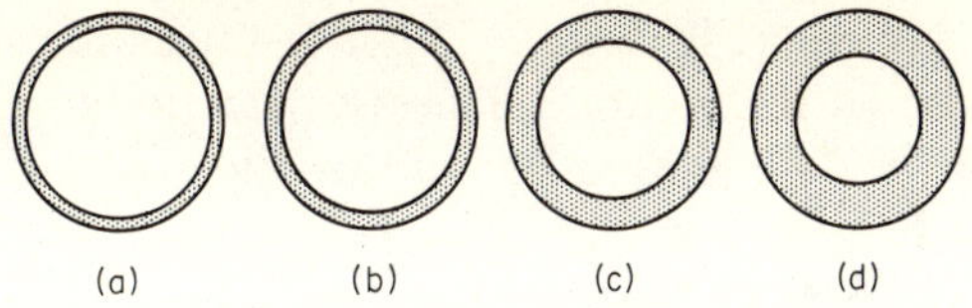


FIG. 6-1 *For any schedule pipe, the O.D. remains the same, but I.D. decreases as the wall thickness increases. (Courtesy Hydraulics & Pneumatics.) (a) Schedule 40 (standard). (b) Schedule 80 (extra heavy). (c) Schedule 160. (d) Double extra heavy.*

have burst pressures of approximately 1200 psi, and should only be used in return lines or other applications in which the maximum pressure is less than 300 psi including any surge pressure that may be present.

Forged-iron or steel fittings have much higher working and burst-pressure ratings and are compatible for use with extra heavy pipe in high-pressure applications (1500 to 2000 psi systems). The fittings shown in Figs. 6-4 to 6-14 are forged-steel fittings which can be used in high-pressure circuits to 5000 psi.

### Pipe Threads

Two types of pipe threads have been standardized in the United States. One is the American (or National) Standard Taper Pipe Thread (NPT) and the other is the Dryseal American (or National) Taper Pipe Thread (NPTF).

On the American Standard (NPT) threads, sealing is done by interference contact between thread flanks (Fig. 6-2). When these threads are applied, sealing compound or sealing tape must be used to keep fluid from leaking through the spiral clearance in the root of the thread.

On the Dryseal American Taper Pipe Thread (NPTF), the roots and crests of the thread engage before the flanks touch (Fig. 6-3*a*). When the thread is wrench tightened, the crests are crushed, creating a seal (Fig. 6-3*b*). JIC (Joint Industrial Conference) hydraulic standards for hydraulic plumbing specify that all pipe threads must be NPTF. These threads do not require a sealing compound or tape, but it is common practice to use one for lubrication and to ensure the seal. The cost of an oil leak in the system is far greater than the expense of a sealer. Oil leaks are unsightly and may present a fire hazard.

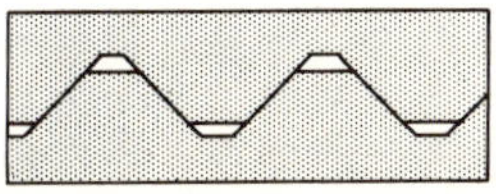

FIG. 6-2 *American Standard Pipe Thread (NPT).*

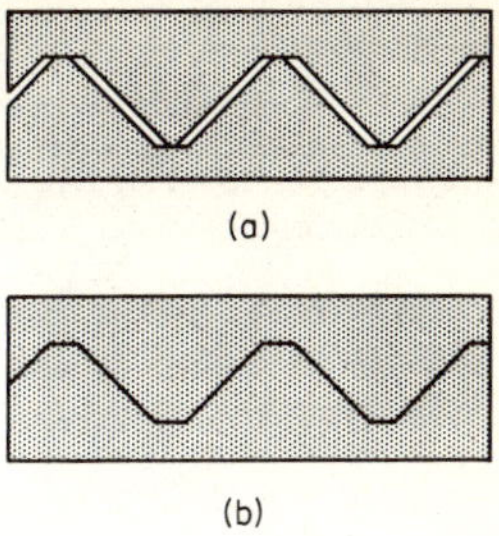


FIG. 6-3 *(a) Dryseal hand tight. (b) Dryseal wrench tight.*

NPT and NPTF threads are interchangeable, but in order to give an effective seal, male and female NPTF threads must be used together.

### Pipe Fittings (Hydraulic Steel or Forged)

The following terms are used throughout the list of fittings:

*Male,* which means the external or outside threaded portion of the fitting.

*Female,* which means the internal or inside threaded portion of the fitting.

*Port,* which is any opening in a valve, pump, or cylinder to which a fitting must be connected.

STREET ELBOW (Fig. 6-4*a*). This is a 90-deg fitting with a male thread on one end and a female thread on the other end. It is used to bring a line in parallel to the port (Fig. 6-4*b*).

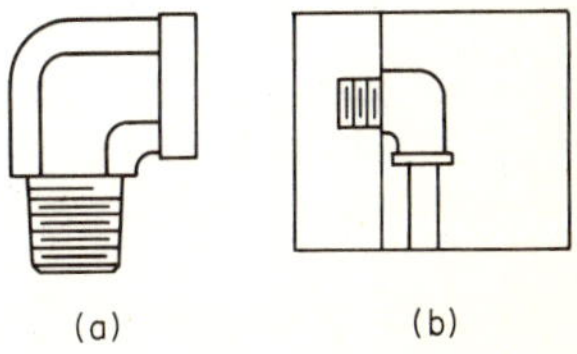


FIG. 6-4 *90-deg street elbow.*

45-DEG STREET ELBOW (Fig. 6-5*a*). This is the same as a street elbow, except that the line is brought in at a 45-deg angle from the port (Fig. 6-5*b*).

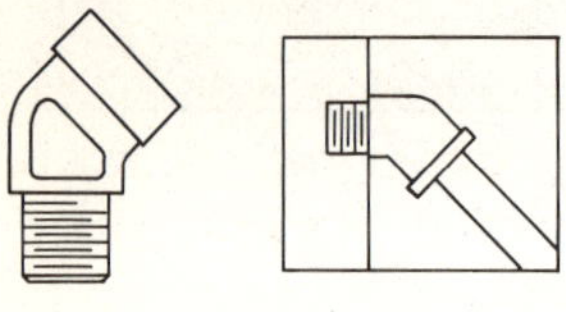

(a) (b)

FIG. 6-5 *45-deg street elbow.*

ELBOW (Fig. 6-6*a*). This is a 90-deg fitting with both ends female. It is used when the port connection and the line going into the fitting are male (Fig. 6-6*b*).

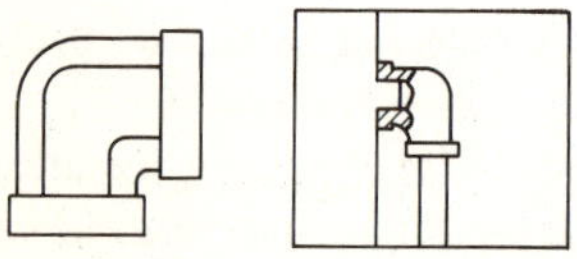

(a) (b)

FIG. 6-6 *90-deg female elbow.*

TEE (Fig. 6-7*a*). This fitting has three female ends. It enables the user to bring three lines together (Fig. 6-7*b*).

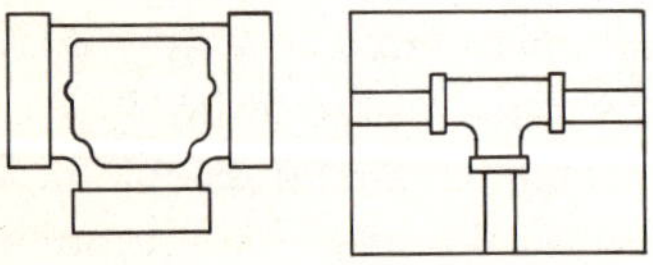

(a) (b)

FIG. 6-7 *Female tee.*

CROSS (Fig. 6-8*a*). This fitting has four female ends. It enables the user to bring four lines together (Fig. 6-8*b*).

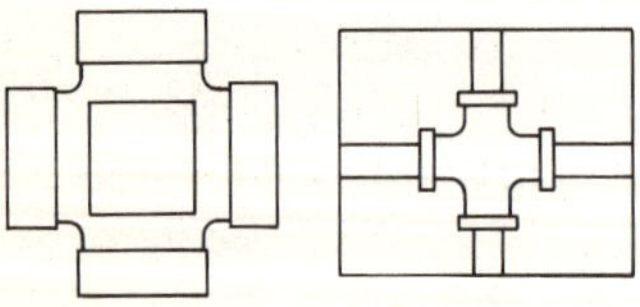

(a) (b)

FIG. 6-8 *Female cross.*

NIPPLE (Fig. 6-9*a*). This is a fitting having two male ends. The user can connect two female lines to this fitting (Fig. 6-9*b*).

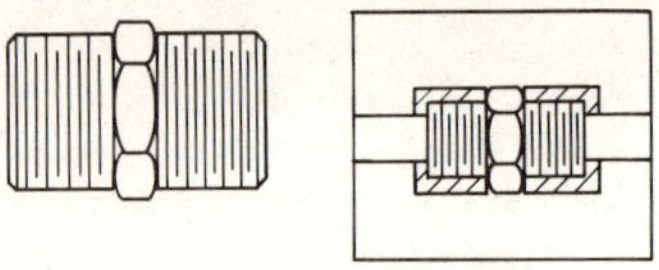

(a) (b)

FIG. 6-9 *Nipple.*

CONNECTOR (Fig. 6-10*a*). This fitting has two female ends. The user can connect two male lines to this fitting (Fig. 6-10*b*).

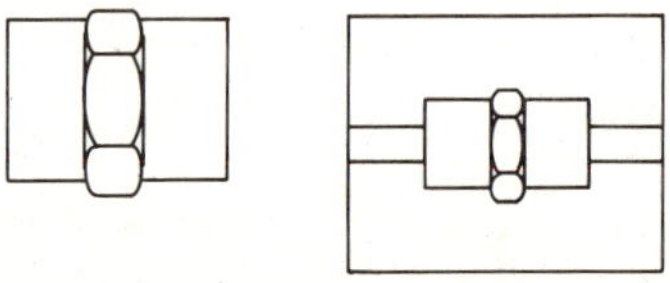

(a) (b)

FIG. 6-10 *Coupling.*

REDUCER BUSHING (Fig. 6-11*a*). This fitting has a male thread on the outside and an internally drilled female thread. It enables the user to reduce the port size or any other fitting size at any female end (Fig. 6-11*b*).

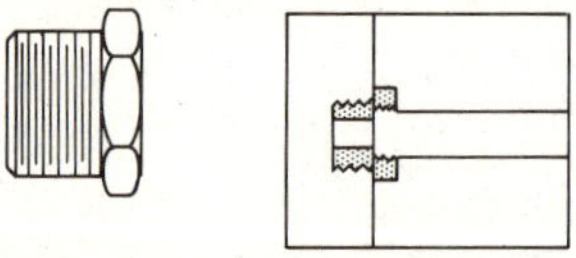

(a) (b)

FIG. 6-11 *Adapter—female to female.*

REDUCING CONNECTOR (Fig. 6-12*a*). This is essentially the same as a reducer bushing except that the user is able to reduce the male end of a port connector or any other fitting (Fig. 6-12*b*).

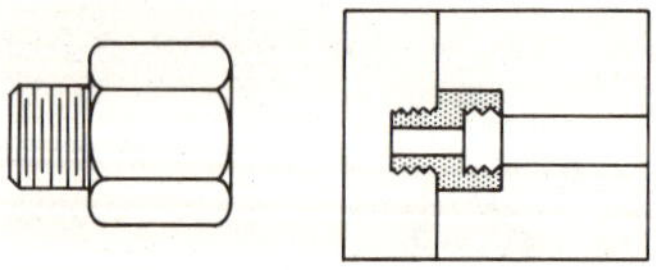

(a) (b)

FIG. 6-12 *Adapter—male to female.*

 

FEMALE UNION (Fig 6-13*a*). This fitting is the same as a connector except that it allows the user to attach the two female parts separately to two male lines and then connect the union together (Fig. 6-13*b*).

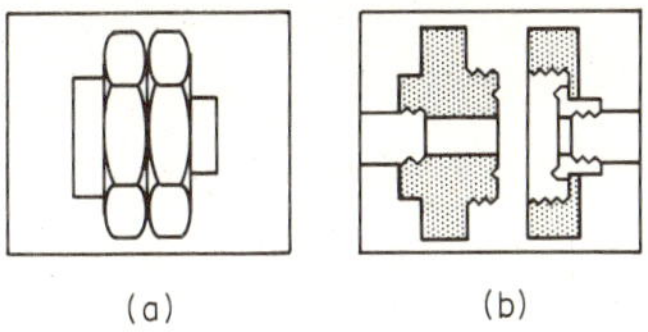


FIG. 6-13 *Female union.*

PLUG (Fig. 6-14). This fitting enables the user to seal off either a male or female port.

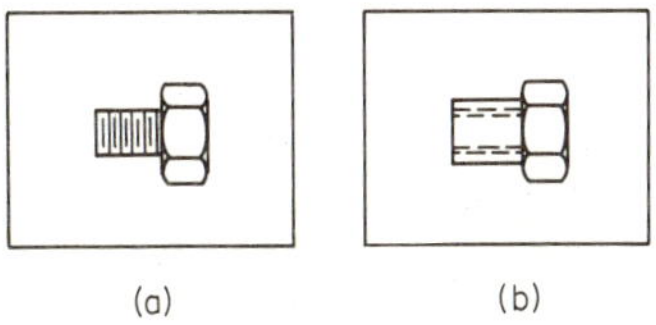


FIG. 6-14 *Plug.*

## HYDRAULIC TUBING

As system pressures increased, it became necessary to develop a method of plumbing hydraulic systems with some material other than iron pipe, since iron pipe does not have the ability to safely withstand high pressure in hydraulic systems, particularly those which have appreciable surge pressures. Hydraulic steel tubing was developed for this purpose. Unlike iron pipe, it can be readily bent or flared with production equipment in the shop or with hand tools in the field. Because it can be bent, such tubing requires fewer fittings, thus eliminating some sources of leakage. Flow restrictions are reduced, compactness is obtained, appearance improved, and maintenance cost reduced. Hydraulic tubing can withstand higher burst pressures than iron pipe, which gives a greater safety factor to the system (see Table 6-2).

System pressures, fluid velocity, flow, and the type of fluid used are some of the major factors which determine the kind and size of tubing a system will require. There are two types of hydraulic tubing presently being used—seamless tubing and electric welded tubing. Both are made of high ductile, soft annealed carbon steel approved in JIC specifications for application in hydraulic equipment.

### Tubing Measurement

The nominal dimension of tubing is given as a fraction in inches or a dash number, both of which refer to the outside diameter of the tube. The dash number represents the outside diameter of the tube in sixteenths of an inch (see Table 6-3). For example, a ½-in. tube would be designated by dash size No. –08, which indicates that the tube's O.D. is $\frac{8}{16}$, or ½, in. The approximate I.D. would be $\frac{13}{32}$ in.

### *Terms Used in Determining Requirements for Tubing*

BURST PRESSURE. This is the amount of pressure applied on the material which would cause it to burst (fail). For example, a ½-in. tube having a burst pressure of 8000 psi would fail when required to hold anything over 8000 psi.

WORKING PRESSURE. This is the amount of pressure under which the system normally operates. For example, if a system is operating for most of the time at 1000 psi, the working pressure is 1000 psi.

SAFETY FACTOR. This represents the margin of safety between the working pressure and the possible failure of the system. It is expressed in ratio terms. For example, a system operating at 2000 psi and having a required safety factor of 4 to 1 would have a burst pressure not less than 8000 psi. That is,

$$\text{Working pressure (2000)} \times \text{safety factor (4:1)} = \text{burst pressure (8000 psi)}$$

TABLE 6-2 BURST PRESSURES OF HYDRAULIC TUBING

| SEAMLESS TUBING | | | WELDED TUBING | | |
|---|---|---|---|---|---|
| *Tube O.D.* | *Wall thickness* | *Burst pressure, psi* | *Tube O.D.* | *Wall thickness* | *Burst pressure, psi* |
| ¼ | 0.035 | 14,000 | ¼ | 0.035 | 14,000 |
| | 0.035 | 9,330 | | 0.035 | 9,330 |
| ⅜ | 0.049 | 13,100 | ⅜ | 0.049 | 13,100 |
| | 0.065 | 17,335 | | 0.065 | 17,335 |
| | 0.035 | 7,000 | | 0.035 | 7,000 |
| ½ | 0.049 | 9,800 | ½ | 0.049 | 9,800 |
| | 0.065 | 13,000 | | 0.065 | 13,000 |
| | 0.042 | 6,720 | | 0.035 | 5,600 |
| ⅝ | 0.049 | 7,840 | ⅝ | 0.049 | 7,840 |
| | 0.065 | 10,400 | | 0.065 | 10,400 |
| | 0.035 | 4,665 | | 0.049 | 6,530 |
| ¾ | 0.049 | 6,530 | ¾ | 0.065 | 8,660 |
| | 0.065 | 8,660 | | 0.095 | 12,665 |
| | 0.065 | 7,430 | | 0.049 | 6,720 |
| ⅞ | 0.095 | 10,855 | ⅞ | 0.065 | 7,430 |
| | | | | 0.095 | 10,855 |
| 1 | 0.065 | 6,500 | 1 | 0.065 | 6,500 |
| | 0.095 | 9,500 | | 0.095 | 9,500 |
| 1¼ | 0.095 | 7,600 | 1¼ | 1.09 | 8,720 |

TABLE 6-3 EXAMPLES OF DASH NUMBERS FOR HYDRAULIC TUBING

| *Dash size no.* | *Tubing O.D.** | *Approximate I.D.†* |
|---|---|---|
| –04 | ¼ | 3/16 |
| –05 | 5/16 | ¼ |
| –06 | ⅜ | 5/16 |
| –08 | ½ | 13/32 |
| –10 | ⅝ | ½ |
| –12 | ¾ | ⅝ |
| –14 | ⅞ | ¾ |
| –16 | 1 | ⅞ |
| –20 | 1¼ | 1⅛ |
| –24 | 1½ | 1⅜ |
| –32 | 2 | 1 13/16 |

* Specifications of tubing O.D. increase in increments of: 1/16 in. from ⅛ in. through ⅜ in.; ⅛ in. from ½ in. through 1 in.; ¼ in. from 1 in. through 1½ in.; ½ in. from 1½ in. through 2 in.

† I.D. varies with wall thickness of tubing.

PROOF PRESSURE. This is a test measurement for determining the reliability of the material. It is in a ratio of 2 to 1 with the working pressure. For example, if proof pressure were 2000 psi, the working pressure would be 1000 psi.

$$\text{Proof pressure (2:1)} = \frac{\text{proof pressure (2000)}}{\text{working pressure (1000)}} \text{ psi}$$

### Tubing Wall Thickness

The thickness of the tubing wall is related to the pressures under which the system will operate (i.e., working pressure). JIC standards for pressure tubing are as follows:

1. When the system is operating at pressures from 0 to 1000 psi, the tubing must have an 8 to 1 safety factor. This means that on a 500-psi operating system all tubing

 

must have a burst pressure of no less than 4000 psi.

2. When the operating pressure of the system is from 1000 to 2500 psi, the safety factor for the tubing must be 6 to 1. On a 1500-psi operating system the tubing must have a burst pressure of no less than 9000 psi.
3. For systems operating at above 2500 psi, the safety factor must be 4 to 1. On systems operating at 3000-psi working pressure, all tubing must have a burst pressure of no less than 12,000 psi. Flow rate (in gallons per minute) is based on maximum flow of 15 ft/sec (this applies to Tables 6-4 and 6-5).

### Tubing Materials (Other than Steel)

Copper or plastic tubing may be used in hydraulic or pneumatic systems operating at low pressures (up to 150 psi).

TABLE 6-4 JIC PIPE AND TUBING SPECIFICATIONS FOR 0 TO 1000 PSI*

| *Flow rate, gpm* | *Valve size* | *Pipe schedule* | *Tubing O.D.* | *Wall thickness* |
|---|---|---|---|---|
| 1 | ⅛ | 80 | ¼ | 0.035 |
| 1.5 | ⅛ | 80 | 5/16 | 0.035 |
| 3 | ¼ | 80 | ⅜ | 0.035 |
| 6 | ⅜ | 80 | ½ | 0.049 |
| 10 | ½ | 80 | ⅝ | 0.049 |
| 20 | ¾ | 80 | ⅞ | 0.072 |
| 34 | 1 | 80 | 1¼ | 0.109 |
| 58 | 1¼ | 80 | 1½ | 0.120 |

* The safety factor is 8 to 1.

TABLE 6-5 JIC PIPE AND TUBING SPECIFICATIONS FOR 1000 TO 2500 PSI*

| *Flow rate, gpm* | *Valve size* | *Pipe schedule* | *Tubing O.D.* | *Wall thickness* |
|---|---|---|---|---|
| 25 | ¼ | 80 | ⅜ | 0.049 |
| 6 | ⅜ | 80 | ⅝ | 0.095 |
| 10 | ½ | 80 | ¾ | 0.120 |
| 18 | ¾ | 80 | 1 | 0.148 |
| 30 | 1 | 80 | 1¼ | 0.180 |
| 42 | 1¼ | 160 | 1½ | 0.220 |

* The safety factor is 6 to 1.

COPPER TUBING. Copper tubing is used extensively in the automotive industry for fuel lines and low-pressure hydraulic lines. It is also used almost exclusively in refrigeration. It is generally best suited for low-pressure applications where there is a minimum amount of surge present, or where the material conveyed is not compatible with iron.

PLASTIC TUBING. Plastic tubing is found in many pneumatic or low-pressure hydraulic systems. Plastic tubing for fluid applications is made from three basic materials: nylon, polyethylene, and vinyl (polyvinyl chloride).

Unreinforced nylon tubing is used at low hydraulic pressures. Nylon tubing has good abrasion and impact resistance, is not affected by fluids, and is good for temperatures from −100 to +225°F. It is generally recommended for pressures up to 250 psi. Compared with metals or other plastics, nylon has a high strength per unit weight; however, temperature and moisture conditions affect its burst strength. An increase in temperature or moisture content lowers the burst strength of the tube.

Polyethylene tubing is another form of plastic tubing used in fluid applications. It is used for liquids, and for air and gas lines, as well as in pneumatic applications. It is more flexible and has a lower initial cost than nylon tubing, but it has a lower burst pressure and a narrower temperature range (−60 to +175°F).

Vinyl tubing (polyvinyl chloride) was especially formulated for low-pressure pneumatic applications. It has a greater flexibility than polyethylene tubing and can be produced in larger sizes, but has the disadvantage of having a lower operating temperature range (+10 to +100°F).

### Clamping of Tubing

Tubing must be supported securely by evenly spaced clamps (Table 6-6). Care must be exercised in order not to damage tubing at points of contact with clamps.

TABLE 6-6 SUPPORT CLAMP SPACING

| *Tube O.D., in.* | *Approx. tubing length between supports, ft* |
|---|---|
| 1/4, 5/16, 3/8, 1/2 | 3 |
| 5/8, 3/4, 7/8 | 4 |
| 1 | 5 |
| 1 1/4, 1 1/2, 2 | 7 |

## Tube Fittings

Unlike iron pipe, which is threaded on the outside, tubing is not threaded, but instead is connected with fittings. Except for welded joints, all tube fittings connect to the tubing with a union which lets the tubing remain stationary while the fitting is tightened. The important difference among various types of fittings is in the method of sealing.

There are two methods of connecting tubing with tube fittings—by *flared* or *flareless* fittings. The latter are also known as compression fittings.

FLARED CONNECTIONS. With this type of connection, the tubing end is flared and a matching surface on the fitting is tightened to it (Fig. 6-15*a*). Fittings of this type are made with a 45- or 37-deg flare angle.

The 45-deg flare fitting is used for low-pressure work and in automotive service (for example, air brakes and fuel-oil lines). The Society of Automotive Engineers (SAE) specifies the 45-deg fitting for these purposes. Generally, these fittings are brass connected to copper tubing. They are two-piece reusable fittings consisting of the fitting body and a matching nut, with the flare end of the tubing in between the two.

The 37-deg flare fitting is used for high-pressure hydraulic applications. Specifications for it are given by the JIC. It provides a longer sealing service and less flare distortion and is primarily used with hydraulic steel tubing. This fitting is a three-piece reusable fitting which consists of the fitting body, a sleeve which matches against the flare end, and a nut which pushes the sleeve against the flare and holds the fitting body (Fig. 6-15*b*). Since the sleeve is used to match against the flare, a shorter nut than that used with the 45-deg fitting can be used; the seal is between the flare and the fitting body. This method allows tube bends closer to the fitting, which is then less affected by vibration.

FLARELESS FITTINGS. Flareless, or compression, fittings are attached to the tube by compressing a metal ring (sleeve or ferrule) or elastomeric seal around the tube (Fig. 6-15*c*). Fittings using the metal ring are sealed by forcing this ring to bite into, or onto, the tube. Some compression fittings do not bite into the tube, such as those used with copper or plastic tubing. However, the large tube contact area of the metal ring, or sleeve, controls tube distortion and holds the fitting in place.

Elastomeric sealing compression fittings seal with an O ring, rubber bushing, or teflon ring in place of a metal seal. The soft seals provide a cushion and permit limited vibration. Pressure ratings are lower for this type of seal.

In general, all of the flareless fittings operate by compressing some material against the tube; the variations depend on whether the material bites into the tube or is sealed by a compressed surface area. The advantages of compression fittings are the fact that the tubing requires no flare, thus elimi-

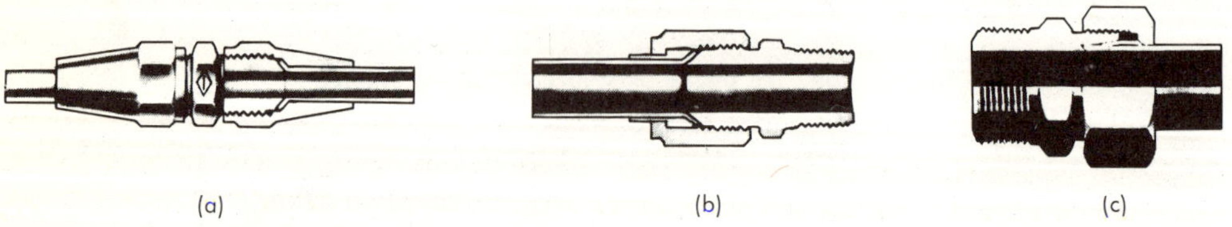


FIG. 6-15 *(a) Flared two-piece. (b) Flared three-piece. (c) Flareless compression three-piece.*

 

nating work, and the fact that the tubing can be bent to a smaller radius. The disadvantage is that the metal ring (sleeve) is not reusable. Once it has been applied to the tube, it is distorted, and a new sleeve is required in order to reuse the fitting.

## HYDRAULIC HOSES

At the same time that iron pipe and tubing were being developed and improved for plumbing hydraulic systems, pressure hose was also being developed for this purpose.

The primary advantage in the use of hose rather than tubing or pipe lies in the flexibility of hose. Neither iron pipe nor tubing is flexible; thus motion within systems using them is precluded. Also, hose can withstand vibration in a system far exceeding that which can be withstood by a rigid connection.

Where parts of the system move in relation to one another, hose is the only application that can be used, and even when the parts do not move, hose will absorb whatever vibration that may occur.

In addition to its flexibility, hose is easier to install and take apart in hydraulic systems. Unlike tubing or pipe, it does not have to be bent with tools, and extra fittings do not have to be installed wherever a turn is made. The disadvantages of using hose are its initial cost, its inability to withstand abrasion, and the fact that it is subject to weathering.

### Hose Construction

Air or hydraulic hose is constructed of an inner tube, a reinforcement, and an outer protective cover (see Fig. 6-18).

TABLE 6-7 THREAD SIZES FOR FITTINGS INSERTED INTO FLARED TUBING*

| PIPE SIZE I.D., IN. | SAE O-RING BOSS† | TUBE O.D., IN. | DASH NO. | THREAD SIZE | |
|---|---|---|---|---|---|
| | | | | *JIC 37-deg flare* | *SAE 45-deg flare* |
| 1/8 | 5/16–24 | 1/8 | –02 | 5/16–24 | 5/16–24 |
| | 3/8–24 | 3/16 | –03 | 3/8–24 | 3/8–24 |
| 1/4 | 7/16–20 | 1/4 | –04 | 7/16–20 | 7/16–20 |
| | 1/2–20 | 5/16 | –05 | 1/2–20 | 1/2–20 |
| 3/8 | 9/16–18 | 3/8 | –06 | 9/16–18 | 5/18–18‡ |
| 1/2 | 3/4–16 | 1/2 | –08 | 3/4–16 | 3/4–16 |
| | 7/8–14 | 5/8 | –10 | 7/8–14 | 7/8–14 |
| 3/4 | 1 1/16–12 | 3/4 | –12 | 1 1/16–12 | 1 1/16–12‡ |
| | 1 3/16–12 | 7/8 | –14 | 1 3/16–12 | 1 1/16–14‡ |
| 1 | 1 5/16–12 | 1 | –16 | 1 5/16–12 | |
| 1 1/4 | 1 5/8–12 | 1 1/4 | –20 | 1 5/8–12 | |
| 1 1/2 | 1 7/8–12 | 1 1/2 | –24 | 1 7/8–12 | |
| 2 | 2 1/2–12 | 2 | –32 | 2 1/2–12 | |

* Pipe thread size (column 1) is for the thread either on the fitting or the port where the fitting is to be inserted. SAE O-ring-boss thread (column 2) is for either the fitting or the port where the fitting is to be inserted. Both the JIC and SAE (column 3) thread sizes are for flared-tube end. This is the size of the fitting nut and flared fitting end, or hose fitting and female swivel end. (Refer to Fig. 6-16 for all flared ends on tube fittings.

† On the SAE O-ring boss (JIC 37-deg flare and SAE 45-deg flare), the number after the thread size refers to the number of threads per inch.

‡ These are the only thread sizes that are not the same for both the JIC and SAE. The others are interchangeable.

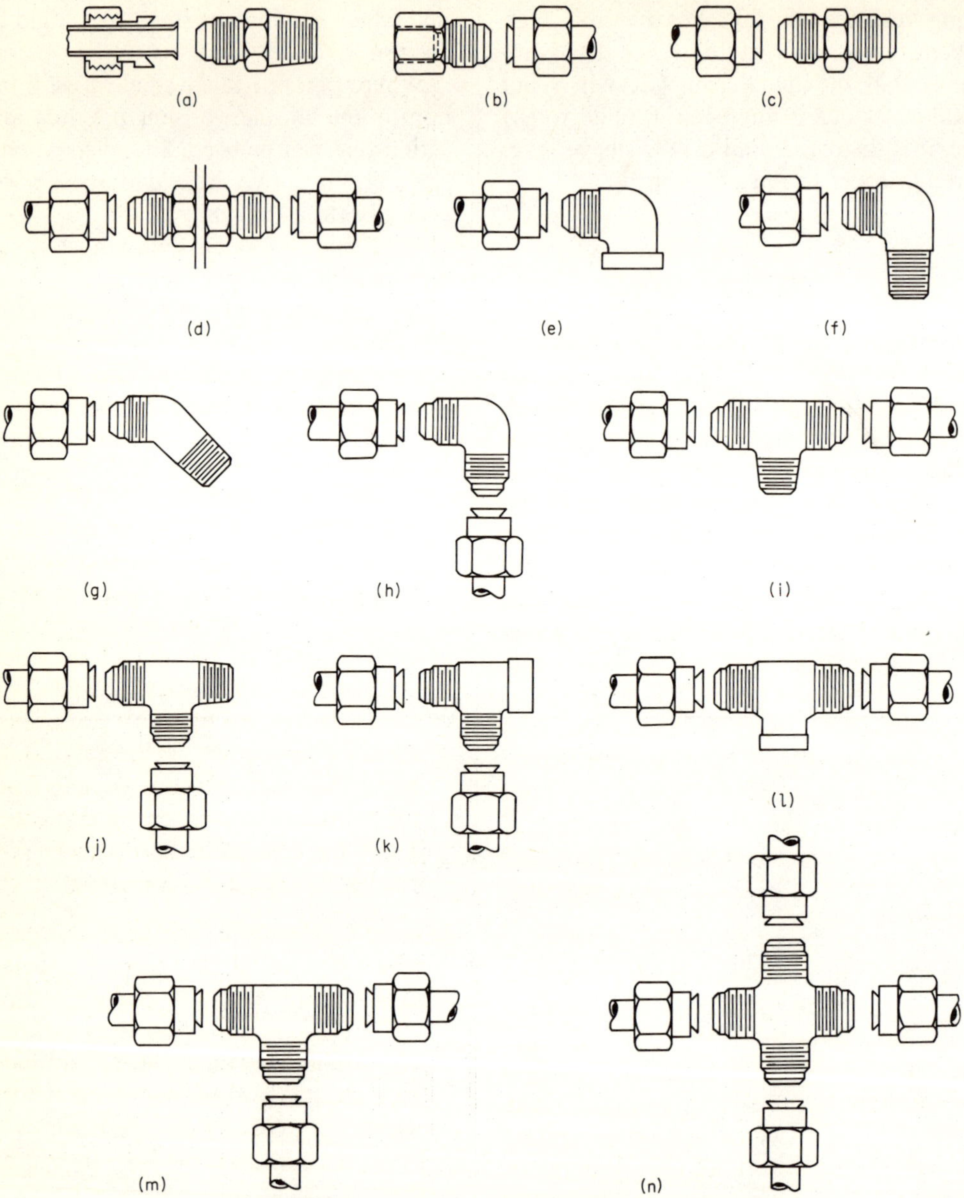


FIG. 6-16 *(a) Male straight connector. (b) Female straight connector. (c) Union–tube to tube. (d) Bulkhead union–tube to tube. (e) Female elbow. (f) Male elbow. (g) Male 45-deg elbow. (h) Union elbow. (i) Male branch tee. (j) Male run tee. (k) Female run tee. (l) Female branch tee. (m) Union tee. (n) Cross.*

THE INNER TUBE. The inner tube is the portion of the hose which carries the hydraulic fluid being conveyed in the system. The rubber compound used in the inner tube determines its compatibility with the fluids conveyed and its operating-temperature range. Hydraulic-hose inner tubes used for conveying fluids are generally made of one of the following:

1. *Buna N* is a synthetic rubber which can be used with petroleum oils but not with

 

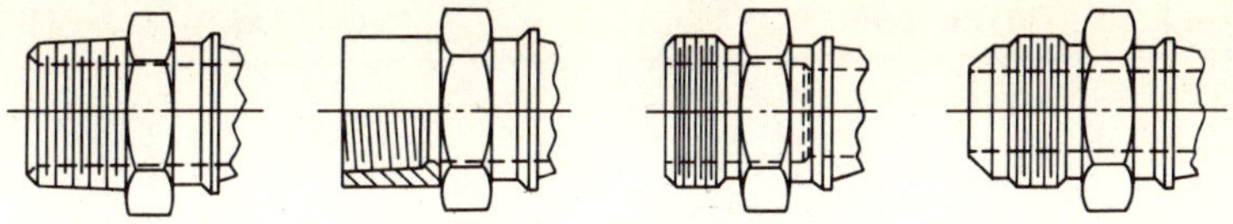


FIG. 6-17 *Variety of port ends which can be combined with flare or flareless connectors. (a) Male pipe thread. (b) Female pipe thread. (c) Male O-ring straight thread (SAE Boss). (d) Male 37-deg JIC straight-thread flare.*

phosphate-ester-based nonflammable hydraulic oils. It has the greatest resistance to swelling when used with petroleum oils. This rubber compound will not absorb the oil; this would decrease the inside diameter of the inner tube, thus causing a restriction to the flow of oil. It is the type commonly used in hydraulic hoses and gasoline pump hoses (see Table 6-8). It generally has a temperature range of $-40$ through $+200°F$, although at additional cost it can be compounded so that its temperature range is increased to $-65$ through $+300°F$. It has very poor resistance to abrasion and weathering; therefore, while it makes an excellent inner tube for hydraulic hose, it is an unsatisfactory outer cover.

2. *Neoprene* is a synthetic rubber compound developed by the Du Pont Chemical Corporation. Like Buna N, it is oil resistant. Unlike Buna N, however, it is not resistant to aromatic hydrocarbons (gasoline) and, therefore, is not used in gasoline pump hoses. It does have an excellent resistance to abrasion and has good characteristics. It is used in many hydraulic oil service conditions. When used with oils, moderate, but not excessive, swelling will occur with time. It makes an excellent outer cover where there are oily conditions existing on the outside. It can be used as the inner tube in a hose.

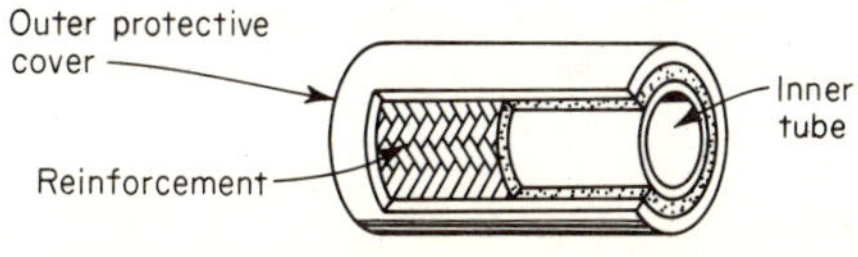


FIG. 6-18 *Component parts of hose.*

3. *Natural rubber* or *GRS* (simulated natural rubber compound) is used in air lines where there is little or no oil present, and it can also be used satisfactorily with hydraulic fluids which have a water glycol base. It is not oil resistant and will swell to over 20 percent of its original size if used with hydraulic fluids which have a petroleum base. It has excellent abrasion resistance. This resistance is better than that of Neoprene, and natural rubber makes an excellent outer protective cover for hydraulic hose except where outside oil conditions are present.

4. *Butyl rubber* is a synthetic compound which has no oil resistance. It deteriorates rapidly when in contact with petroleum-based fluids. It is, however, the only rubber available which is compatible with the phosphate-ester-based nonflammable hydraulic fluids. (When this rubber is used as an inner tube in hydraulic hose, a green outer cover indicates that it should be used only with phosphate-ester-based fluids.)

THE REINFORCEMENT. The reinforcement is part of the hose (see Fig. 6-18) which is placed over the inner tube and which provides the structural strength for the pressure applied on the hose. It is commonly called the braid of the hose. There are five types which are commonly used in hydraulic hose: rayon, synthetic yarn, nylon, wire, and spiral wire.

TABLE 6-8 INNER-TUBE COMPOUNDS AND FLUID CHARACTERISTICS*

| *Fluid* | *Neoprene* | *Buna N* | *Butyl* | *Natural rubber or GRS* |
|---|---|---|---|---|
| Acetylene | C | $C^2$ | B | C |
| Air | A | A | A | A |
| Air (hot) | C | C | A | C |
| Alcohols, aliphatic | B | A | A | A |
| Ammonia gas | B | B | B | B |
| Ammonia liquid (anhydrous) | B | B | B | $C^2$ |
| Butane | A | A | D | $C^2$ |
| Diesel fuel | B | $B^2$ | D | D |
| Ethylene glycol (antifreeze) | B | A | A | A |
| Freon 12 | C | $C^2$ | C | D |
| Freon 22 | C | C | D | D |
| Fuel oil | B | A | D | D |
| Gasoline, refined | B | A | D | D |
| Houghto safe—600 series (hydraulic fluid-water glycol base) | A | A | A | A |
| Houghto safe—1000 series (hydraulic fluid-phosphate ester base) | C | D | A | D |
| Hydraulic fluid, petroleum base | B | A | D | D |
| Hydraulic fluid, water glycol base | A | A | A | A |
| Hydraulic fluid, phosphate ester base | C | D | A | D |
| Hydraulic fluid, chlorinated hydrocarbon | D | D | D | D |
| Kerosene | B | A | D | D |
| Lubricating oils, petroleum base | B | A | D | D |
| Lubricating oils, diester base | D | B | D | D |
| Mineral oil | B | A | D | D |
| Petroleum oils, refined | B | B | D | D |
| Skydrol series | D | D | A | D |
| Steam | D | D | C | A |
| Water | A | A | A | A |
| Irus 902 hydraulic fluid (water-oil emulsion) | A | A | D | D |

* Key:
A—Excellent; little or no swelling
B—Good; swelling or softening is moderate
C—Fair; conditional service may be expected
D—Unsatisfactory; not recommended

Rayon braid (Fig. 6-19) is the standard type used for low-pressure hydraulic applications. The burst pressures on hose with rayon-braid reinforcement range from 1000 psi in I.D. sizes of 3/16 to 3/8 in., down to 300 psi in I.D. sizes of 3/4 in.

The pressures used with rayon-braid hose can be increased by adding more layers of rayon, although only two layers of rayon braid are commonly used in hydraulic hose; this provides a 1¼ in. I.D. hose with up to 700 pounds burst pressure. An advantage to this hose is its flexibility. The minimum

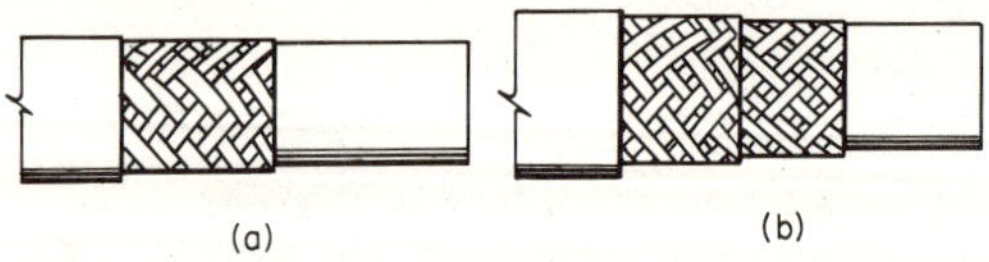


FIG. 6-19 *(a) One-rayon braid. (b) Two-rayon braid.*

 

TABLE 6-9 RAYON HOSE PRESSURE AND BENDING RADIUS

| SIZE | | PRESSURE | | BEND RADIUS |
|---|---|---|---|---|
| *Hose I.D., in.* | *Hose O.D., in.* | *Recommended maximum working pressure, psi* | *Minimum burst pressure, psi* | *Minimum bending radius, in.* |
| 3/16 | 7/16 | 250 | 1000 | 2½ |
| ¼ | 17/32 | 250 | 1000 | 2½ |
| 5/16 | 19/32 | 250 | 1000 | 3 |
| ⅜ | 21/32 | 250 | 1000 | 4 |
| ½ | ⅞ | 200 | 800 | 4 |
| ⅝ | 1 1/32 | 100 | 400 | 5 |
| ¾ | 1⅛ | 75 | 300 | 6 |
| ¼ | 19/32 | 300 | 1200 | 3 |
| ⅜ | 23/32 | 300 | 1200 | 4 |
| ½ | 29/32 | 250 | 1000 | 5 |
| ¾ | 1¼ | 250 | 1000 | 6 |
| 1 | 1½ | 200 | 800 | 8 |
| 1¼ | 1 13/16 | 175 | 700 | 10 |

bending radius is smaller than that of other hydraulic hoses.

High-burst rayon is used with a specific high-tensile rayon-braid fabric which can withstand much higher pressures because it has a closer weave (Table 6-10). (Also see Fig. 6-20.)

TABLE 6-10 HIGH-BURST RAYON HOSE PRESSURE AND BENDING RADIUS

| SIZE | | PRESSURE | | BEND RADIUS |
|---|---|---|---|---|
| *Hose I.D., in.* | *Hose O.D., in.* | *Recommended maximum working pressure, psi* | *Minimum burst pressure, psi* | *Minimum bending radius, in.* |
| 3/16 | ½ | 1250 | 5000 | 3 |
| ¼ | 9/16 | 1250 | 5000 | 3 |
| ⅜ | ¾ | 1125 | 4500 | 4 |
| ½ | 15/16 | 1000 | 4000 | 5 |
| ¾ | 1¼ | 750 | 3000 | 6 |
| 1 | 1½ | 550 | 2250 | 8 |
| 1¼ | 1¾ | 375 | 1500 | 10 |

TABLE 6-11 BRAIDED YARN PRESSURE AND BENDING RADIUS

| SIZE | | PRESSURE | | BEND RADIUS |
|---|---|---|---|---|
| *Hose I.D., in. (±1/64″)* | *Hose O.D., in.* | *Recommended maximum working pressure, psi* | *Minimum burst pressure, psi* | *Minimum bending radius, in.* |
| ¼ | ½ | 250 | 1000 | 3 |
| ⅜ | ⅝ | 250 | 1000 | 3 |
| ½ | ¾ | 250 | 1000 | 5 |
| ⅝ | 29/32 | 250 | 1000 | 6 |
| ¾ | 1 1/32 | 250 | 1000 | 7 |

Synthetic yarn is a loose-weave fabric which is used on low-pressure hydraulic lines (Table 6-11). Its characteristics are similar to rayon-braid hose with one exception—it has the ability to hold a coupling internally. The yarn is woven such that it strives to maintain a consistent position under pressure and acts like a Chinese finger puzzle. Hose having this type of braid is commonly called "push-on" hose because a fitting can be pushed into it, and the yarn will grab the fitting and hold it (Fig. 6-21).

Single wire braid is the common reinforcement used in hose where high working pressures are involved (Table 6-12). Wire has a much higher tensile strength than rayon braids or synthetic yarns, and therefore has the ability to withstand higher pressures.

FIG. 6-20 *SAE 100R3.*

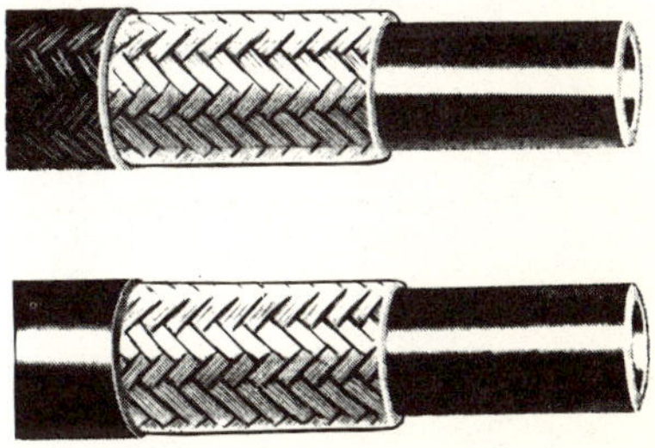

FIG. 6-21 *Push-on hose.*

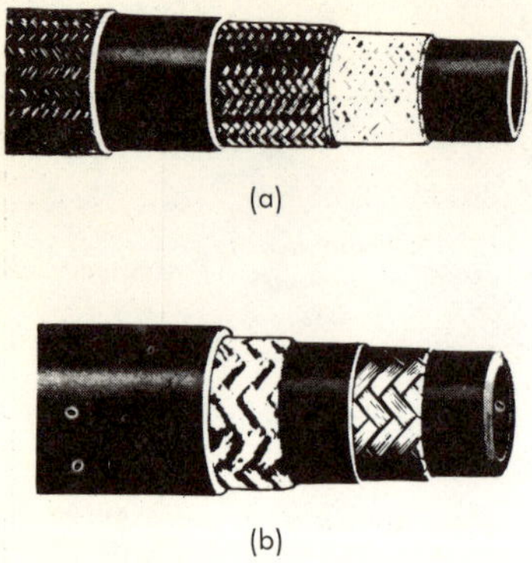

FIG. 6-22 *(a) SAE 100R5. (b) SAE 100R1.*

As with rayon braids, wire braids are used with either one- or two-layer construction. One wire braid reinforcement is referred to as single wire braid hose and is commonly used in systems operating at 1500 psi where flow volume permits a hose I.D. of 5/8 in. or less. SAE specifications for one wire braid hose are SAE 100RI (for hose with a rubber cover) and SAE 100R5 (for hose with a cotton yarn cover). (See Fig. 6-22.)

TABLE 6-12 SINGLE WIRE BRAID PRESSURE AND BENDING RADIUS

| SIZE | | PRESSURE | | BEND RADIUS |
|---|---|---|---|---|
| *Hose I.D., in.* | *Hose O.D., in.* | *Recommended maximum working pressure, psi* | *Minimum burst pressure, psi* | *Minimum bending radius, in.* |
| 3/16 | 33/64 | 3000 | 12000 | 3 |
| 1/4 | 37/64 | 2500 | 10000 | 3 3/8 |
| 5/16 | 43/64 | 2250 | 9000 | 4 |
| 13/32 | 49/64 | 2000 | 8000 | 4 5/8 |
| 1/2 | 59/64 | 1750 | 7000 | 5 1/2 |
| 5/8 | 1 5/64 | 1500 | 6000 | 6 1/2 |
| 7/8 | 1 15/64 | 800 | 3200 | 7 3/8 |
| 1 1/8 | 1 1/2 | 625 | 2500 | 9 |
| 1 3/8 | 1 3/4 | 500 | 2000 | 10 1/2 |
| 1 13/16 | 2 7/32 | 350 | 1400 | 13 1/4 |
| 3/16 | 1/2 | 3000 | 12000 | 4 |
| 1/4 | 5/8 | 2750 | 11000 | 4 |
| 5/16 | 11/16 | 2500 | 10000 | 5 |
| 3/8 | 25/32 | 2250 | 9000 | 5 |
| 1/2 | 29/32 | 2000 | 8000 | 7 |
| 5/8 | 1 1/32 | 1500 | 6000 | 8 |
| 3/4 | 1 3/16 | 1250 | 5000 | 9 1/2 |
| 7/8 | 1 5/16 | 1125 | 4500 | 11 |
| 1 | 1 1/2 | 1000 | 4000 | 12 |
| 1 1/4 | 1 13/16 | 625 | 2500 | 16 1/2 |
| 1 1/2 | 2 1/16 | 500 | 2000 | 20 |
| 2 | 2 5/8 | 375 | 1500 | 25 |

High-burst-pressure hose (Table 6-13) is constructed with a two wire braid layer construction. This is referred to by the Society of Automotive Engineers as SAE 100R2 (Fig. 6-23). The second layer of wire braids boosts the working pressure of the hoses by as much as 2000 lb in the smaller 1/4 in. and 3/16 in. I.D. sizes. There are two types of SAE 100R2, type A and type B, as shown in Fig. 6-23. Both have synthetic rubber outer covers and the same pressure rating; however, the construction of the reinforcement in SAE 100R2 type B has a spiral-wrapped inner-reinforced layer instead of braided wires.

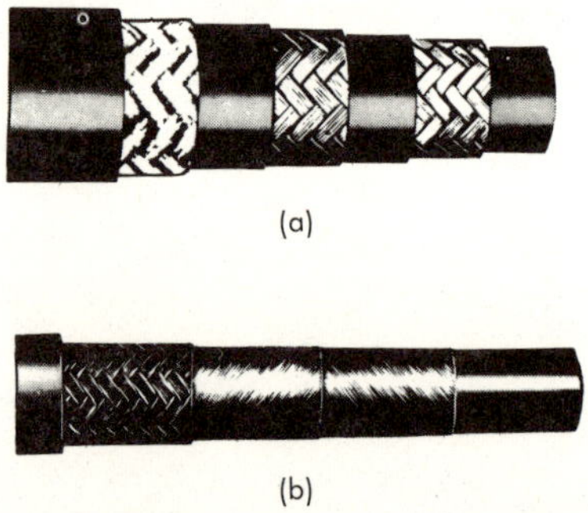

FIG. 6-23 *(a) SAE 100R2 type A. The cotton yarn pictured as the outer layer in the SAE 100R2 type A is not used as reinforcement, but is used to improve the adhesion between the outer rubber cover and the reinforcement. (b) SAE 100R2 type B.*

Spiral wire reinforcements are wires which are laid out in layers on the inner tube and are not braided. This design affords high pressure ratings, flexibility, and better impulse life. In addition, heavier wires can be used, which further increase the pressure ratings under which the hose can be worked. Figure 6-24 shows four and six spiral wire braid construction. Table 6-14 indicates their pressure ratings. Hoses with all spiral

 

TABLE 6-13 TWO WIRE BRAID PRESSURE AND BENDING RADIUS

| SIZE | | PRESSURE | | BEND RADIUS |
|---|---|---|---|---|
| *Hose I.D., in.* | *Hose O.D., in.* | *Recommended maximum working pressure, psi* | *Minimum burst pressure, psi* | *Minimum bending radius, in.* |
| 3/16 | 5/8 | 5000 | 20000 | 4 |
| 1/4 | 11/16 | 5000 | 20000 | 4 |
| 5/16 | 3/4 | 4250 | 17000 | 5 |
| 3/8 | 27/32 | 4000 | 16000 | 5 |
| 1/2 | 31/32 | 3500 | 14000 | 7 |
| 5/8 | 1 3/32 | 2750 | 11000 | 8 |
| 3/4 | 1 1/4 | 2250 | 9000 | 9 1/2 |
| 7/8 | 1 3/8 | 2000 | 8000 | 11 |
| 1 | 1 9/16 | 2000 | 8000 | 12 |
| 1 1/4 | 2 | 1625 | 6500 | 16 1/2 |
| 1 1/2 | 2 1/4 | 1250 | 5000 | 20 |
| 2 | 2 3/4 | 1125 | 4500 | 25 |
| 1/4 | 11/16 | 5000 | 20000 | 4 |
| 3/8 | 27/32 | 4000 | 16000 | 5 |
| 1/2 | 31/32 | 3500 | 14000 | 7 |
| 3/4 | 1 1/4 | 2250 | 9000 | 9 1/2 |
| 1 | 1 9/16 | 2000 | 8000 | 12 |
| 1 1/4 | 2 | 1625 | 6500 | 16 1/2 |
| 1 1/2 | 2 1/4 | 1250 | 5000 | 20 |
| 2 | 2 3/4 | 1125 | 4500 | 25 |

wire braid construction are generally found in the larger I.D. sizes ranging from 3/4 to 2 in.

Nylon braided construction can be used under higher working pressures than can rayon braid and does not have the scissoring effect that wire braids have. At the present time it is primarily used in smaller I.D. sizes, and especially with light mobile equipment or in power-line applications where a nonconductor of electricity is required.

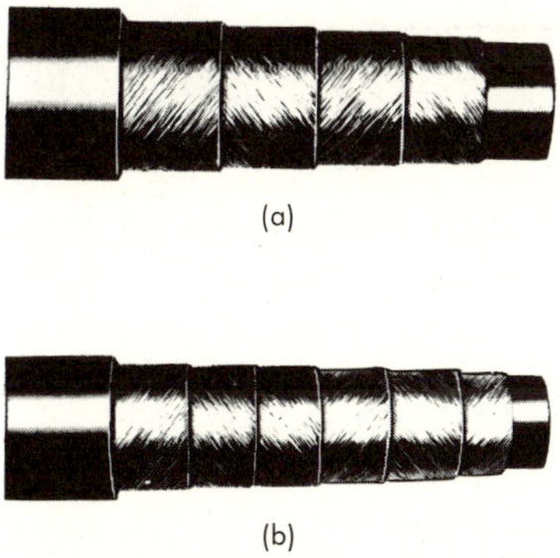

FIG. 6-24 *(a) Four-spiral and (b) six-spiral wire braid construction.*

TABLE 6-14 FOUR- AND SIX-SPIRAL WIRE BRAID PRESSURE AND BENDING RADIUS

| SIZE | | PRESSURE | | BEND RADIUS |
|---|---|---|---|---|
| *Hose I.D., in.* | *Hose O.D., in.* | *Recommended maximum working pressure, psi* | *Minimum burst pressure, psi* | *Minimum bending radius, in.* |
| 3/4 | 1 7/16 | 5000 | 20000 | 10 |
| 1 | 1 3/4 | 4000 | 16000 | 12 |
| 1 1/4 | 2 | 3000 | 12000 | 18 |
| 1 1/2 | 2 1/4 | 2500 | 10000 | 22 |
| 2 | 2 25/32 | 2500 | 10000 | 27 |
| 3/4 | 1 9/16 | 6250 | 25000 | 10 |
| 1 | 1 29/32 | 5000 | 20000 | 12 |
| 1 1/4 | 2 5/32 | 3500 | 14000 | 18 |
| 1 1/2 | 2 13/32 | 3000 | 12000 | 22 |
| 2 | 2 31/32 | 3000 | 12000 | 27 |

THE OUTER PROTECTIVE COVER. The outer protective cover of the hose protects the reinforcement from corrosion, abrasion, and other damage from weather or because of service conditions. As mentioned previously, Neoprene or synthetic GRS rubber are commonly used for this cover because they have excellent abrasion- and weather-resistant characteristics. Neoprene does have an advantage over synthetic GRS rubber because it has a resistance to swelling and softening caused by oil. GRS rubbers, however, have a greater abrasion resistance. Ambient conditions will dictate the type of outer cover best suited.

Cotton yarn is also used as a protective cover in the case of SAE 100R5 (one-wire braid hydraulic hose) and on one type of synthetic yarn push-on hose. Cotton-covered SAE 100R5 hydraulic hose has an advantage in that in putting on a reusable hydraulic hose fitting in the field, the cover does not have to be cut down to the wire (skived) to provide a

TABLE 6-15 MEASUREMENT OF SAE 100R5 COTTON-COVERED HOSE

| *Dash size no.* | *Hose I.D., in.* | *Tube size O.D., in.* | *Actual I.D., in.* |
|---|---|---|---|
| -03 | 3/16 | | |
| -04 | 1/4 | 1/4 | 3/16 |
| -05 | 5/16 | 5/16 | 1/4 |
| -06 | 3/8 | 3/8 | 5/16 |
| -08 | 1/2 | 1/2 | 13/32 |
| -10 | 5/8 | 5/8 | 1/2 |
| -12 | 3/4 | 3/4 | 5/8 |
| -16 | 1 | 1 | 7/8 |
| -20 | 1 1/4 | 1 1/4 | 1 1/8 |
| -24 | 1 1/2 | 1 1/2 | 1 3/8 |
| -32 | 2 | 2 | 1 13/16 |

holding grip for the coupling. However, it has a much shorter life because it is less resistant to abrasion and corrosion.

Polyethylene plastic is also used as a protective outer cover, primarily in hose which has a nylon-braid construction. It has a much greater resistance to oil, abrasion, and weathering than rubber, but it deteriorates rapidly in solvents and has a tendency to kink when it is not under pressure.

### Hose Application

Selection of hydraulic hose depends on the working pressure and surge pressures of the system in which it is being installed, the hydraulic or pneumatic fluid which is being conveyed, the temperature under which the system will operate, outside conditions, and the volume of flow which is required.

Hose is generally measured by its inside diameter and dash number. SAE 100R5 cotton-covered hose is the one exception; it is measured according to tube O.D. size (see Table 6-15).

Flow characteristics are determined through the use of a nomographic chart which enables the user to determine the I.D. hose required as listed in Table 6-16.

This chart (Table 6-17) is provided for rapid determination of hose size.

*Example:* At 10 gal/min, determine hose assembly size within recommended velocity range.

*Solution:* Put a straight edge at 10 gal/min on the left column and at 10 ft/sec (the center of the recommended velocity range) on the right column. Then the hose size in the center column nearest to the straight edge is the proper size. In this example, 3/4 in. is the proper size.

For suction lines, use recommended velocity for intake lines, and follow the same procedure.

The length of a hose assembly must provide sufficient length to prevent bending strain from localizing directly back of the couplings. (See, for example, the diagrams

TABLE 6-16 SIZE REQUIREMENTS FOR HYDRAULIC HOSE

| I.D. OF HOSE, IN. | B CONSTANT FOR STRAIGHT PORTION INCLUDING COUPLING, IN. | MIN. A, IN. | MINIMUM OVERALL LENGTH, IN. | |
|---|---|---|---|---|
| | | | *Fig. 16-24* | *Fig. 16-25* |
| 3/16 | 10 | 8 | 23 + 1/2T | 23 + T |
| 1/4 | 10 | 8 | 23 + 1/2T | 23 + T |
| 3/8 | 10 | 10 | 26 + 1/2T | 26 + T |
| 1/2 | 12 | 14 | 34 + 1/2T | 34 + T |
| 3/4 | 14 | 19 | 44 + 1/2T | 44 + T |
| 1 | 16 | 22 | 51 + 1/2T | 51 + T |
| 1 1/4 | 18 | 32 | 68 + 1/2T | 68 + T |
| 1 1/2 | 20 | 44 | 87 + 1/2T | 87 + T |
| 2 | 20 | 48 | 95 + 1/2T | 95 + T |

 

TABLE 6-17 FLOW CHARACTERISTICS NOMOGRAPHIC CHART

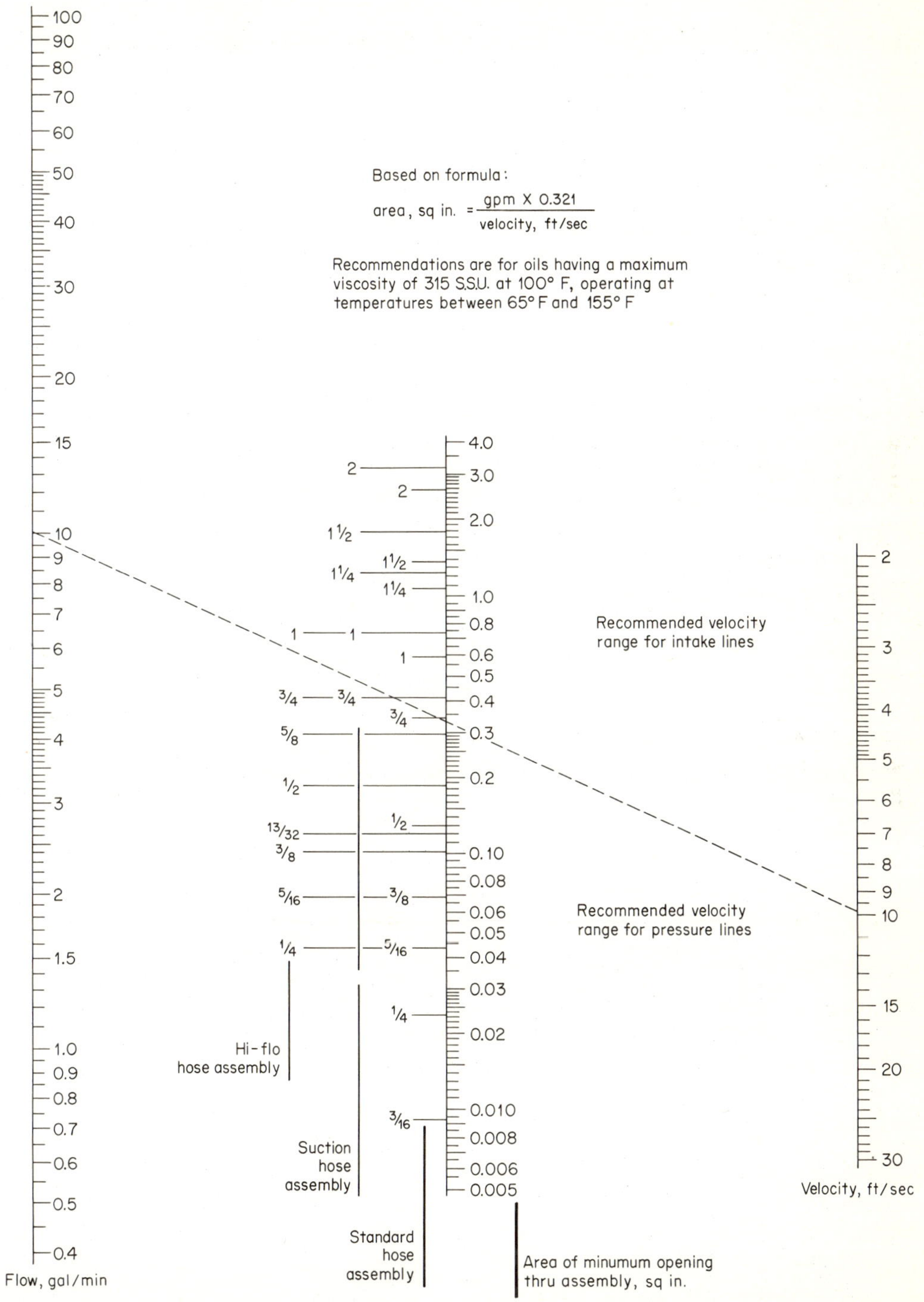

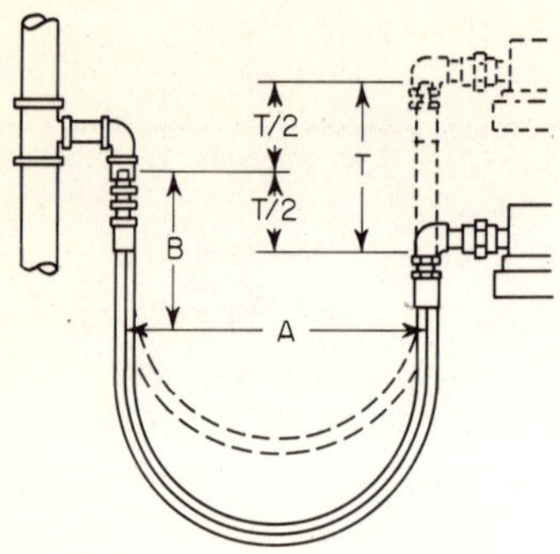


FIG. 6-25 *Hose installation.*

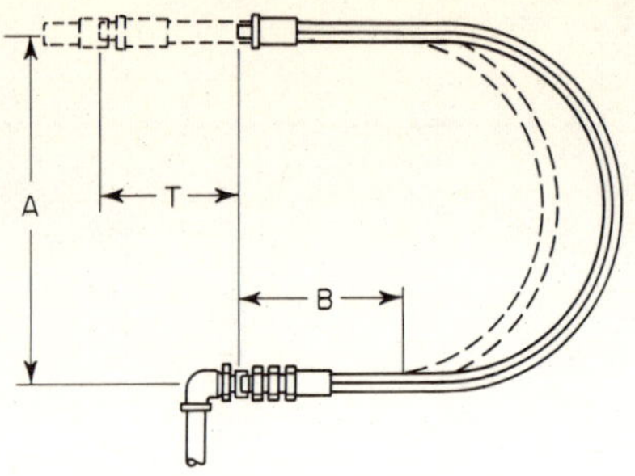


FIG. 6-26 *Hose installation.*

in Figs. 6-25 and 6-26.) Measurement B allows for a straight section of hose beyond the coupling to prevent concentration of bending. T designates the amount of travel. A indicates the smallest diameter to which hose should be bent.

Hose assemblies are measured from their overall length—the extreme end of one coupling to the extreme end of the other. Where elbow couplings are used, measurement should be made to the centerline of the sealing surface of the elbow end. The method of measurement should be specified. The following are some length tolerances for hose assemblies:

| *Length* | *Tolerance* |
|---|---|
| Up to 12 in. | ±1/8 in. |
| Above 12 in. | ±3/16 in. |
| Above 18 in. | ±1/4 in. |
| Above 36 in. | ±1 percent of length measured to nearest 1/8 in. |

### Hose Installation

There are seven rules for hose installation which, if followed, will provide longer service life with less maintenance problems.

1. Allow enough slack to avoid kinking the hose. Only the hose is flexible—not the fittings—so the hose run should be long enough to bend freely with machine action.
2. Avoid a taut run of hose, even where the end connections do not move in relation to each other. Pressure tends to bulge the hose and also shorten it, so allow some slack to avoid strain.
3. Use fittings to avoid unnecessarily long loops. This cuts down the length of line and makes a neater installation.
4. Avoid twisting the hose either during installation or during machine operation. Twisting weakens the hose and loosens the connections.
5. Design hose runs to avoid rubbing and abrasion. Clamps are often helpful to keep hoses away from moving parts; sometimes an extended end fitting will do the job. But if neither of these work, use a hose guard. This can be either a wire spring or a flat armor spring. These are available from hose suppliers especially for this purpose. Not only does the guard protect against abrasion, but it also relieves the strain caused by flexing at the coupling.

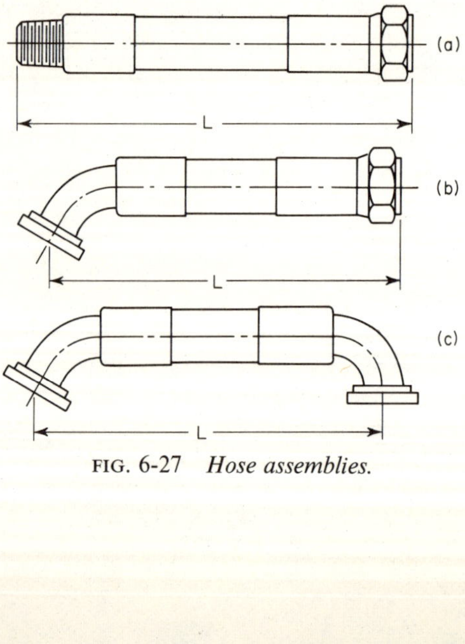


FIG. 6-27 *Hose assemblies.*

 

6. Keep hoses away from hot parts. High temperatures damage the outer cover and shorten hose life. If you can't route the hose away from the heat, insulate.
7. Follow manufacturer's specifications on minimum bend radius. Bend radius depends on hose construction, size, and pressure. It is specified for nonflexing applications at working pressures, and must be reduced for flexing. At lower pressures a tighter bend is permissible. By using the curve you can determine a factor to multiply rated bending radius according to flex angle.

## HOSE COUPLINGS

Hydraulic hose couplings were developed at the same time that hose usage was being applied to fluid power systems. Hose, like tubing, cannot be threaded and therefore requires a fitting before it can be attached to any other part of the system.

All couplings for hose (with the exception of push-on fittings, which will be discussed later) have three basic component parts: an insert, an outer shell, and a coupling end, which will hold the fitting to the hose and to the unit to which it is attached. Figure 6-28 illustrates these component parts.

### The Insert

The insert, or shank, of the coupling has barbs which act like fishhooks and grip the hose tube from the inside. Pressure in the hose pulls it away from the coupling, and therefore the barbs face the coupling end to grip the hose as it tries to pull away from the coupling. The insert and the coupling end are drilled through to allow fluid to pass through it.

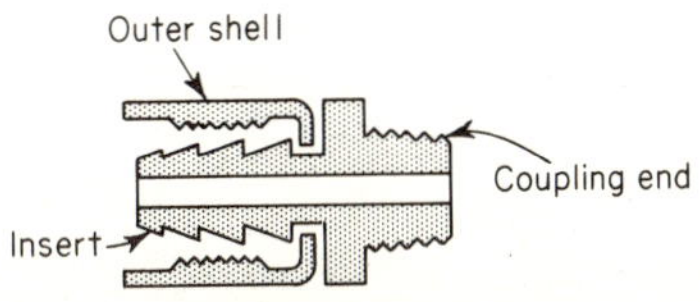


FIG. 6-28 *Hose fitting.*

### The Outer Shell

The outer shell pushes down on the insert, which clamps the hose between itself and the insert. This shell can be a complete unit, but in some cases (as in automobile radiator hoses) it is nothing more than a clamp applying the external force to the insert. For higher pressures, the outer shell and the inserts are increased in thickness and length to give a larger holding area.

### The Coupling End

The coupling end is the part of the fitting which is attached to the component, e.g., a pump, valve, or cylinder port. There are many varieties of coupling ends used today.

MALE PIPE COUPLING END. This type (Fig. 6-29) is found on over 50 percent of all hydraulic hose applications today. It is used in mobile and industrial hydraulic equipment. Male pipe coupling with NPTF Dryseal threads is an effective sealer at lower pressures. However, in order to install a hose assembly onto a piece of equipment, either the hose or component part (cylinder or motor) must be turned into the port and wrench-tightened. Frequently a swivel union adapter must be used in areas where the component parts cannot be turned. In this instance the hose is turned into one port, and the other end uses a swivel female NPTF adapter.

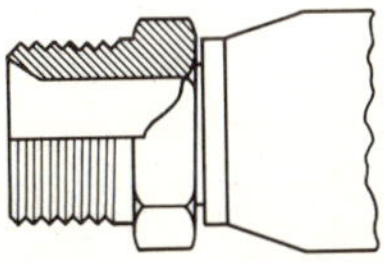

FIG. 6-29 *Male pipe.*

JIC FEMALE SWIVEL. This type (Fig. 6-30) is a 37-deg coupling end. It can be furnished in straight ends, and with 90- and 45-deg bends. This fitting is found on all hydraulic applications. The advantages in using the 37-deg female swivel fitting are that the female end can be turned into a 37-deg adapter (as shown under hydraulic tube fitting in this section) without turning the entire hose. It also has a metal-to-metal flared seat which, when it is tightened, provides a leakproof seal. These features are advantageous especially when longer and heavier hoses are needed to make connections.

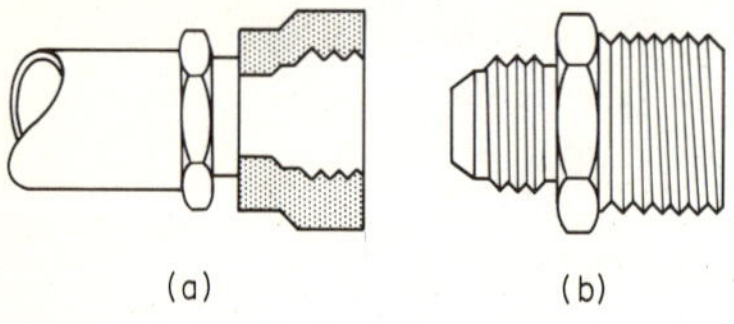


FIG. 6-30 *(a) JIC female swivel. (b) 37-deg adapter.*

45-DEG SAE SWIVEL. This swivel (Fig. 6-31) is essentially the same as the 37-deg flare JIC swivel, except that the flare set is at a 45-deg angle instead of a 37-deg angle. This fitting is used primarily on automotive applications. With the exception of the –06 and –12 tube sizes, the threads are interchangeable and a 45-deg flare fitting can be used with a 37-deg adapter or, in cases of extreme emergency, a 37-deg fitting can be used with a 45-deg adapter.

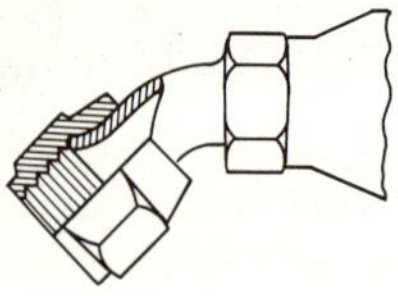

FIG. 6-31 *45-deg SAE swivel.*

SPLIT-FLANGE SAE BOLT ENDS. These (Fig. 6-32) are found primarily on heavy-duty machinery and particularly on large-size hose lines (1-in. to 3-in. I.D. hose sizes). This fitting seals with an O ring, which acts as a

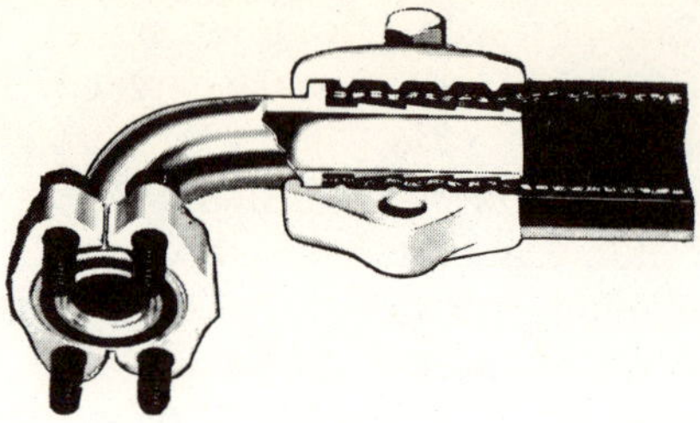

FIG. 6-33 *Split-flange SAE four-bolt ends.*

gasket under pressure with the four bolt flanges threaded into the port. The advantages of this fitting are that they provide high-pressure leakproof seals and can be installed in a confined area where access to the components is limited. The technician can position the flange ends and bring the split halves up to it, using a socket wrench to install the bolts. This coupling end can be furnished with straight ends, or bent to any degree required (Fig. 6-33). Figure 6-34 gives standard dimensions for split-flange assemblies.

MALE O-RING STRAIGHT-THREAD FITTING. This fitting (Fig. 6-35), also known as an O-ring boss fitting, also seals with an O-ring gasket. However, it is not necessary to use flanged halves and bolts to attach the end of the straight-thread fitting to a port, as it is with the flanged fitting. The fitting has straight threads which thread into the port, and the O ring is tightened against the body of the receiving component by a washer and a locknut which are directly behind the O ring. These are available in straight, 90-deg, and 45-deg bends.

INVERTED FLARE FITTINGS (Fig. 6-36). These are used for low pressures and are found primarily on automotive equipment.

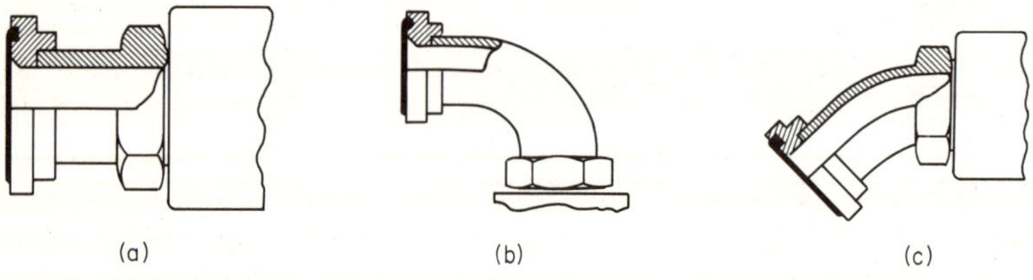


FIG. 6-32 *(a) Split-flange straight. (b) Split-flange 90-deg elbow. (c) Split-flange 45-deg elbow.*

 

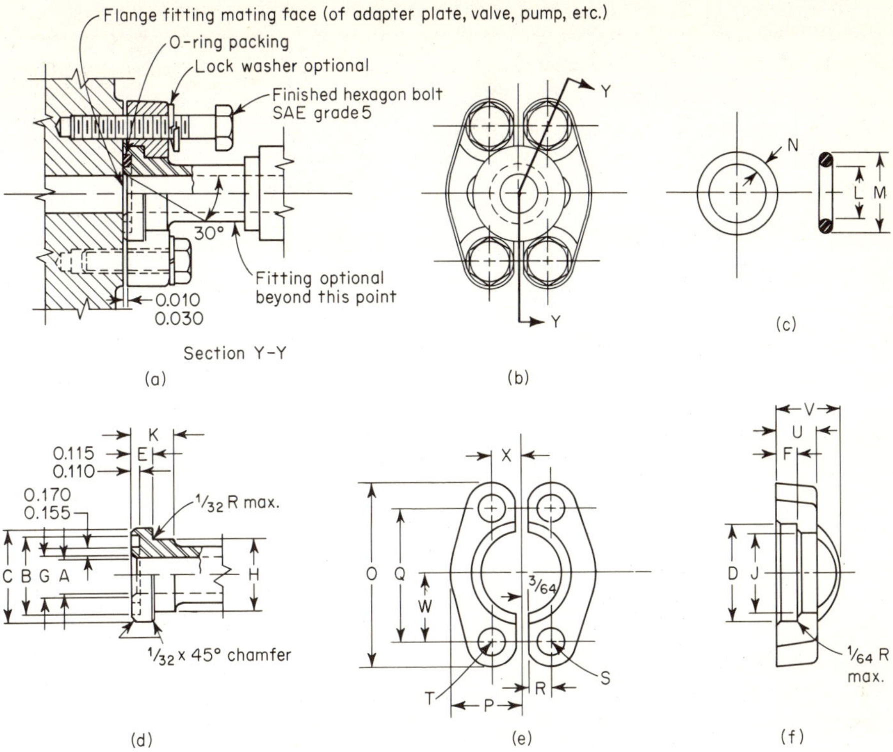


FIG. 6-34 *(a,b,c) Flanged connections. 6-33c O-ring dimensions shown for reference only. (d) Flanged head. (e,f) Split-flange clamp half. (These are dimensions for split-flange assemblies in accordance with SAE Handbook.)*

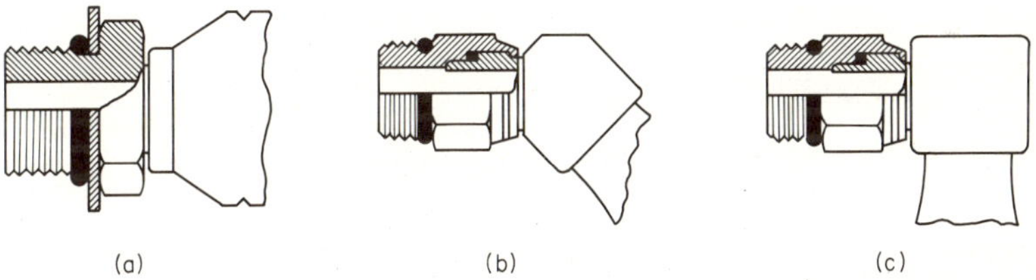


FIG. 6-35 *(a) Straight-thread O-ring seal. (b) 45-deg elbow, straight-thread, O-ring seal. (c) 90-deg elbow, straight-thread, O-ring seal.*

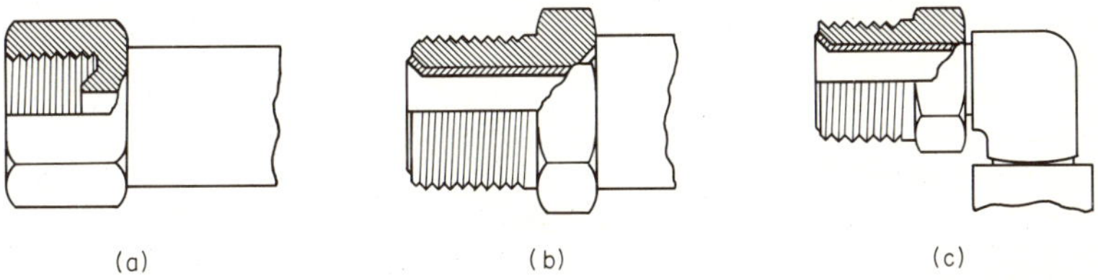


FIG. 6-36 *(a) Female inverted flare, fixed. (b) Male inverted flare, swivel. (c) Male elbow, inverted flare.*

TABLE 6-18 DIMENSIONS OF HYDRAULIC FLANGED CONNECTIONS

| *Flange size* | *A max.* | *B* | *C ±0.010* | *D ±0.010* | *E* | *F* | *G* | *H max.* | *J ±0.010* | *K* | *L I.D.* | *M O.D.* | *N dia.* |
|---|---|---|---|---|---|---|---|---|---|---|---|---|---|
| 1/2 | 1/2 | 1.005 | 1 3/16 | 1 7/32 | 0.270 | 0.250 | 9/16 | 15/16 | 31/32 | 1/2 | 3/4 | 1 | 1/8 |
| | | 1.000 | | | 0.260 | 0.240 | | | | | | | |
| 3/4 | 3/4 | 1.255 | 1 1/2 | 1 17/32 | 0.270 | 0.250 | 13/16 | 1 1/4 | 1 9/32 | 9/16 | 1 | 1 1/4 | 1/8 |
| | | 1.250 | | | 0.260 | 0.240 | | | | | | | |
| 1 | 1 | 1.565 | 1 3/4 | 1 25/32 | 0.320 | 0.300 | 1 1/16 | 1 1/2 | 1 17/32 | 9/16 | 1 5/16 | 1 9/16 | 1/8 |
| | | 1.560 | | | 0.310 | 0.290 | | | | | | | |
| 1 1/4 | 1 1/4 | 1.755 | 2 | 2 1/32 | 0.320 | 0.300 | 1 5/16 | 1 45/64 | 1 47/64 | 9/16 | 1 1/2 | 1 3/4 | 1/8 |
| | | 1.750 | | | 0.310 | 0.290 | | | | | | | |
| 1 1/2 | 1 1/2 | 2.125 | 2 3/8 | 2 13/32 | 0.320 | 0.300 | 1 9/16 | 1 63/64 | 2 1/64 | 5/8 | 1 7/8 | 2 1/8 | 1/8 |
| | | 2.115 | | | 0.310 | 0.290 | | | | | | | |
| 2 | 2 | 2.500 | 2 13/16 | 2 27/32 | 0.380 | 0.360 | 2 1/16 | 2 29/64 | 2 31/64 | 5/8 | 2 1/4 | 2 1/2 | 1/8 |
| | | 2.490 | | | 0.370 | 0.350 | | | | | | | |

| *Flange size* | *O* | *P ±1/32* | *Q ±0.010* | *R ±0.010* | *S Rad.* | *T Dia. hole* | *U* | *V* | *Hex bolt SAE grade 5* | *W* | *X* |
|---|---|---|---|---|---|---|---|---|---|---|---|
| 1/2 | 2 1/8 | 55/64 | 1 1/2 | 19/64 | 5/16 | 11/32 | 1/2 | 3/4 | 5/16–18 x 1 1/4 lg | 3/4 | 11/32 |
| 3/4 | 2 9/16 | 63/64 | 1 7/8 | 25/64 | 11/32 | 13/32 | 9/16 | 7/8 | 3/8–16 x 1 1/4 lg | 15/16 | 7/16 |
| 1 | 2 3/4 | 1 7/64 | 2 1/16 | 15/32 | 11/32 | 13/32 | 9/16 | 7/8 | 3/8–16 x 1 1/4 lg | 1 1/32 | 33/64 |
| 1 1/4 | 3 1/8 | 1 25/64 | 2 5/16 | 35/64 | 13/32 | 15/32 | 9/16 | 7/8 | 7/16–14 x 1 1/2 lg | 1 5/32 | 19/32 |
| 1 1/2 | 3 11/16 | 1 37/64 | 2 3/4 | 21/32 | 15/32 | 17/32 | 5/8 | 1 | 1/2–13 x 1 1/2 lg | 1 3/8 | 45/64 |
| 2 | 4 | 1 55/64 | 3 1/16 | 51/64 | 15/32 | 17/32 | 5/8 | 1 1/32 | 1/2–13 x 1 1/2 lg | 1 17/32 | 27/32 |

Generally they are used in conjunction with copper tubing.

MALE SWIVEL FITTING (Fig. 6-37). This type is similar to the male pipe end, with the exception that the fitting will swivel into the port so that the hose assembly or component part does not have to be turned. Sealing is done in two parts—on the thread ends and with an O ring inside the fitting.

PUSH-ON COUPLINGS. Unlike other hydraulic couplings, these have only two basic components—the barbed insert and the coupling end (Fig. 6-38). As described in the section on hose, the yarn braid in the hose acts like a Chinese finger puzzle. When pressure is applied in the hose and it wants to push away from the insert, the yarn grabs the barbed insert. As the pressure increases and the hose pushes away harder, the yarn grabs the fitting tighter. A plastic cap acts as a protector for the end of the hose and prevents hydraulic fluid from attacking the braid.

### Methods of Fastening Couplings

There are two basic types of couplings used for hoses—those that are permanently attached and those that are reusable.

PERMANENTLY ATTACHED COUPLINGS (Figs. 6-39 and 6-40). These are put on by pressure and cannot be reused if the hose has to be replaced. They are pressed on by the manufacturers of hydraulic couplings or by special machinery in the field. Manufac-

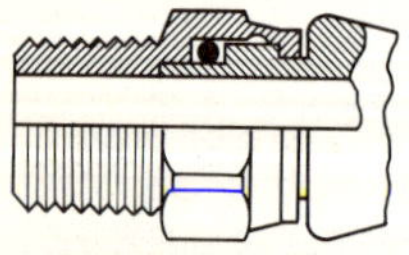

FIG. 6-37 *Male pipe swivel, O-ring seal.*

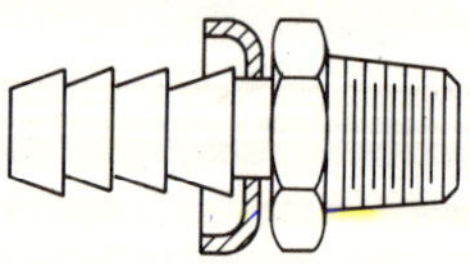

FIG. 6-38 *Push-on coupling.*

 

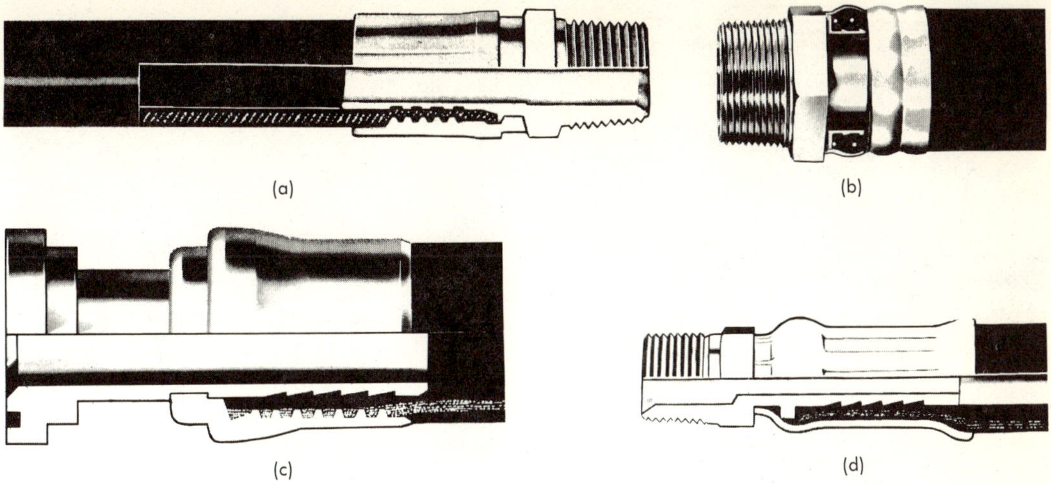


FIG. 6-39 *(a,b,c) Crimped-on factory press. (d) Crimped-on portable press.*

turers of hydraulic equipment generally use hose assemblies with these couplings; in some instances, however, they will use the reusable fitting. There are two methods of applying permanently attached couplings—by crimping or by swedging.

In the method of *crimping* pressure is put on the outer shell of the coupling to a preset amount. A press, which has eight dies that circle around the coupling, forces the dies down onto the shell. Even with variances in wire tolerances, the press will push the shell down until it hits the wire and then will put a predetermined amount of pressure on the shell so that it imbeds the serrations firmly into the wire. With this method of applying the còupling, the assembler can gauge the optimum amount of pressure required to give the maximum holding ability. By setting the pressure, he can run large numbers of assemblies and still be assured of maximum reliability for each assembly. All major manufacturers of hydraulic fittings use this method for coupling permanently attached assemblies.

*Swedging* permanently attached assemblies (Fig. 6-40) consists of forcing a fitting, which has been put on a hose, through a fixed orifice (generally two dies). The fixed orifice, with the force applied to the fitting, forces the shell down onto the reinforcement of the hose. This method is not as reliable as the crimping method because tolerances on the hose cannot be taken into account. If the hose tolerances vary from length to length, different dies must be used for each assembly. (For example, if the hose O.D. were on the high side of the tolerance, the coupling would be pushed down too far and would crack the wires, and if the hose O.D. were on the low side of the tolerance range, the coupling would not be pushed down far enough to give maximum holding ability.) This method is satisfactory where lower pressures are encountered with a minimum of surge, and where dependability is not critical.

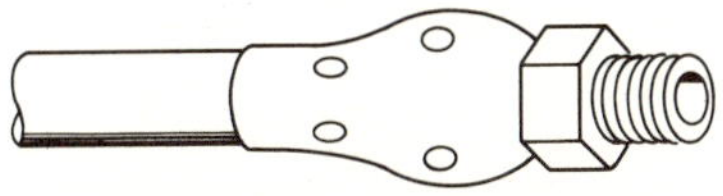

FIG. 6-40 *Swedged-on.*

REUSABLE COUPLINGS. Unlike permanently attached couplings, these can be unscrewed or unclamped and used again after the hose has worn out. There are four basic types of reusable couplings—screwed-on, clamp, compression, and push-on.

A screwed-on reusable coupling depends upon mechanical holding ability between a threaded insert and a threaded outer shell (Fig. 6-41). The outer shell is screwed onto the hose first and then the insert is threaded inside it.

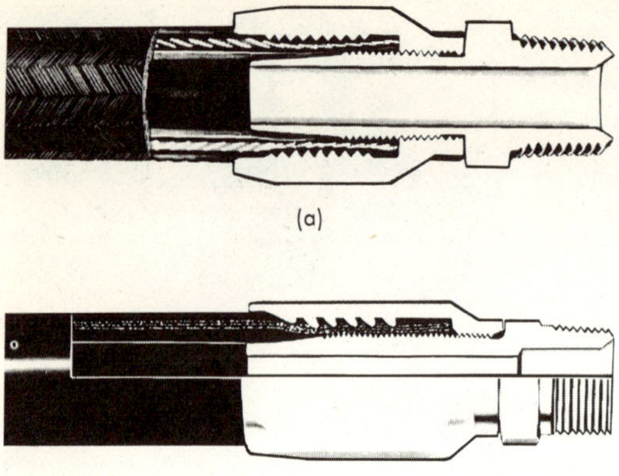

(a) (b)

FIG. 6-41 *(a) Screwed-on type reusable for cotton-covered SAE 100R5 hose. (b) Screwed-on type reusable for rubber-covered hose.*

The clamp-type reusable coupling gets its holding ability from two split halves of the outer shell which are bolted onto each other over the insert (Fig. 6-42).

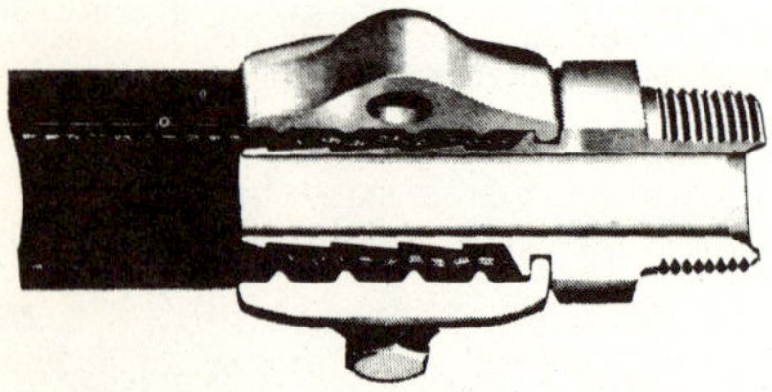

FIG. 6-42 *Clamp type, reusable.*

The compression-type reusable coupling operates in the same manner as the compression-tube fitting. When the nut is tightened onto the body of the coupling, it forces the sleeve down onto the hose (Fig. 6-43).

The push-on-type coupling is reusable because when the user wants to replace a hose line, he cuts the hose from the fitting and installs a new hose (Fig. 6-44).

## ADAPTERS

Adapters are designed to be used with fluid power hose assemblies. They contribute to ease of assembly and installation, improved appearance, and afford compactness. There are three types in common use today—the swivel adapter union, the 37-deg flared adapter, and the steel-pipe adapter.

### Swivel Adapter Union

The swivel adapter union (Fig. 6-47) has a swivel female end and a male Dryseal NPTF thread or O-ring-boss straight thread on the other end. The male Dryseal NPTF thread or O-ring-boss straight thread screws into the port connection, and the swivel female end is attached to a male NPTF coupling end of the hose assembly. The male hose couplings which are fastened into the swivel adapters have a 30-deg flared seat on the inside (see, e.g., Fig. 6-41). The swivel adapter has a corresponding 30-deg seat on the female end. When connected, these two surfaces provide a metal-to-metal seal. The following types of swivel adapter are currently available for use in hydraulic plumbing.

STRAIGHT SWIVEL ADAPTER UNION (Fig. 6-48). This fitting is primarily used for connecting a hose assembly where the cou-

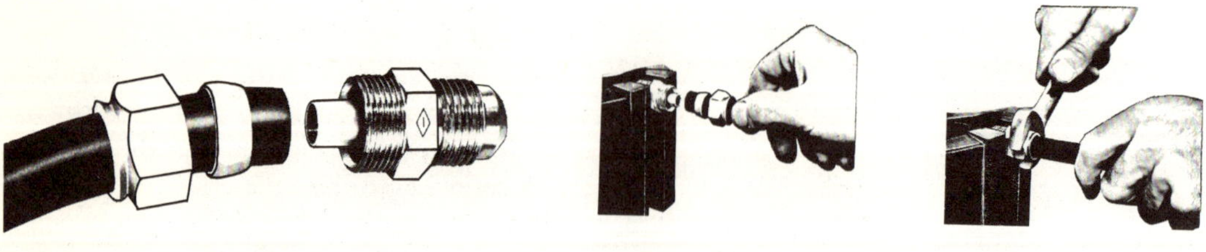

(a) (b) (c)

FIG. 6-43 *(a) Slide nut and sleeve over hose and oil inside of hose and coupling threads. (b) Compression type, reusable. (c) Push hose over insert to bottom of cavity in coupling body. Tighten nut.*

 

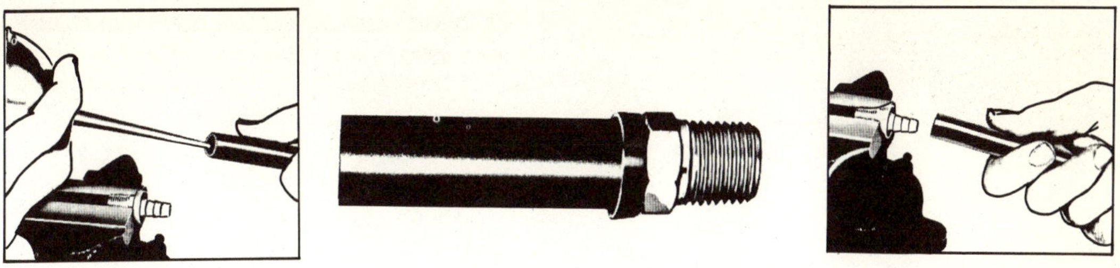

(a) (b) (c)

FIG. 6-44. *(a) Oil inside of hose and outside of nipple. (b) Push-on type, reusable. (c) Push hose over barbs as far as it will go. End should be inside of protective cap.*

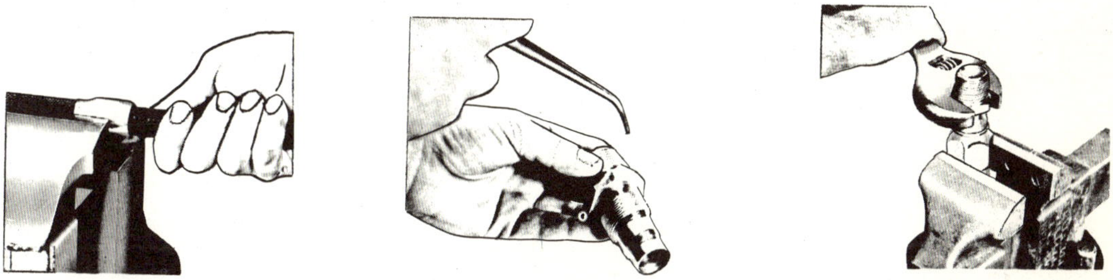

(a) (b) (c)

FIG. 6-45 *For two rayon-braid, cotton-covered, and nonskive rubber-covered hose. (a) Clamp coupling socket in vise. Turn hose counter-clockwise into coupling socket until end of hose is seated against bottom of socket. (b) Oil threads on insert and inside of hose generously. (c) Screw insert into coupling socket, turning insert clockwise, and tighten completely.*

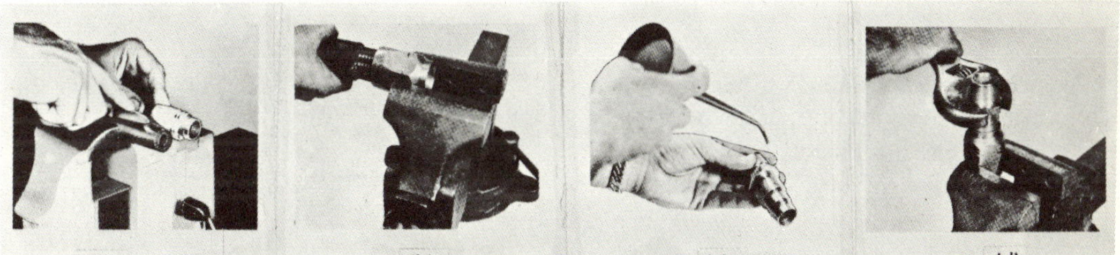

(a) (b) (c) (d)

FIG. 6-46 *Methods of applying reusable fittings. Reusable couplings for single and double wire braid and two spiral wire, one wire braid rubber-covered hose. (a) These hoses must be skived (stripped of rubber cover) before applying coupling. To strip hose, cut around it, down to the wire braid. (An Imperial-Eastman tube cutter is ideal for this operation.) Slit lengthwise, lift flap, and pull off with pliers. Be careful not to fray or damage wire braid. Length of hose to be stripped can be determined by placing coupling shell next to hose and measuring from hose end of shell to large notch. (b) Place shell in vise and screw hose counterclockwise into shell until it is inserted to full depth. (c) Oil threads on insert and inside of hose liberally. Use grease instead of oil for larger sizes. (d) Screw insert into shell, turning clockwise. (Courtesy of Imperial-Eastman.)*

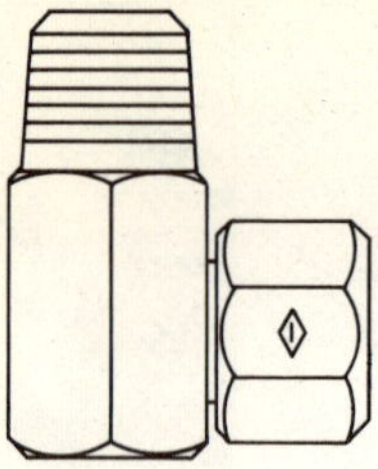

FIG. 6-47 *Swivel adapter union.*

plings are male pipe ends and the component parts (valves, pumps, cylinders, etc.) are mounted so that they cannot be turned.

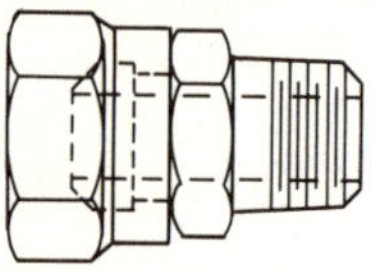

FIG. 6-48 *Straight swivel adapter union.*

STRAIGHT SWIVEL ADAPTER UNION WITH MALE O-RING STRAIGHT THREAD (Fig. 6-49). This fitting is the same as the straight swivel adapter union except the O-ring thread is designed for ports which have compatible threads.

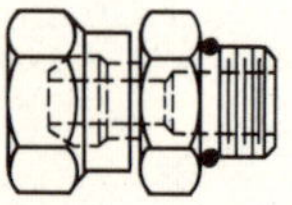

FIG. 6-49 *Straight swivel adapter union with male O-ring straight thread.*

90-DEG SWIVEL ADAPTER ELBOW (Fig. 6-50). This fitting is designed for use where a 90-deg elbow is needed to conserve space and maintain the hose line in a straight plane.

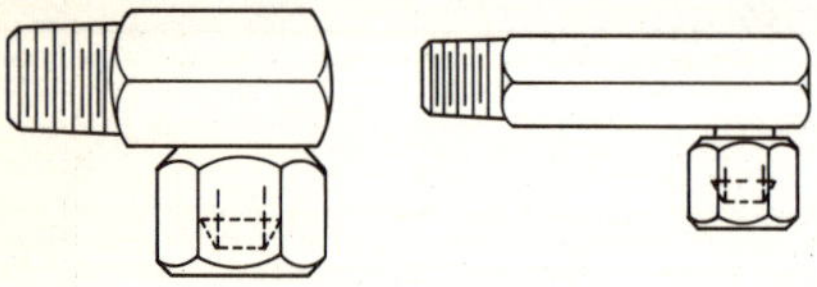

(a) (b)

FIG. 6-50 *(a) Standard elbow. (b) Long 90-deg elbow.*

90-DEG O-RING BOSS ELBOW (Fig. 6-51). This is the same as the 90-deg swivel adapter elbow except that the port thread is for O-ring thread instead of male pipe thread.

45-DEG ELBOW STRAIGHT PIPE AND O-RING BOSS THREAD (Fig. 6-52)

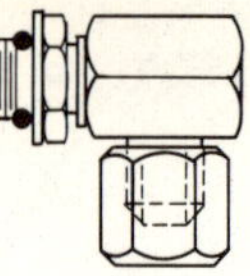

FIG. 6-51 *O-ring boss elbow.*

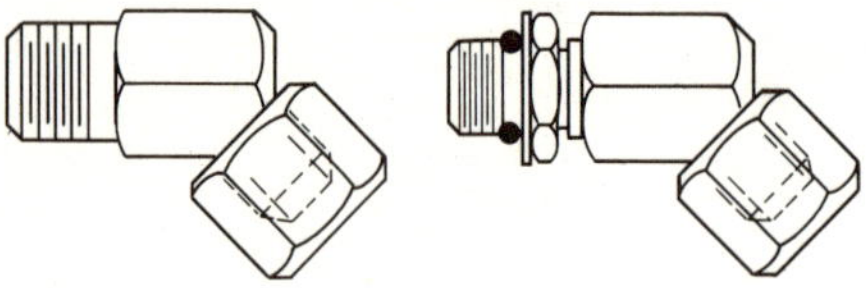

(a) (b)

FIG. 6-52 *(a) 45-deg elbow straight pipe and O-ring boss thread. (b) 45-deg elbow straight pipe and O-ring boss thread.*

SWIVEL ADAPTER UNIONS WITH FEMALE PIPE THREADS FOR MALE PORTS (Fig. 6-53).

Other types of swivel adapter unions are shown in Fig. 6-54.

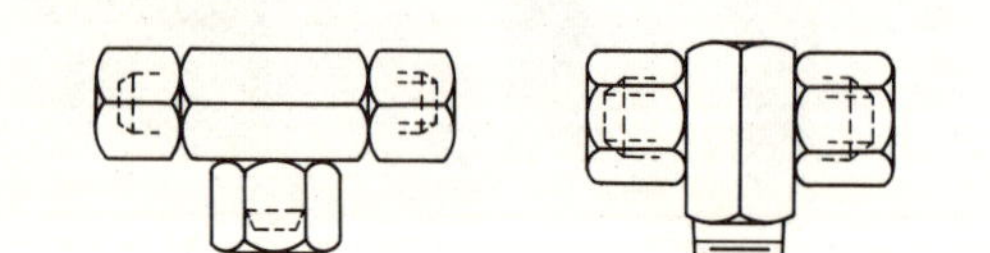

(a) (b) (c)

FIG. 6-53 *Swivel adapter unions with female pipe threads for male ports.*

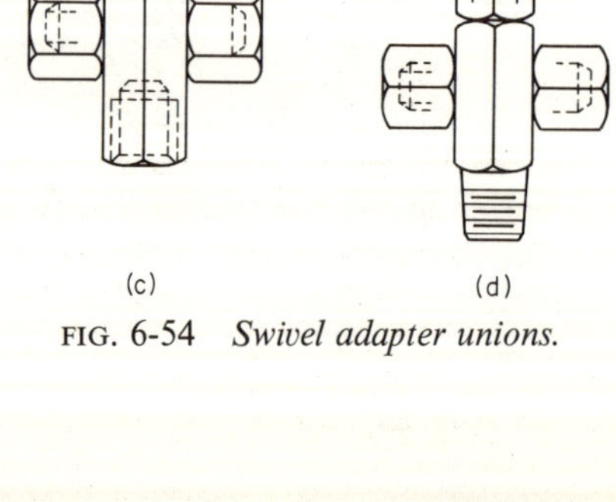

(a) (b)

(c) (d)

FIG. 6-54 *Swivel adapter unions.*

 

**The 37-deg Flare Adapters**

Because 37-deg flare adapters have a larger flare seat than do swivel adapter unions, they make an effective seal at higher pressures. These adapters are designed for usage with hose assemblies which have 37-deg JIC female coupling ends, whereas swivel adapter unions are designed for use with the hose assemblies which have male NPTF coupling ends (Fig. 6-55). 37-deg flare adapters, like swivel adapter unions, are available with male O-ring-boss straight threads where they are required. The type used is dictated by the type of thread at the component.

Some 37-deg flare adapters in common use in industry today are depicted in Fig. 6-56.

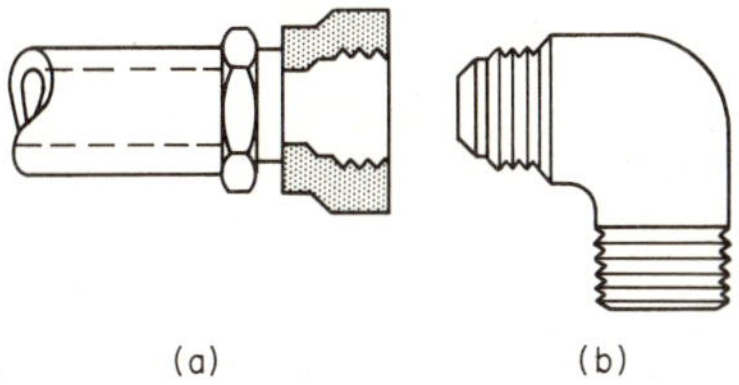


FIG. 6-55 *(a) JIC female coupling end. (b) JIC male elbow.*

**Steel Pipe Adapters**

These adapters are essentially like the iron-pipe fittings mentioned in the first part of this chapter; however, they consist of steel instead of iron pipe and are capable of withstanding pressures up to 5000 psi. All the configurations available in iron pipe are also available in steel. Steel adapters can withstand higher pressures, and they afford greater compactness.

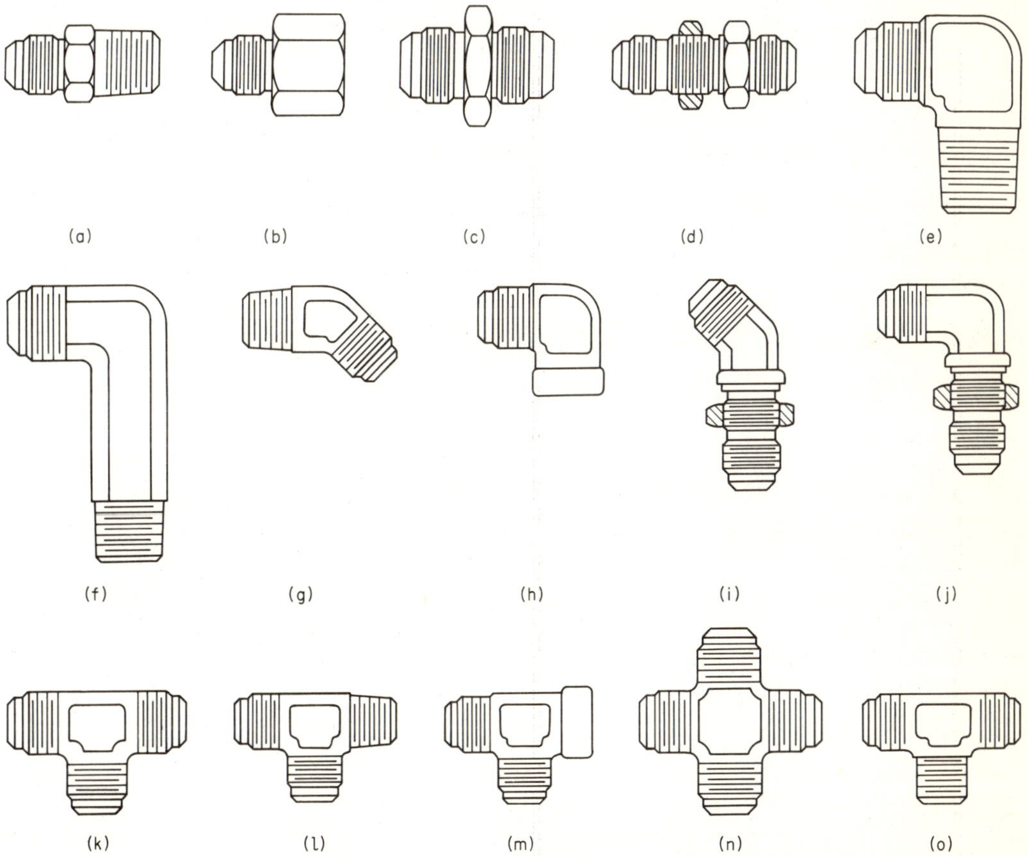


FIG. 6-56 *37-deg flare adapters which are in common use in industry today.*

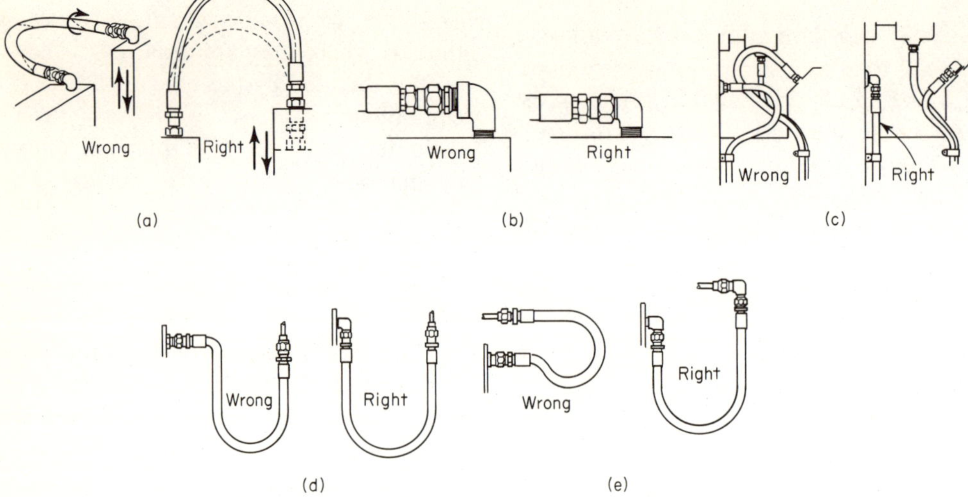


FIG. 6-57 *Adapter application. (a) To prevent twisting and distortion, hose should be bent in the same plane as the motion of the boss to which hose is connected. (b) To reduce number of pipe thread joints, shorten assembly line, and improve appearance, use proper hydraulic adapters instead of pipe fittings. (c) Obtain direct routing of hose through use of 45-deg and 90-deg adapters. Improve appearance by avoiding excessive hose length. (d) Avoid sharp twist or bend in hose by using proper angle adapters. Increase hose life by preventing breaks at sharp bends. (e) Where the radius falls below the required minimum, an angle adapter should be used to avoid sharp bends in hose.*

## PROBLEMS

### Iron Pipe

**1.** How are pipe sizes specified?

**2.** (*a*) What are the two pipe threads which have been standardized in the United States?

(*b*) Are these two threads interchangeable?

**3.** If a length of ½-in.-I.D. schedule 40 iron pipe is connected to a length of ½-in.-I.D. schedule 80 iron pipe, will there be a reduction in flow? Why?

**4.** What effect does the use of a sealing compound or sealing tape have on a Dryseal thread?

**5.** Can iron-pipe fittings found in hardware stores be used on high-pressure systems? Why?

 

**6.** What is a street elbow and how is it used?

**7.** Is it possible to connect two elbows without another fitting? Why?

**8.** Which fitting would be used to connect four lines together?

**9.** What is the difference between a nipple and a connector?

**10.** What three factors determine the pipe size to be used in a hydraulic system?

**11.** If you had a female port size on a cylinder which was ½-in. I.D. and you wanted to connect a ⅜-in. I.D. male line to it, which type of fitting would you use?

**12.** When is iron pipe generally used in a hydraulic system?

### Tubing and Tube Fitting

**13.** How is hydraulic tubing specified?

**14.** What size would a –10 size tube represent? Would this dimension represent the outside or inside diameter?

**15.** If a –6 size tube were to be used in a 1000- to 2500-psi system, what should be the tube's rated burst pressure? What wall thickness should be used?

**16.** What two types of steel hydraulic tubing are recommended under JIC specifications?

**17.** Name two types of tubing other than hydraulic steel sometimes found in hydraulic systems.

**18.** Why should tubing be clamped to the frame?

**19.** Name the two styles of tube fittings used to connect hydraulic tubing and describe the primary difference between the two.

**20.** Can two flared tubes be connected together with a tube fitting? What type of fitting would be used?

**21.** If you had to make a line by running the tubing parallel to the frame, and you did not want to flare the tubing, which type of fitting would you use? Why?

**22.** Name two types of threads which can be found on flareless or flared fittings other than pipe threads.

### Hydraulic Hose and Fittings

**23.** What is the primary advantage of using hydraulic hose instead of tubing in a hydraulic system?

**24.** What are the three primary parts of a hose?

**25.** What inner-tube compound has (*a*) the greatest oil resistance?

(*b*) The greatest abrasion resistance?

**26.** What is the standard operating range for most hydraulic hoses?

**27.** Can phosphate ester oils be used in a Buna N tube hose, and, if not, what type of rubber should be used?

**28.** Name two types of reinforcements commonly used in hydraulic hose.

**29.** Under extreme flexing conditions, which is the best type of hose to be used, SAE 100R2 type A or type B?

**30.** What are the important factors in determining what type of hose should be used in a given application?

**31.** What is the I.D. of SAE 100R5 hose in a –08 or one-half size? ______

**32.** Name three of the seven rules for hose installation.

**33.** How are hose assemblies measured?

**34.** Using the nomographic chart, determine what size I.D. hose should be used where the flow required in gallons per minute is 15 and the velocity range for the pressure lines is 15 ft/sec. ______

**35.** What are the three basic components of a hydraulic hose coupling?

 

**36.** Name four types of coupling ends available.

**37.** What are the two basic types of fittings used for hydraulic hoses?

### Adapters

**38.** Why are adapters used with hose assemblies?

**39.** What are three types of adapters used in plumbing hydraulic applications?

**40.** Can a length of pipe with a threaded male end be screwed into the female swivel end of a swivel adapter and give a seal?

**41.** What end must a hose assembly have to use 37-deg flare adapters?

**42.** Are steel-pipe adapters the same as pipe fittings?

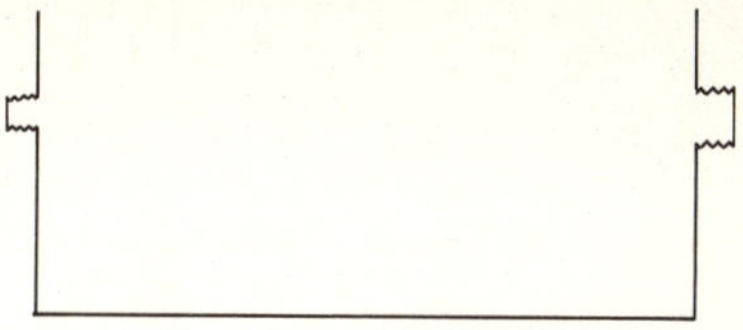

FIG. 6-58 *Hose installation problem.*

**43.** The port on the left side of Fig. 6-58 is tapped for ½-in. pipe thread, and the port on the right side is tapped for ¾-in. pipe thread. Neither port section is movable. The flow requirements are for a ½-in.-I.D. (8 gpm) line, and the pressure on the line will be 1000 psi. Describe how you would connect these two ports with (*a*) steel tubing and (*b*) hydraulic hose. (On steel tubing a straight line cannot be made between the two because the tubing must be bent to allow it to be put into the fittings. If it is too long, it will be impossible to install. If it is too short, there will be no material available for flaring; or, in the case of flareless fitting, there will be insufficient length of tube for the ferrule to bite into.) Describe and list the fittings to be used as well as the size tubing and hose.

FIG. 6-59 *Hose and tube installation problem.*

**44.** Suppose that there is an obstruction between the two ports shown in Fig. 6-58. Describe how you would connect these ports with hydraulic tubing and hose (Fig. 6-59).

**45.** Fig. 6-60 is a plumbing design for a hydraulic unit. Describe how you would clean up this design and shorten the hose lengths by the use of 37-deg flare adapters and 37-deg female swivel ends on the hoses. (Name the 37-deg flare adapters to be used for each line.)

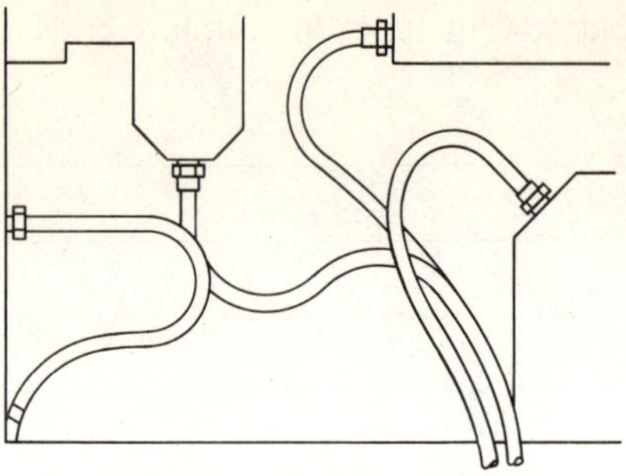

FIG. 6-60 *Hose installation.*

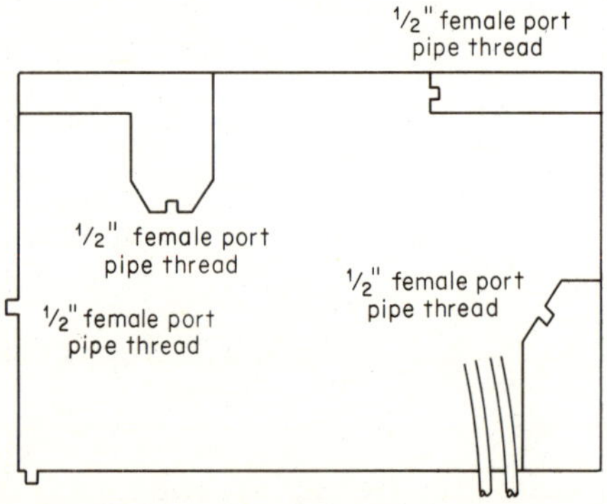


FIG. 6-61 *Revised hose installation.*

**46.** Which of the six ways shown in Fig. 6-62 are correct for hydraulic hose connections? Why?

 

NAME .............................................. DATE .............. CLASS ...........

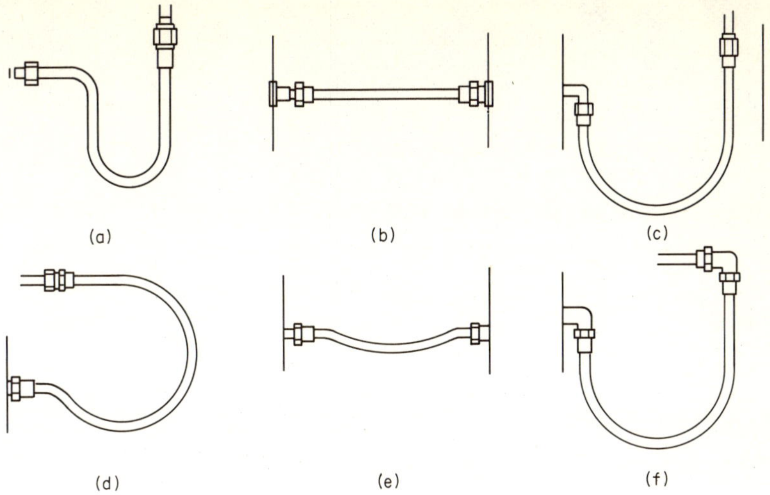


FIG. 6-62 *Hose connections.*

# 7 PUMPS

The pump is the heart of the hydraulic system. It converts work applied to its drive shaft into either kinetic or potential energy in the fluid. This energy, measured as pressure and volume, is converted back into useful work on an object or load when transmitted to the working element of the system. Pumps are either of the positive-displacement or the non-positive-displacement type.

*Positive-displacement* pumps are those whose pumping volume changes from maximum to minimum during each pumping cycle. That is, the pumping element expands from a small to a large volume and is then contracted to a small volume again. These pumps have the following advantages.

1. They can generate high pressure.
2. They are relatively small.
3. They have relatively high volumetric efficiency.
4. There is relatively small change in efficiencies throughout the pressure range.
5. They have great flexibility of performance; that is, a given pump can operate over a wide range of speeds and under varying pressure requirements.

Nearly all pumps used in fluid power are positive-displacement pumps.

*Non-positive displacement* pumps are those whose volume does not change during the pumping cycle. They rely on other means of imparting energy to the fluid. Most common of these are centrifugal pumps, in which change in linear velocity of the pumping element (and of the oil being pumped) applies centrifugal force to the fluid. Other types are propeller, jet, and aspirator pumps. The advantages of this kind of pump are:

1. There is a low initial cost.
2. There is minimum maintenance cost.
3. They can be operated quietly.
4. They are capable of handling almost any type of fluid—for example, sludge and slurries.

Since they are not capable of withstanding high pressures (usual maximums are in the vicinity of 250 psi), they are of little use in the fluid power applications described here.

Positive-displacement pumps are either *reciprocating* or *rotary*. Reciprocating pumps are essentially cylinders operating as a pump. Because their size, shape, and required driving mechanism are awkward, they are used only for specialized applications. (Such pumps are capable of developing high pressure if the reciprocating principle is used in a rotary form. A pump using this principle is known as a piston pump.)

Rotary pumps are the most common in oil hydraulics. Although there are many varieties of rotary pump, we shall concentrate our study on three principal categories, and

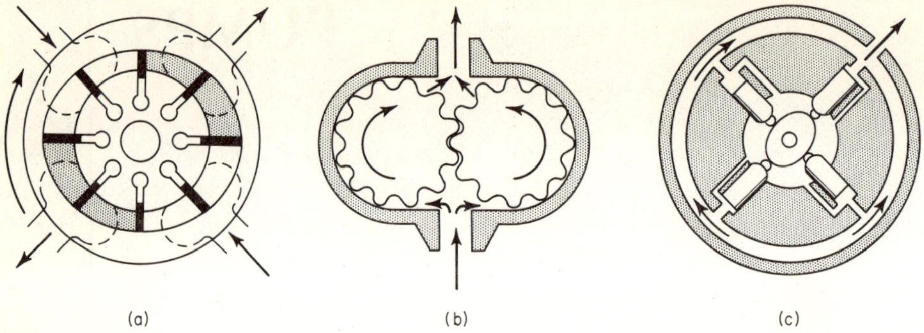


FIG. 7-1 *Positive-displacement pumps. (a) Vane. (b) Gear. (c) Radial piston.*

mention the others only when appropriate. The three prime types are: (1) gear; (2) vane; and (3) piston (see Fig. 7-1). Although they overlap to some degree in both performance and application, they are completely different in structure and usually serve quite different functions.

## GEAR PUMPS

The functions and operating principles of the various parts of the gear pump will be described in greater detail than the parts of the other types of pumps. Thus discussion of the gear pump and a description of its parts will serve as an introduction to all pumps.

### Elementary Structure

The basic gear pump consists of two meshed gears, a case or housing surrounding the gears, and two covering pieces that enclose the ends of the gears. Each gear is mounted on a shaft which is supported on bearings in the end covers. One of these shafts, the drive shaft, extends out through one of the covers and is coupled to the prime mover.

Two ports—an intake and a discharge—must be provided. Their exterior position on the pump is relatively unimportant, but they are most often located on opposite sides of the gear case. It is absolutely essential, however, that inside the pump these ports open directly on opposite sides of the mesh point of the gears. For best performance, the ports should be as large, straight, and unobstructed as possible (Fig. 7-2).

### Operating Principle

The pumping action occurs as follows. A tooth space of one gear is filled by a tooth of the other gear. As the meshed gears rotate, one tooth space after another is evacuated, and the resulting spaces become, in turn, vacuums. Atmospheric pressure forces oil into the tooth space from the port located at that point. This tooth space full of oil is carried around the periphery of the gear until the teeth again mesh; at this time the oil is forced out of the space by the meshing tooth and flows out of the port located at that side of the mesh point (Fig. 7-3). The elimination of this space prevents

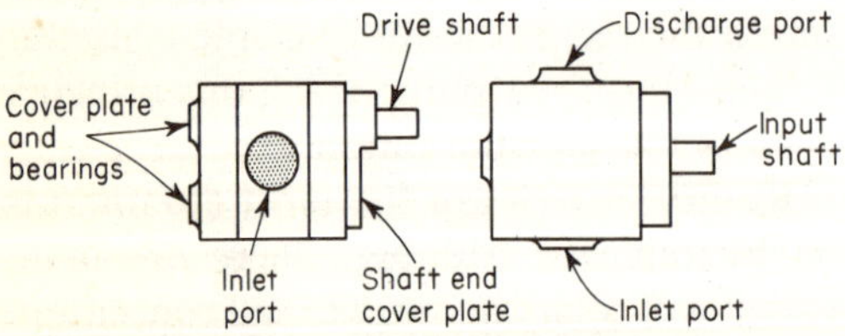


FIG. 7-2 *Pump configurations.*

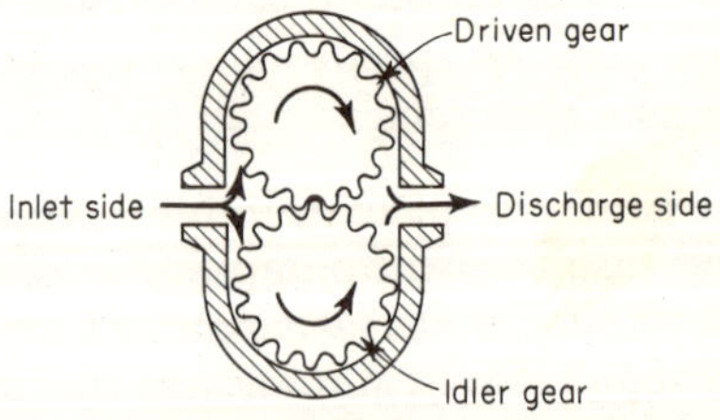


FIG. 7-3 *Pump cross-section.*

 

oil from crossing over to the intake side, and thus the pump dispels one toothful of oil per tooth per revolution. Close-fitting covers prevent fluid passage over the end. The line contact of the gear teeth on one another prevents flow through the mesh, and the close-fitting housing prevents flow back around the periphery.

The fact that the gear teeth can force fluid out of the tooth spaces against resistances placed downstream is the essence of the pumping concept and is the measure of success of the gear pump. Without a restriction downstream, it can be seen that the pump would merely transfer fluid from one side to the other at zero pressure. Positive pressure is created only when a restriction is encountered.

#### Position in the Field

Of all hydraulic pumps, gear pumps are by far the most common. This is probably because they are simple in design, are rugged in construction, are the least sensitive to dirt and other contaminants in the fluid, and are the easiest to manufacture.

Gears are rugged devices in themselves. Since the pumping action is confined to the teeth (that is, the solid part of the gear circle is not involved), the shafts can be relatively large, and large bearings can be used. In short, massive support can be given the components. In addition, only the simple elements described above are required—no complex or delicate components are necessary.

This simple design and construction permits gear pumps to be manufactured with standard machine tools; no unique or complicated tooling is required. This fact no doubt explains why there are so many manufacturers of gear pumps and why so many companies make their own.

Vane and piston pumps require a certain minimum, or threshold, precision in order to work well enough to justify their manufacture. Gear pumps do not; or at least we can say that the minimum precision required is so negligible that it is of little practical concern. Thus, if a certain job needs only a low-performance pump, it is easy to produce a cheap gear pump which will meet the needs of such a job (an example of this is the oil pump used in an automobile engine). On the other hand, very precisely built gear pumps that will meet very demanding performance requirements are also manufactured.

#### Design and Manufacturing Requirements

The performance level of the pump is primarily determined by the precision of its manufacture. Parts must be manufactured to close-tolerance fit specifications, and these close tolerances must be preserved while the pump is operating. Obtaining the proper fits (initially) requires fine quality of materials, accurate machines, and careful craftsmanship. Preserving these close fits during operation is the constant concern of engineers. Close tolerance is maintained by designs which are structurally sound and which employ balanced pressures.

A perfect pump would be one with zero clearance between all parts. Although this goal is practically impossible, moving toward it theoretically is the object of continuing research and development. Working clearances should be the smallest possible that can be obtained while maintaining an oil film across the ends of the gear, providing unobstructed rotation of the gear teeth, and allowing for thermal expansion. These factors vary with the size and speed of the pump, its materials, the pumping medium, and the operating environment. Clearances in current practice cover a range of from 0.001-in. end and diametral clearances on small pumps (1-in.-dia. gears, for example) to 0.004-in. to 0.005-in. end clearance and 0.006-in. diametral clearance on pumps with 3½-in.-dia. gears.

Surface finish, parallelism, concentricity, and absolute location of mating parts are as important to pump performance as clearance, and, indeed, are necessary for achieving properly minute clearances. The finer the surface finish, the more accurately the measured dimensions approach the absolute dimensions. The more nearly parts are made perfectly square or parallel, for example, the less allowance need be made for misalignment.

Thus we see that care and quality in machining are the foundation of successful pump manufacture. This applies to all hydraulic equipment and will be emphasized repeatedly throughout this book. The finest quality machine tools and most refined machining techniques are common in hydraulic equipment. It is fair to say that no other industry utilizes finer machining techniques or produces higher-quality products than the fluid power equipment industry.

### Gears

Gears commonly used in industry are comprised of 14½- or 20-deg involute teeth. The primary requirement of such power transmitting gears is that they must roll without sliding on the pitch lines of the teeth. This permits longer life. However, as the number of teeth is reduced, the roots of the teeth are undercut and the teeth are weakened. The minimum number of teeth which should be cut on a 14½-deg gear is about 16; if there are fewer, the gear will not be strong enough.

In pump gears, the rolling of the gears without sliding is of secondary importance. More important is the amount of oil carried in each tooth space and the strength of the teeth. In a gear of any given diameter, the teeth and their intermediate spaces become larger as the number of teeth is reduced. This means that as the number of teeth is reduced, the amount of oil carried per space (or per revolution) increases. However, when conventional 14½-deg involute teeth are used (and they are used in some pumps) and the number of teeth is reduced to 12 or less, the resultant undercutting of the teeth makes them structurally weak and leaves unproductive "pockets" on each side of the mating gear tooth when the gears are meshed (see Fig. 7-4). Thus when gears of this structure are used, a certain amount of oil always remains in the tooth space, which reduces

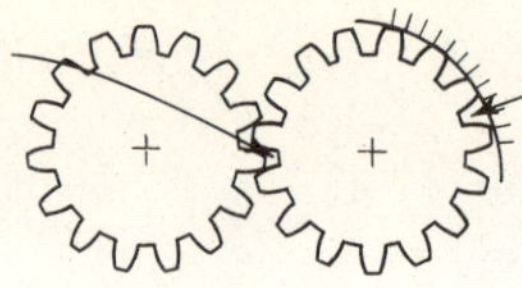


FIG. 7-4 *Pump gears, set 14½.*

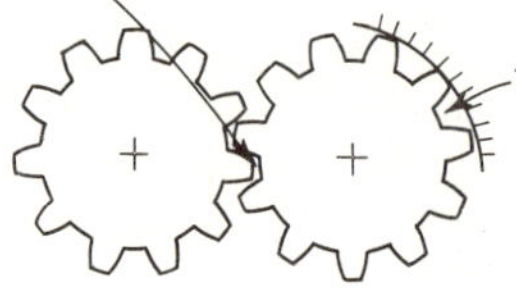


FIG. 7-5 *Eleven-tooth gears.*

the usable volume of oil the gears can carry.

If the involute angle is increased, the gear teeth become more triangular. If such gears are used, the undercutting will be substantially reduced, and the number of teeth can be less than 12. The teeth remain thick and strong although the roots and unproductive "pockets" are reduced. This type of gear is now predominant in pumps, and 28- to 28½-deg involutes are currently in favor. Although many pumps are made with 9- and 10-tooth gears, most have from 11 to 15 (see Fig. 7-5).

Although most pump gears are straight spur gears, a few helical gear pumps are made (Fig. 7-6). The advantages to helical gears are that because they mesh with a wiping or slicing action, rather than a shearing action, they are stronger and quieter. However, they are more expensive to manufacture. The helical angle must be carefully chosen so that the entire gear tooth lies *between* the inlet and discharge ports (which, you recall, lie on opposite sides of the mesh, only one gear tooth apart), or the gear teeth will short-circuit the pump. That is, when

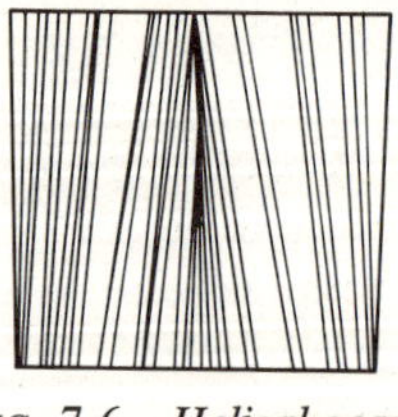

FIG. 7-6 *Helical gears.*

 

the gear teeth are completely meshed, one end of the tooth may be in the suction area and the other end in the pressure area; then the oil will flow down the clearance space at the root of the tooth space from the pressure area to the suction area. This problem would probably not be troublesome if only one size of pump were built. But most manufacturers build a line or series of pumps using the same diameter gears in a variety of widths. In this case, either the helical angle or port position would have to be changed for each gear thickness, and the advantages of mass production would disappear.

Length and thickness, or diameter, are important factors in gear selection. Running clearance at each end of the gear allows a certain internal loss of oil from the theoretical volume. This loss is influenced by the diameter of the gear and will remain reasonably constant regardless of the length. If this loss remains constant, it will be a greater percentage of small-capacity pump than it will of a larger. Thus, the thinner the gear, the smaller the capacity and the greater percentage of loss. Consequently, there is a logical minimum thickness below which the loss becomes too large a percentage to be practical. Further, the drag of the wear plates is a function of the gear diameter, and is not dependent on gear length. As the gears become narrower, the percentage of power input necessary to overcome drag increases; the mechanical efficiency drops. Conversely, as the gear length is increased, a point can be reached at which the bearing loads become excessive for the chosen side. Also, the capability of the fluids to flow into the suction side and completely fill the tooth spaces diminishes as the length increases.

Therefore, in a pump or family of pumps, the diameter to be used sets certain practical limits to the thickness that can be used. Although there is no set practice governing these limits, it is a rather accepted rule of thumb that optimum gear size is where the thickness of the gear is approximately 75 to 100 percent of the diameter of the gear. Because fewer problems occur with slightly thinner gears, a family of pumps usually covers gear lengths from 50 or 60 to 100 percent of diameter.

#### Wear Plates and Pressure Balancing

From fluid mechanics we know that:

1. Pressure acts equally in all directions.
2. Pressure exerted upon one point or area of a fluid body immediately translates to all portions of body and acts with equal intensity throughout the body.
3. Pressure always seeks to move fluid from a high to a low pressure point.
4. Pressure acting upon an area becomes a force.
5. Equal pressures acting in direct opposition to one another cancel or neutralize each other.

Pressure balancing is the proper utilization of these facts. As a pump operates, it carries oil from a point of less-than-atmospheric pressure (vacuum) to one of high pressure (discharge). This action creates a pressure gradient around the perimeter of the gear. If we plotted the gradient ($p$ versus $l$), the curve would resemble that in Fig. 7-7.

In accordance with the fluid properties enumerated above, the pressure acts against the confining surfaces to distort them. When the pressures are relatively low, the strength of the pump materials is sufficient to resist; but as the pressure increases, the materials measurably distort. We can say, then, that this distortion is the result of the unbalanced pressure force. If and whenever this force can be balanced (neutralized), the distortion is eliminated.

Wear plates are placed at the ends of the pump gears between the gears and the cover plates. The first function of the wear plates is to provide contact surfaces of superior bearing quality. Most wear plates are bronze, but aluminum is also used. Since cast iron is an excellent bearing material, many pumps have no wear plates and use the ground surfaces of the cover plates as bearing surfaces. (These pumps, however, have no

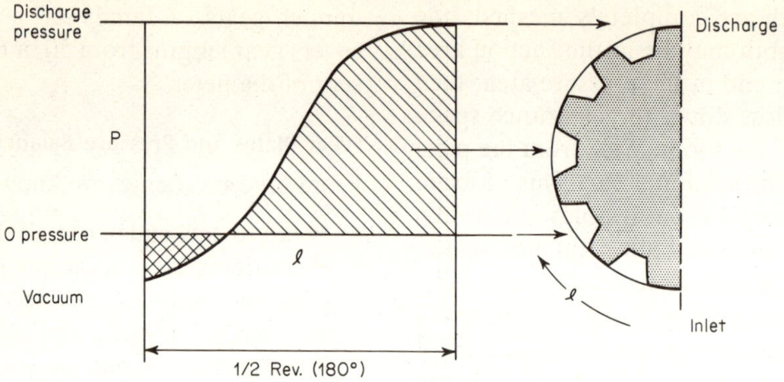


FIG. 7-7 *Pump gear pressure gradient.*

capability for pressure balancing; this is discussed below.)

Through clearances *around* the wear plates and holes (such as gear-shaft holes) *through* the wear plates, oil finds its way between the wear plates and the cover plates. This oil exerts pressure against the wear plates; in accordance with our second fluid mechanics "law," this pressure is the maximum of the system. In the area of the discharge port, the pressures on both sides of the wear plate are equal; the wear plate is in neutral atmosphere and remains flat (the fifth "law"). In the area of the inlet port, however, the high pressure on the "underside" is substantially greater than the pressure on the gear side. This creates an unbalanced force which deflects the plate against the gear. In a relatively short time, the gear teeth will mill away the bearing surface, increasing the clearance across the ends of the gears and reducing the pump's performance (Fig. 7-8).

If the areas of varying pressures are isolated, pressures on both sides of the plate can be neutralized and deflection of the wear plate can be eliminated. This is done by sealing off areas of differing pressures under the wear plates so that fluid and pressure transfer from one area to another cannot occur. The sealed areas can then be vented to the area in the gear chamber directly above so that the pressures on both sides of the wear plate become substantially equal.

Although this pressure balancing cannot exactly follow the pressure gradient of the pump chamber, any pressure separation is helpful, and reasonable approximations can be made. Since the inlet area is the most

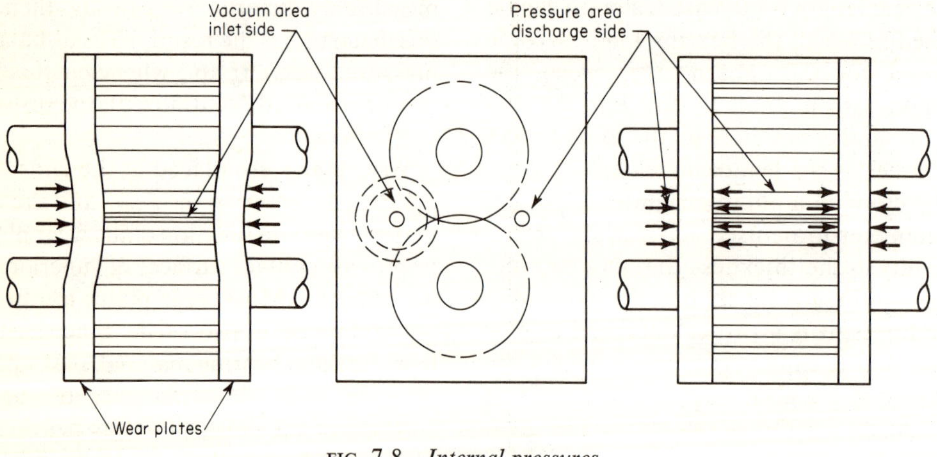


FIG. 7-8 *Internal pressures.*

 

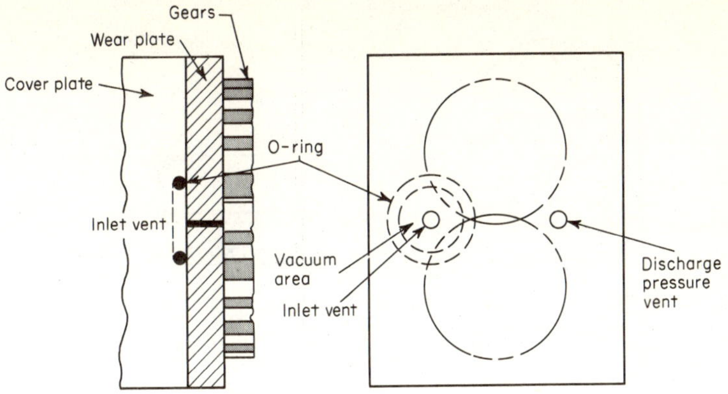


FIG. 7-9 *Pressure balancing.*

critical, elementary pressure balancing consists of merely sealing the areas beneath the wear plates at the intake port (Fig. 7-9). More elaborate balancing separates the entire area under the gears into many small areas, each one vented to the gear chamber directly opposite. This balancing provides a reasonably neutral atmosphere for the entire wear plate, allowing it to remain flat. This provides a uniform bearing surface for the gears, maintains design clearances, improves operating efficiency, and extends service life.

The question might logically be asked, "What about the distortion of the cover plates?" The cover plates do distort, but since the pumping chamber is confined by the wear plates, which remain flat, the effect on pump performance is negligible. This is not to say that the distortion is of no concern. Since the bearings are carried in the end plates, distortion affects their alignment. However, this is an engineering problem entirely separate from pressure balancing and beyond the scope of this text. Figures 7-10 to 7-12 show the three principal methods of pump pressure balancing commonly used.

#### Performance Ranges

Gear pumps are classed as medium-pressure pumps. At this time, most gear pumps operate from 1000 to 1500 psi under continuous duty and up to 2000 psi intermittently. Shaft speeds range from 1500 to 1800 rpm. Gallonages are unlimited, being governed by the sizes of the gears. Whereas many control pumps have outputs so small they are measured in gallons per hour, pumps in common use today on familiar earth-moving machines have 100 to 150 gpm capacities.

The speed range matches current electric motor and crankshaft speeds, and the pressures adequately satisfy most power requirements. However, increased speeds and pres-

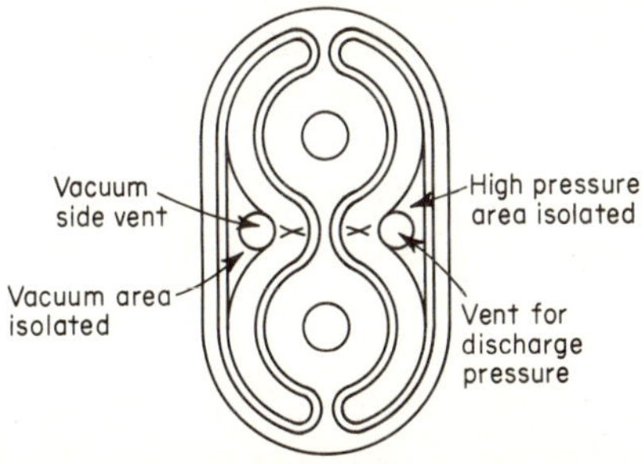


FIG. 7-10 *Type of pressure balancing.*

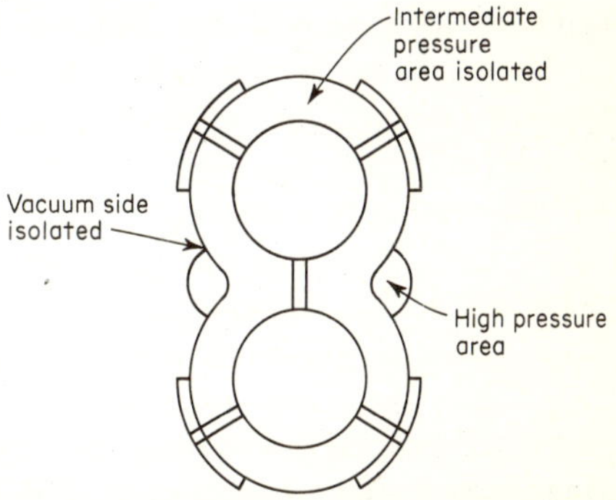


FIG. 7-11 *Another system of pressure balancing.*

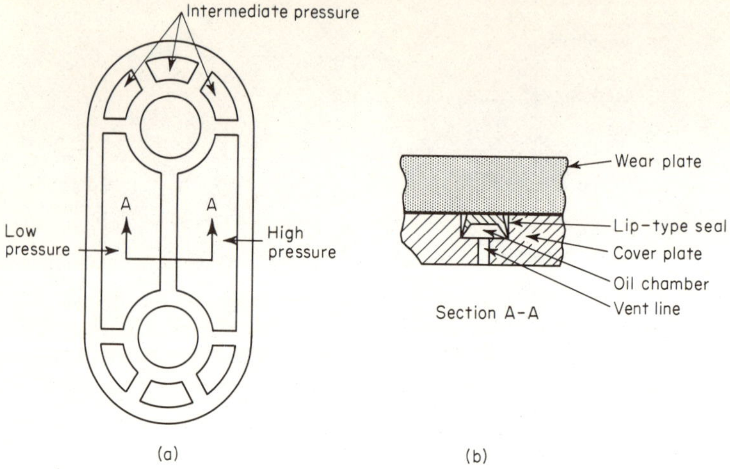


FIG. 7-12 *(a) Another method of pressure balancing. (b) Cross-section of venting chamber.*

sures offer certain advantages, notably in size and weight reductions. These advantages are very important to the aircraft industry. Spurred, no doubt, by aircraft, other industries are also demanding higher performance pumps. To meet this demand, gear pumps have been made available in the 2000- to 2500-psi range with shaft speeds from 2000 to 3000 rpm.

Upgrading the performance has required closer fitting and finer finishes, but is mostly the result of improved wear-plate design. Pressure balancing, as discussed previously, maintains pump-chamber clearance by minimizing distortions, reducing wear, and reducing wear-plate drag. As stated earlier, preserving the designed shape and fit of the pump chamber is the key to elevated performance.

Metallurgical improvements have also contributed to increased performance ratings. Higher-strength castings, particularly in the gear housings, are required to maintain body stiffness.

It is interesting to note that at the present time most high-pressure gear pumps use bronze bushings rather than antifriction bearings. Because bushings offer area support to the gear shaft, they are capable of carrying higher bearing loads than are roller bearings, which make line contact with the shaft. However, bushings require a film of oil lying between the shaft and the bearing surface. Since maintenance of this film is a function of shaft speed, it is not a coincidence that increased pump pressure is accompanied by increased shaft speeds.

Increased speeds also contribute to increased operating pressures by increasing the relative viscosity of oil. Viscosity is a measure of the resistance of oil molecules (in an oil film, for example) to slide relative to one another within a specified time period. The higher the viscosity, the greater the resistance to sliding. In a sense, the shear strength of oil is a function of viscosity, although shear strength is a static concept. As the shaft speed is increased, the time interval between the closing and subsequent opening of the gear teeth at the mesh point is reduced. The relative viscosity of the oil is increased and the capability of leakage past the mesh point is reduced. Hence, higher pressures are obtainable.

It must be emphasized that not all high-pressure pumps use bushings; many use roller bearings. However, most operate at speeds above 2000 rpm and have vented bearings and unidirectional seals.

### Shaft Seals

Three types of shaft seals are used: (1) pressure; (2) wiper; and (3) rotary. In every case the seal is located as far out along the shaft as possible to give the shaft maximum lubrication.

Pressure seals are of the O-ring type. They are seated in a groove machined in the housing containing the shaft. The seal groove dimensions are such that a compression or squeeze is put on the O ring when the shaft is installed. This squeeze provides an initial seal on the shaft. Pressure, applied down the shaft, packs the seal tightly against the far wall of the groove and the shaft surface. The higher the pressure, the tighter the seal packs against the groove and shaft. Deformation of the seal prevents oil from getting past the shaft or past the far wall of the groove (see the section on sealing devices).

The pressure type (O ring) seal is one of the best available, and it is also the least expensive. The shaft must have a ground finish, and the groove must not only be cut to close dimensions, but must have well finished surfaces, free of chatter marks and rings. A seal whose inside diameter is 5 to 10 percent larger than the shaft diameter should be used. In most instances, this will be the next largest seal available from the stock list. (For example, for a 1-in.-dia. shaft, a 1¹⁄₁₆-in.-I.D. seal should be used.)

The disadvantage of the pressure-type seal is that it involves heavy mechanical drag on the shaft, causes excessive wear on both shaft and seal, and generates high heats. In a light-duty, low-cost pump, a pressure seal is quite acceptable; but under high-pressure, continuous-duty conditions, seal life is too short and shaft attrition too great.

The wiper-type seal is a synthetic rubber or leather sealing member encased in a steel shell. The sealing member has a spring-loaded sharp lip which rides on the shaft. The seal lip is faced into (against) the pressure and wipes the oil in. The seal is press-fitted into a countergore machined into the outboard face of the shaft housing (or in a seal retainer on the outboard end of the housing). (See, for example, Fig. 7-13.)

The shaft must have a ground surface, and the lip of the seal must be smooth and continuous. The spring loading on the shaft is light; wear on shaft and seal is light but definitely present, particularly in dusty atmosphere. Effect of dirt on the shaft is minimized in many applications by installation of a double-lipped seal, whose second lip wipes dirt out. This type of seal is capable of withstanding only 35 to 50 psi pressure; consequently fluid must be bled off from behind the seal. Undoubtedly the most commonly used shaft seal, it is more expensive than pressure seals but less expensive than rotary seals. Machining the counterbore is a routine operation, and the cost is nominal. Seal life is relatively long and shaft attrition low. However, bleed-off provisions are sometimes difficult to make, and they add to machining cost even when simple. Unless special check-valve-type bleed passages are used, pumps using wiper seals are unidirectional.

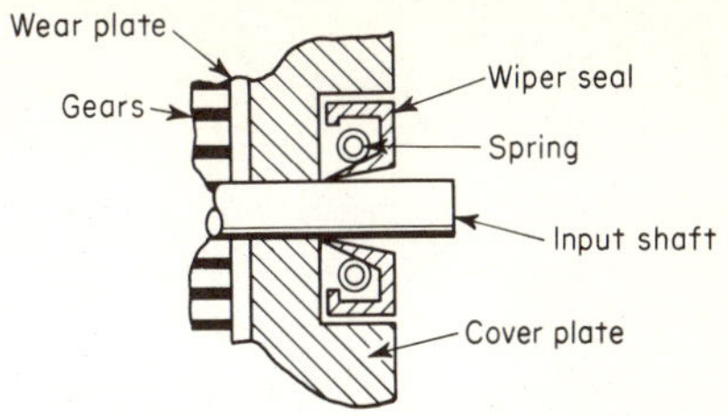

FIG. 7-13 *Stationary shaft seal.*

Rotary seals, also known as mechanical seals, seal by maintaining intimate (bearing) mechanical contact between the two primary elements of the assembly. The assembly consists of a stationary element, a rotating element, and a preloading spring (Fig. 7-14). The stationary element is mounted in the housing as far toward the outboard end as practical; the rotating element is mounted on the shaft. A static seal (usually an O ring) surrounds the stationary element, preventing oil from leaking out between the housing and the stationary element. Similarly, a static seal between the shaft and the rotating element prevents leakage down the shaft from behind the seal. A spring mounted behind one of the elements holds the two in direct bearing contact. Figure 7-14 shows the spring behind the rotating member, but

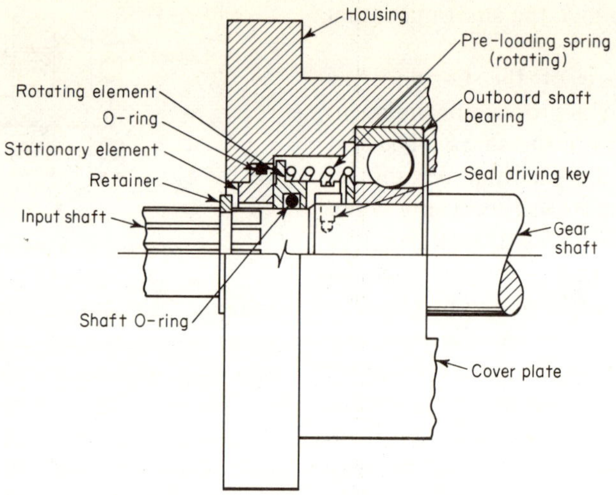


FIG. 7-14 *Rotating shaft seal.*

it must be pointed out that many designs have the spring behind the stationary element.

The materials of the two contacting elements must be of excellent bearing combination. Usually one of the pieces is hardened steel and the other is bronze. The surface of the steel piece is lapped; the bronze surface is very smoothly machined.

As it rotates, the shaft carries the rotating element with it. As it rotates, the rotating element is held in direct bearing contact with the stationary element. At zero and low pressures, this contact is maintained by the spring; at higher pressures, the pressure maintains the contact. As previously mentioned, static seals prevent leakage down the shaft and around the hole in the housing. Dynamic sealing is provided by the bearing contact of the two contact surfaces, and prevents oil leakage from the inboard pressure area to the outboard end of the shaft.

Other factors, such as unit bearing load of the surfaces, unbalanced pressure force, pressure gradient across the surfaces, and lubrication of the surfaces, must be considered in designing these seals, but they will not be discussed in this text.

The mechanical seal is generally regarded as the most effective shaft seal. Shaft attrition is light. Since it relies on pressure for its success, pumps with this seal can be bi-rotational. This seal is less affected by ingress of dirt from the outboard end of the shaft than any other type. This is particularly important in applications in which the outboard end of the shaft is exposed to heavy dirt accumulation, such as in construction equipment. However, the mechanical seal is far more expensive initially than other types and is also more expensive to install, since it requires special preparation of both shaft and housing.

### Performance Data and Pump Ratings

Performance data of pumps are shown in either graph or chart form. The charts show the delivery of the pump at various shaft speeds and pressure levels. Typical performance data are shown in Figs. 7-15 and 7-16.

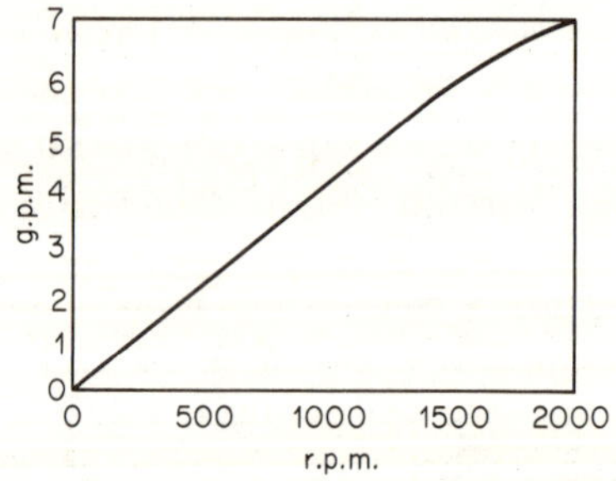


FIG. 7-15 *Typical pump performance data.*

 

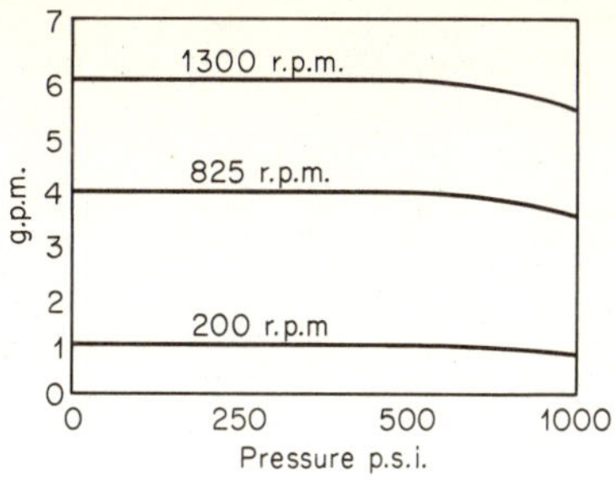


FIG. 7-16 *Pump performance curves.*

Pressure is measured in pounds per square inch (psi); volume is measured in gallons per minute (gpm); and shaft speeds are measured in revolutions per minute (rpm). Very often horsepower input is also measured. By comparing the changes of delivery at different pressures, one can note the effect of pressures and shaft speed on delivery.

Output of the pump is the sum total of the number of toothfuls of oil (see Figs. 7-4 and 7-5) carried by the gears per minute. It is apparent that the faster the gears turn, the more oil there is delivered. We can say with reasonable accuracy that output is proportionate to shaft speed (e.g., if the speed is doubled, the output will double).

The primary use of performance data is in the selection of a pump to meet the requirements of some application. This applies to determining the speed and pressure at which a certain size pump should be operated. Also, since most manufacturers make a family of pumps rather than a single size, the performance data of all pumps considered together allows the customer to select the pump size that best meets all his operating criteria.

The pump *rating* is a "shorthand" notation for the performance data. The rating denotes the output of the pump at some pressure and shaft speed (usually those for which the pump was designed) set by the manufacturer. The most common "rating" shaft speeds are 1200 rpm and 1500 rpm; the most common pressure is 1000 psi. Thus, a pump rated at 10 gpm would be one which produces about 10 gpm at the rated speed and pressure.

**Venting**

To prevent unwanted pressure from building up, certain areas of the pump are vented (bled off) to some point of low pressure. Such areas are shaft-wiper seals (Fig. 7-14), wear plates (Fig. 7-9), and the enclosed end of the drive shaft. Venting is accomplished by connecting the points of high pressure to low pressure through suitable passages either external or internal.

*External Venting*

A passage is drilled from the critical area to an area outside the pump at which a tube or hose connects to a tank (or to a return line). Oil entering the critical area flows out of the passage and back to the tank. Pressure in the critical area is thus usually prevented from exceeding atmospheric pressure; occasionally it may be slightly higher than atmospheric to overcome the resistance to flow in the bleed line. Since the critical area is usually the drive shaft wiper seal, the bleed line is most often connected to the pump in the vicinity of the drive shaft.

Pumps using this method of venting can be birotational, since the venting action has no relation to direction of shaft rotation. Although effective, this method is usually objected to because of the difficulties in running the vent line back to the tank (Fig. 7-17).

*Internal Venting*

A passage, or combination of passages, is arranged inside the pump to vent critical

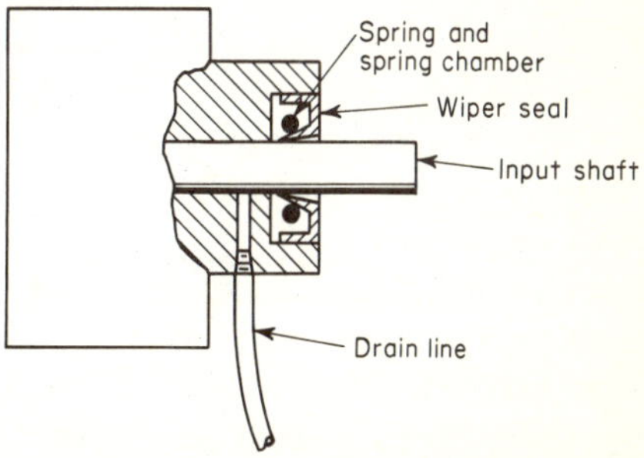


FIG. 7-17 *External drain system.*

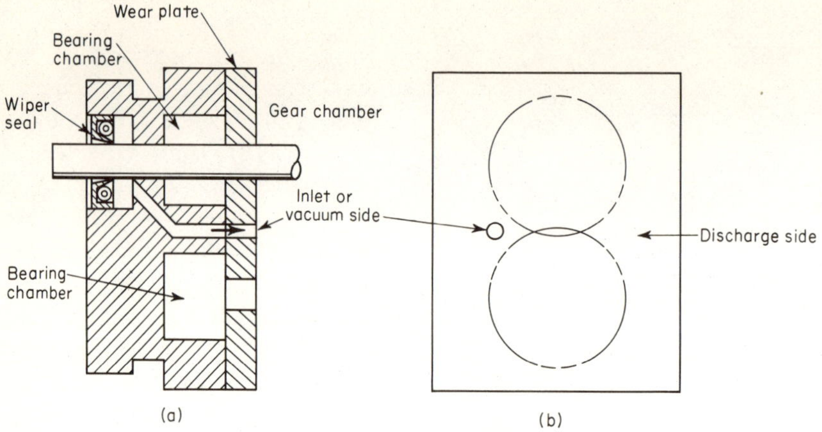


FIG. 7-18 *Internal venting.*

areas back to the inlet side of the pump. As the pump operates, the inlet side of the gear chamber creates a less-than-atmospheric pressure in the entire venting circuit. Oil entering the vented areas flows immediately to the inlet side of the pump and runs through the pumping cycle again.

The simplest form of internal venting is merely a hole drilled in the front housing directly from the inlet side of the gear chamber to the shaft area (Fig. 7-18). Because of casting configurations a single drilled hole is not always possible. Therefore, the seal chamber is often cored out (enlarged) to provide a place to drill into. In other cases, *two* intersecting drilled holes—one from the gear chamber down, the other from the seal chamber up—may be required to make a suitable passage.

Somewhat more complex is the practice of venting from the seal chamber through a passage to the idler gear bearing hole in the front housing, through the hollow idler gear shaft to the idler bearing hole in the rear housing, and finally through a passage to the inlet side of the gear chamber (Fig. 7-19). This method of venting is most common when areas additional to the seal chamber, such as wear plates and shaft ends, are to be vented. There is, however, a definite advantage to venting through the rear cover; with this method the direction of pump rotation can be changed by reversing the rear cover.

Internally vented pumps are usually unidirectional, and the correct direction of rotation must be used if the venting is to occur successfully. If rotation is in the wrong

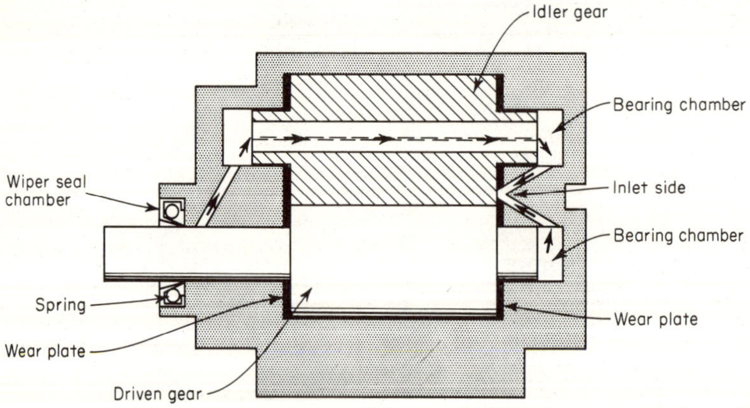


FIG. 7-19 *Internal venting through idler gear.*

 

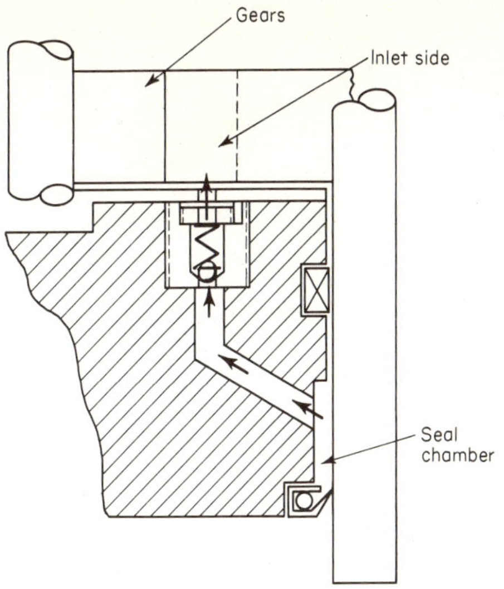


FIG. 7-20 *Birotational pump venting.*

direction, the venting circuit will be pressurized, and, inevitably, the front seal will leak severely. Often it will be blown out of its counterbore.

Birotation can be arranged by providing venting passages to both sides (inlet and pressure) of the gear chamber. The passages are most often made in the drive shaft cover plate, one on each side of the mesh point. These passages are then fitted with check valves, which permit flow when the side is being used as the inlet side, but which close under pressure, that is, when the side is being used as the pressure side (Figs. 7-20 and 7-21). In Fig. 7-21 the arrow shows the direction of flow when this side of the pump is the inlet side. Upon reversal of the pump, the check valve prevents the pressure from entering the sealed chamber. Regardless of the direction of rotation, the check valves will vent correctly, Although effective, such

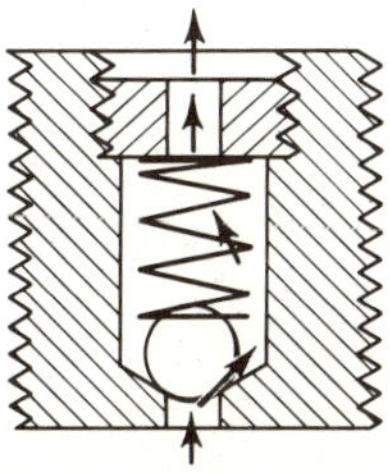

FIG. 7-21 *Internal check and flow direction.*

check-valve venting is expensive and in most instances unnecessary, since most pumps operate in only one direction throughout their operating lives.

## VANE PUMPS

The main parts of a vane pump are the rotor, the vanes, and the cam ring. These pieces are surrounded by front (shaft) and rear cover castings which contain the inlet and discharge ports, the drive shaft, and mounting provisions. Originally, vane pumps were what is today referred to as "unbalanced." Since the development of the "balanced" design, the "unbalanced" has become obsolete for all except the variable-volume vane pumps (which are discussed in Section 14).

As in gear pumps, the exterior location of the ports is of little importance. However, most vane pumps carry one port in the front casting and one in the rear. The interior position of the ports is extremely critical, however, since they are an important part of the pump timing.

The drive shaft extends from the front housing and, when coupled to a prime mover, rotates the pumping element. The drive shaft is the only shaft in the pump, since the pumping mechanism consists of a single rotating element. In many designs, the shaft extends through the pumping chamber and is supported on bearings fore and aft. Just as common, however, are designs which carry the shaft entirely in the front housing with the aft end protruding into the pumping chamber only to drive the mechanism.

The *rotor,* as its name implies, is the rotating element. A short cylindrical piece with radial slots cut in it, the rotor is usually driven by the drive shaft through a splined connection. Although the number of radial slots is determined by each manufacturer, and varies with the size of the rotor (larger rotors often having more slots), the most common numbers of slots seem to be 10 and 12.

The *vane* is a flat rectangular piece as long

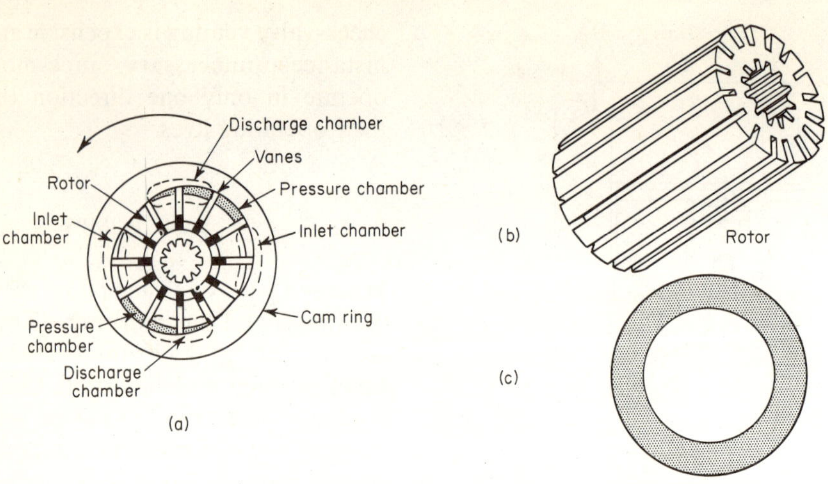


FIG. 7-22 *(a) Vane pump assembly. (b) Rotor. (c) Cam ring.*

as the rotor is wide; it is thick enough to fit closely in the slot provided, yet is able to slide freely. Each radial slot carries a vane. The end of the vane which extends beyond the periphery of the rotor is crowned, so that it presents a curved surface to the cam surface (Fig. 7-22).

The *cam ring* is the surrounding housing for the rotor and vanes. The word "cam" is derived from the action of the interior surface of the piece which performs a camming, or actuating, function on the vanes. Broadly described, the cam surface is an oval-shaped hole symmetrically disposed about the center of the rotor. In actuality, it is a carefully calculated series of curves, forming a closed path about the center, which guides the vanes to form the pumping chambers properly. Note (Fig. 7-22) that across its minor axis, the curve of the cam is essentially tangent to the rotor, whereas across its major axis, the cam surface is spaced away from the rotor.

### Principle of Operation

With the rotor set on the drive shaft, the vanes in the slots of the rotor, and the cam ring surrounding the rotor, the shaft is rotated. Rotation imparts centrifugal force to the vanes, moving them radially outward in their slots until their tips (or edges) contact the cam surface. As the rotor turns, the vane tips will traverse the curve of the cam surface. They move away from the rotor surface starting from one of the tangent points; then (at approximately 90 deg of rotor rotation) they turn and return to the rotor surface at the other tangent point (at 180 deg of rotor rotation). As each vane makes this traverse, it slides partially out of the slot in the rotor and back in again. After the cam curve passes center (180 deg of rotation), it again recedes from the rotor, with vane tips following, and returns to the rotor at 360 deg (Fig. 7-23).

It will be noted that as the vanes, following the cam surface (starting again at a tangent point), slide out of their slots, the volume enclosed by the cam surface, the two vanes, and the rotor starts essentially at zero and grows larger. As the space enlarges, it

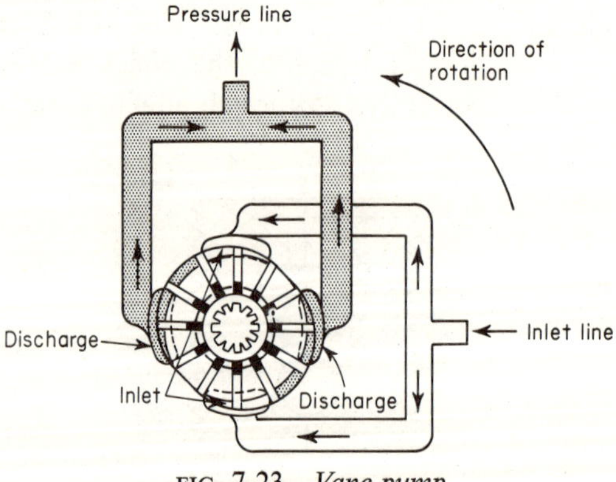


FIG. 7-23 *Vane pump.*

 

is evacuated. Atmospheric pressure in the reservoir forces oil into the vacated area as the cam surface recedes from the rotor; the vane cavity continues to grow, and the intake process continues. When the cam surface turns and becomes roughly concentric to the rotor, the intake process stops, and for a short time the vane cavity remains static. As the cam surface returns toward the rotor, the vanes are forced back into their slots, the vane cavity becomes smaller, and oil is forced out. The discharge process continues until the cam surface is virtually tangent to the rotor; at this point the vane cavity is essentially zero. The intake-discharge cycle requires 180 deg of rotor travel and is repeated exactly in the succeeding 180 deg.

### Port Configuration

The gradual expansion and contraction of the vane cavities produce relatively long intake and discharge cycles. In each case, so long as the vane cavity is changing size, the related port must be open to the cavity. This dictates that each port opening follow the path of the vane cavity; indeed, the shape of the port opening is essentially the plan view of the vane cavity.

### Pressurizing Vanes

Earlier it was stated that the vanes are forced against the cam surface by the centrifugal force that is imparted to the vanes by the rotation of the rotor. This is true at the start-up, before any pressure has built up. As soon as pressure is available, it is fed into the vane slot of the rotor behind the vane and pushes the vane tightly against the cam surface. Thus, although centrifugal force continues to act, it is pressure that maintains the tight contact of the vane against the cam surface (Fig. 7-23). Some designs include light springs behind the vanes to assist the initial (no pressure) sealing contact of vanes to cam.

### Design and Manufacturing Consideration Clearances and Tolerances

As with all pumps, the ideal vane pump would be one with zero clearance between all parts, and the continuing effort by all manufacturers is to approach that ideal. Clearances are reduced to the least possible that will still allow freedom of motion and maintain an oil film on all moving parts.

Close tolerance in fitting the parts is partially accomplished by very fine finishing techniques. Vanes and end-bearing surfaces are commonly lapped, the cam surface is ground, and the ends of the rotor are ground and sometimes lapped. The vane slots pose the most difficult finishing problem, but in most cases, they, too, are ground. Squareness and parallelism are also important to maintenance of close tolerances, since misalignments of parts require additional clearances. Two design characteristics that greatly assist in gaining squareness and parallelism are the following.

1. Rotor, cam ring, and vanes are all flat-sided pieces without projections, which assemble in their flat orientations. This makes these pieces natural for parallel grinding, with the result that they can be held within very close tolerances. [An 0.0002-in. total indicator reading (TIR) is a very common limit.]
2. The rotor is driven through a connection with the shaft (usually splined) which is relatively loose. This allows the rotor to float within the pumping chamber; it is not tipped or otherwise misaligned by the shaft. In other words, the shaft misalignment need not be reflected in misalignment of the rotor.

These two characteristics permit the pump parts to be manufactured with very slight end clearance, yet allow them to operate with remarkable freedom. End clearances vary with pump size and manufacturer, of course, but are generally from 0.0005 in. for small pumps (5 gpm) to 0.002 in. for large pumps (30 gpm).

### Pressure Plate

In addition to providing wearing surfaces, the pressure plates maintain the designed end clearance during operation. Most designs are so arranged that one wear surface is solidly based, and the other is relatively movable. The movable plate is held against the pumping chamber by some form of mechanical preloading, such as a spring. When pressure is developed, it is fed around behind the pressure plate and acts to hold the plate tightly in position and, at the same time, holds the whole pumping chamber tightly in assembly. Thus, it is pressure that, once established, holds the pumping chamber in its designed condition just as it holds the vanes in contact with the cam surface.

Pressure balancing, as practiced on gear pumps, is not done on vane pumps. But the pressure-plate application described above is similar inasmuch as the pressure behind the plate is analogous to that in the pressure portion of the pump chamber.

### Pressure Forces and Strength of Materials

Creation of pressure develops forces within the pumping chamber which can distort or burst the chamber. Since pressure always moves from high to low, the resultant force vector is from discharge toward intake.

In terms of the entire pumping chamber, this means that a pressure force is constantly applied that acts as a force trying to move the rotor away from pressure toward intake. Here we see the value of the "balanced" structure. If we consider only half of the pumping chamber, namely, one suction portion and its corresponding pressure portion, we can see that the force moves the rotor toward suction and also, generally, away from the major axis of the cam ring. This causes high bearing loads and, ultimately, contact of the rotor with cam surface, which results in increased wear. (The above is a thumbnail description of the problems of the "unbalanced" pump.)

In terms of the whole pump again, it is apparent that the pressure forces are balanced and that the net side force against the rotor is zero (Fig. 7-24). The only unbalanced forces within the pump are those attempting to burst the pumping chamber and those against the vanes attempting to move oil from the high-pressure side back to the low-pressure, or inlet, side.

The burst forces are counteracted by high-strength materials, rigid structure, and pressure balancing. The cam ring, if it is an external part (part of the housing), is massive in size and has high strength as well. If it is an internal part, it is supported all around by heavy housing structure. Earlier, we mentioned pressure plates and the pressure balancing applied to them. Additionally, these plates are either thick in themselves or are backed by a thick or rigid structure. These heavy structural pieces are specifically designed to resist distortion from pressure.

Undoubtedly, the most critical strength problem is in the vanes. The geometry and strength requirements of the rotor dictate that the vane slots be narrow. The vanes must, therefore, be thin. However, at the point in the pumping cycle at which the pressure port begins, the maximum pressure of the system is applied against the vane. The oil is attempting to move past the vane to the inlet area. (It should be noted that this cycle, in which there is maximum pressure through the entire pressure portion of

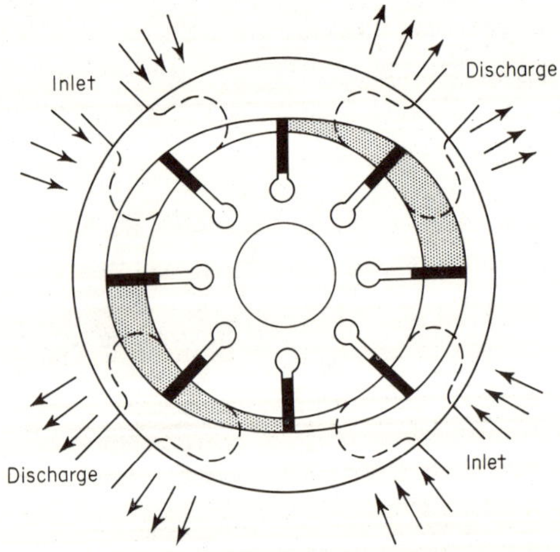


FIG. 7-24 *Pressure on cam ring.*

 

the cycle, differs remarkably from the cycle in a gear pump. There a gradient exists around the periphery of the gear, and because the port covers the entire pressure area of the vane pump, the maximum pressure applies throughout the area.) It is at this point that the vane is most extended from the slot. Thus, a very high cantilever load is applied against the vane; the strength requirement is extreme. Consequently, tool steels, which combine high strength and toughness, are universally used for vanes.

Rotors must also be of high-strength material, since they must support the vanes. With the geometry of the radial slot in mind, it should be noted that the support must come at the smallest part of the rotor—near the center.

High-strength hardened parts are necessary not only because of pressure but because they must be wear resistant. In a vane pump, the vanes come into continuous sliding contact with all other surfaces. Minimum clearance, as in gear pumps, is not enough. Resistance to and reduction of wear are also aided by using extremely fine finishes on all surfaces. Finally, lubrication is important for wear resistance; it is vitally important that oils with good lubricity be used in vane pumps.

#### Performance Ranges

The vane pump is classed as a medium-pressure pump, and the usual operating range is from 500 to 1500 psi. Gallonages are merely a matter of size. Pumps delivering up to 100 gpm exist, but the most common ranges are from 5 gpm to 40 gpm.

Shaft speeds are usually those of electric motors or crankshafts, since these are the most common prime movers. However, the low inertia of the vane structure and long intake and discharge cycles equip the vane pump for high-velocity operation. Shaft speeds up to 4000 rpm are often used. This high-speed capability is important for many applications, since it permits the use of smaller, lighter components without sacrificing the volume of oil delivered. (Note that the volume is approximately proportional to shaft speed. Thus a pump turning at twice the speed of another can be half the size yet deliver the same quantity of oil.)

Since it is important to some applications, it should be mentioned that vane pumps are comparatively quiet and free of pulsations.

As with gear pumps, the interest in higher pressures also affects vane pumps. In recent years, vane pumps rated up to 2500 psi have become available. To accomplish this, changes have been made that strengthen the vanes, reduce tip pressures against the cam surface, and increase the sealing capabilities of the tips. Various approaches to this problem involve the use of: (1) double vanes to obtain laminar strength and increase tip area; (2) hollow vanes to create an oil cushion and reduce tip wear; and (3) internal pressure-compensating vanes to damp out flutter. Additionally, the eccentricity of the cam ring can be reduced; this reduces the cantilever load on the vanes and improves their relative strength. This change in eccentricity reduces the pump volume per revolution. However, many of these higher-pressure pumps operate at increased shaft speeds, which offset the reduced volume. The higher shaft speeds also increase the relative viscosity of the oil, thus improving the oil seal.

To this writer's knowledge, no change has been made in bearings as a result of the increased pressures. The balanced rotor neutralizes bearing loads, and the overall pump design lends itself to the use of ball bearings, which are high-speed bearings anyway.

#### Position in the Field, Efficiencies, Service Life

Vane pumps are used in probably the widest variety of applications of any type. Without question, they are the most common in machine tools and are also used in construction machinery, material handling, and aircraft. This wide acceptance can be attributed to many desirable operating characteristics: wide range of shaft speeds; low noise; negligible bearing load, which extends bearing life; and compact size.

Vane pumps operate in approximately the same range as gear pumps, although they are capable of slightly higher continuous-duty pressures and considerably higher shaft speeds. Their volumetric efficiency is slightly higher than that of gear pumps; it is generally accepted as 85 to 90 percent throughout most of the operating range. Mechanical efficiency is also slightly higher; it is approximately 75 to 80 percent. (Both these efficiencies will drop slightly as pressures and shaft speeds increase.) But throughout their performance range, vane pumps satisfactorily meet the competitive demands of industry.

The most serious service shortcomings of vane pumps are their sensitivity to dirt, their need of high lubricity in the oil, and the fact that high-frequency vibrations often develop in the vanes. All three problems have the same result: deterioration of the surfaces of the pumping chamber and loss of efficiency.

Most of the sealing in on hydraulic systems is done by the vanes. A seal must, of course, be maintained between the rotor and the pressure plates, but the primary barrier between any two vane cavities is the vanes themselves. All vane surfaces are affected because the tips must seal against the cam surface, the ends must seal against the pressure plates, and the sides must seal against the seal in the slots. The sealing must be accomplished by little more than a line.

To be effective, this line must be in continuous, rubbing contact with the bearing surface. A break in the surface, such as a score mark, represents a substantial loss of seal. Score marks on the pressure plates between the discharge and intake ports (caused, for example, by foreign particles becoming lodged under the vanes) are especially damaging, since they short circuit the pump.

Lack of lubricity affects most the tips of the vanes and the cam surface. Oil pressure applied behind the vane exerts a very high unit bearing pressure on the tip of the vane. The vane tip must be lubricated extremely well to effectively resist wear. It is not enough to assume that the pump is full of oil and that, therefore, adequate lubrication exists. Lubricities of oil vary, and the vane pump requires an oil with high lubricity even at elevated temperatures such as 200°F. Inadequate lubrication causes galling between tip and cam surface. The cam surface becomes torn and the vane-tip wear excessive. This wear is self-increasing, and extensive damage quickly results.

High-frequency vibrations of the vanes are the result of many factors, most of them interrelated, such as shaft velocity, pressure variations, and cam curvature, all of which are beyond the scope of this text. The effects of this show on the cam surface as ripples caused by the vane tips. These ripples are also self-increasing, and they ultimately become notches, which are extremely destructive.

## PISTON PUMPS

Piston pumps possess two operational characteristics that establish their position in hydraulics.

1. They are capable of developing high pressures.
2. They readily lend themselves to variable-volume output.

Since only an introductory discussion of piston pumps can be given, this text will concern itself with constant-volume pumps and item (2) will not be discussed.

Constant-volume piston pumps can be divided into three broad categories: those with (1) radial drive; (2) in-line drive; and (3) off-set drive. There are many similarities among these types, since the functional components are substantially the same for each. Although there is some overlapping of performance, however, each of these combinations has a fairly well-defined position.

### Radial Pumps

A typical radial pump consists of a cluster of cylinders disposed radially and spaced evenly about a center, like spokes of a wheel. The center of the circle is the driving element; each spoke is a pumping unit.

The driving element is a cam or eccentric, generally circular in character, which is driven by a shaft in the center. As with all pumps, the shaft extends to the outside and is coupled to the prime mover. The shaft is appropriately supported by bearings. The cam is some suitable shape with alternate lobes and valleys. Each pumping unit is a cylinder, and is comprised of a barrel, a piston, a return spring, and suitable valving.

The pistons are assembled in their respective barrels with the return springs arranged to hold the pistons in constant contact with the cam. The pistons, almost universally, are cylindrical plugs long enough to hold themselves straight and true in their barrels. Most designs have a cam roller attached to the piston to give a rolling contact of the piston with the cam. Note that the discharge check valve in Fig. 7-25 is closed; this prevents high-pressure backflow. It is open on the pumping cycle, which permits oil discharge to the collection ring. The inlet check valve prevents high-pressure oil from returning to the inlet area during the pumping cycle. This valve opens and permits the piston chamber to be filled on the intake stroke.

The valves are located at the end of the cylinders opposite the cam. Most often, the valves are ball checks. There are two checks per cylinder—one intake and one discharge. Each ball check has a ball lightly held against a seat by a spring. The seats are juxtaposed to the cylinder so that one ball opens to inlet and the other to pressure. It should be evident that when one ball is open, the other is closed.

Intake and discharge manifolds surround the cylinder structure. In many designs, the discharge (pressure) manifold is like the rim of a wheel, and all the cylinders (spokes) empty into it. Almost all designs use the kind of housing that encases the cylinders as they do the intake manifold.

In operation, when the shaft is rotated, the cam is rotated. As a lobe contacts a piston, it pushes the piston radially outward in its barrel. As the lobe passes the piston and a valley is presented, the return spring forces the piston radially inward as far down the barrel as the depth of the valley permits. Additional lobes and valleys follow in endless succession, which results in a reciprocating motion of the pistons.

As the pistons move inward from their maximum upstroke position, they evacuate the barrel. This drop in pressure opens the appropriate ball check and sucks oil into the

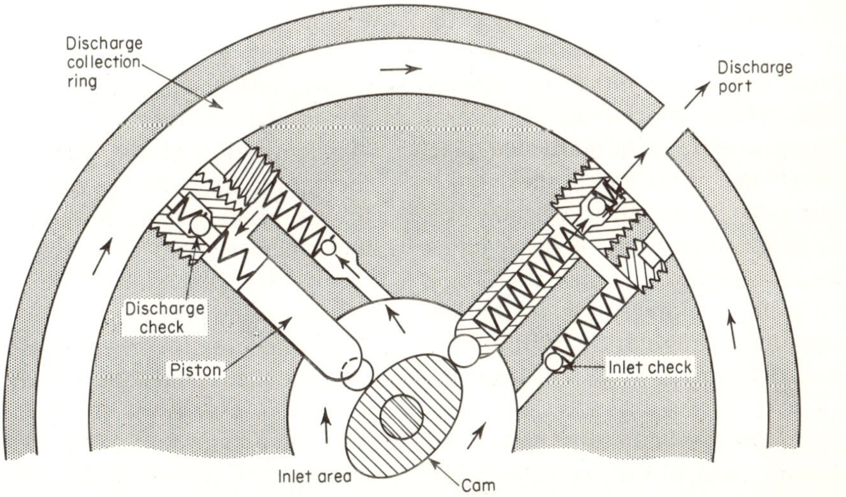


FIG. 7-25 *Radial piston pump.*

barrel. This suction (intake) continues until the piston has reached the bottom of the cam valley. As the lobe carries the piston up (outward), the positive pressure created closes the intake valve and, simultaneously, forces the discharge ball check open. Oil under pressure is forced out into the discharge manifold and, ultimately, to the system.

Pumps of this design usually contain an even number of cylinders, with each cylinder 180 deg away from another. The cam is designed to have pairs of lobes and valleys 180 deg apart so that pairs of opposing cylinders are simultaneously on their pressure strokes. This balances all bearing loads on the shaft.

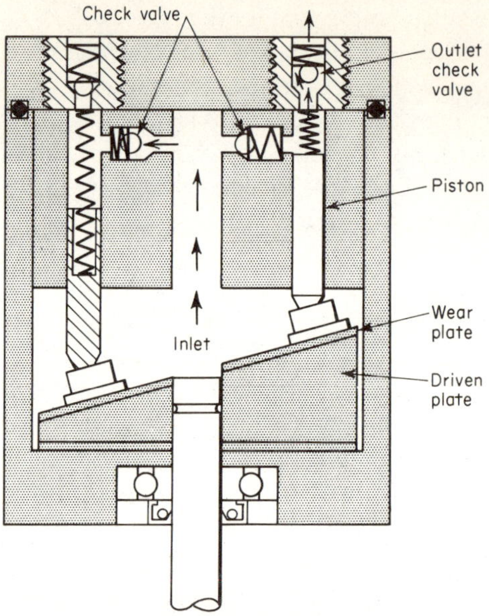

FIG. 7-26 *In-line piston pump.*

### In-line Pumps

The typical in-line pump consists of a cylinder block containing a number of pistons, a drive plate, a drive shaft, and suitable valving.

The cylinder block is a cylindrical piece (usually a cast-iron casting) containing a number of axial holes spaced around the circle of the block. Each hole is a cylinder and contains a piston. This entire cylinder assembly is positioned axially (in-line) with the drive shaft.

The drive shaft is located at one end of the cylinder block and is concentric with it; it extends outside the body to couple to the prime mover. Depending upon the design, the internal end drives either the driving plate or the cylinder block (Fig. 7-26).

The driving plate is a hardened steel washer positioned at the end of the cylinder block where the drive shaft is located. It is set at an angle to the axis of the cylinder block, and is supported at this angle by a truncated cylindrical bearing block behind it. Because of its angular disposition and the motion it imparts to the pistons, the driving plate is referred to as a "swash plate," a "wobble plate," a "nuttating plate," etc.

The pistons, lying in their cylinder bores, are arranged to maintain bearing contact with the driving plate. Again depending upon driving method, this contact is maintained by either spring loading the pistons against the plate or by attaching the pistons to the plate by shoes.

The valving is accomplished at the end of the cylinder block opposite the drive shaft. The valving is either a pair of ball checks for each cylinder—one for intake and one for discharge—or a valve plate containing intake and discharge ports. The type of valving is determined by the driving method.

In some designs, the driving plate is rotated by the drive shaft; in this case, the cylinder block remains stationary. As the driving plate rotates, the pistons, following the plate, are pushed into their cylinder barrels as the "high" portion of the plate moves under them, and are pulled partially out of their barrels as the "low" portion of the plate moves under them. Continuous rotation of the wobble plate imparts a reciprocating motion to the pistons.

In this design, the pistons are usually spring loaded against the wobble plate. The valving is usually of the ball-check type, with the pressure-operated checks opening into a pressure manifold and the intake checks opening into the pump body.

 

In other designs, the drive shaft rotates the cylinder block; the driving (swash) plate remains stationary. The reciprocation of the pistons is identical with that in the other method, since the pistons follow the slanted surface of the swash plate. The pistons are most often held in contact with the swash plate by shoes on the ends of the pistons; the shoes are held in place by a cover ring (Fig. 7-27).

The valving is done by a valve plate with single discharge and intake ports, and is similar to the porting of a vane pump. As the barrel rotates through the low to high portion of the swash plate, the pistons are on their pressure stroke; consequently the pressure port covers this area. As the barrel moves from the high to low portion of the swash plate, the pistons are on their intake stroke; consequently the inlet port covers this area of the cylinder circle. An odd number of pistons is usually used.

### Performance and Position in the Field

Piston pumps are classed as high-pressure pumps. Although pressures of 8000 to 10,000 psi are common, most pumps operate in the

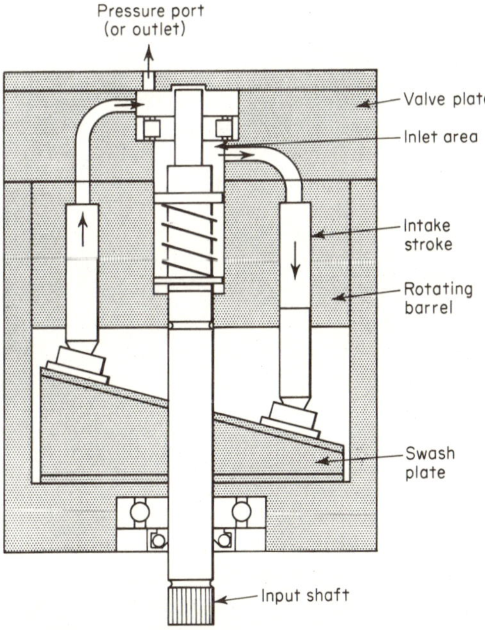


FIG. 7-27 *Rotating cylinder block in-line pump.*

2500 to 4000 psi range. Large-gallonage pumps are rather rare; the vast majority are within 25 gpm, and the most common sizes are 10 gpm or less. Piston pumps are capable of very high shaft speeds with very little loss of efficiency. However, most pumps operate at electric-motor shaft velocities (1750 to 1800 rpm), since these are the most common prime movers.

Piston pumps have the highest volumetric efficiencies of all pumps; 90 to 95 percent is common. Mechanical efficiencies are not necessarily higher, however, since piston pumps require more power to turn.

Since check-valve porting provides positive sealing, the very high-pressure pumps use this type. Valve-plate porting is used in most pumps that have operating pressures of up to approximately 3000 psi. Not only is it satisfactory in this range, but it is much cheaper to manufacture and permits more competitive prices. Since gear and vane pumps are now approaching this range, price is definitely a consideration.

Piston pumps are widely used in machine tools, especially in hydraulic presses, in which high pressures are mandatory. They are also widely used in aircraft, for which small size and light weight are highly desirable. Since they are very sensitive to dirt, they are seldom used for mobile equipment.

Piston pumps must be manufactured to the highest standards of precision if they are to perform satisfactorily at elevated pressures. Cylinder barrels must be honed, and pistons and cam surfaces lapped. Since the pressure forces are high, working parts are highly stressed; hence, hardened alloy steels are used extensively. The design is more complex than other types. All these factors add cost. This results in prices which, as a class, are substantially higher than those for gear or vane equipment.

But it is not cost of the pump alone that makes piston pumps expensive. The entire system must be comprised of components matched to the elevated pressures; hoses and tubes must be stronger, valves must have heavier sections, and cylinders must be stronger. Also, careful filtration is necessary to protect the pump against contaminated oil. Almost all these components are higher priced than are comparable components for lower-pressure systems. Therefore, the total cost of a high-pressure system is usually considerably higher than that of a medium-pressure system. It is this cost, perhaps more than any other factor, which limits the use of piston pumps to those applications in which their special qualifications are highly advantageous.

 

# ACTUATORS

# 8

## MOTORS

Just as an electric motor is essentially the converse of an electric generator, so the hydraulic motor is the converse of the hydraulic pump.

Under the broadest definition of a motor, which includes reciprocating as well as rotating actuators, cylinders can be classed as motors. This classification, however, is usually reserved for cylinders which are so controlled that they reciprocate continuously, thereby performing some cyclical function. There is nothing to distinguish a reciprocating motor from a power cylinder; the difference lies in the valving.

Rotary motors are the type that usually come to mind when the subject is presented. They can be broadly classified as having either nonpositive or positive displacement. Nonpositive-displacement motors are waterwheels and Pelton wheels. Each of these machines converts the kinetic energy of water striking an exposed surface (blade) of the wheel into work at the axle of the wheel. Since these devices are not used in fluid power as we consider it, they will not be discussed in this text.

Positive-displacement motors, like pumps, are one of three types: gear, vane, or piston. Each of these types has the same parts and construction as its pump counterpart, and we need therefore discuss only the operation and functions of motors.

In all cases, oil is pumped into the motor to act on the driving mechanism. The spent oil discharges from the opposite side of the motor and returns to the tank. Work is accomplished by the conversion of the oil pressure into torque on the shaft. Volume produces shaft speed.

Whereas a pump converts torque on the shaft into pressure in the oil, a motor converts pressure in the oil into torque on its shaft.

### Gear Motors

Figure 8-1 shows a pair of motor gears in their housing. Oil enters the motor through the inlet port, travels around the periphery of the gears, and exhausts from the discharge port. As it sweeps around the gears, it rotates them. The volume of oil going into the motor determines the shaft speed. Within reasonable limits, determined by port

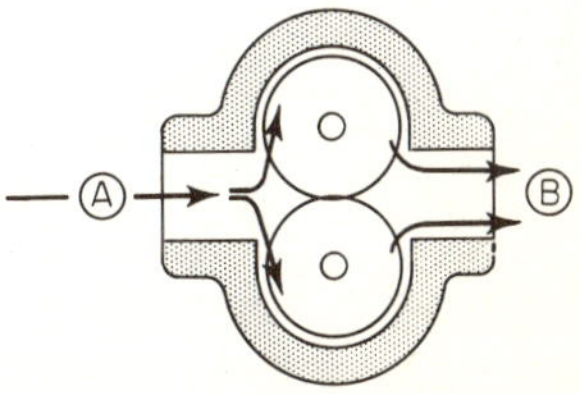


FIG. 8-1 *Motor cross-section.*

sizes and internal restrictions and slippage (efficiency), the shaft speed is directly proportional to the volume of oil impressed across the motor.

Torque is developed by the application of pressure to the exposed areas of the gear teeth. The pressure is a result of resistance to rotation. Figure 8-2 is an enlarged drawing of the teeth at the mesh point. Oil under pressure is applied at point *A*. Point *B* is at low pressure. The pressure differential across the gears, that is from point *A* to point *B*, develops the force which creates the torque on the shaft.

Another way of understanding this is to observe that in all but one tooth space, the pressure acts equally against both sides of the space and so balances out. The exception is at point 1, where side *A* of the tooth space is exposed to high pressure, whereas side *B* is exposed to discharge pressure. This unbalanced pressure force creates the torque on the shaft.

As we have learned in the section on mechanics, $T = Fr$. That is, torque equals a force acting through a radius. From Pascal's law, we know that $F = pA$; that is, force equals unit pressure times exposed area. Combining the two, we have

$$T = pAr$$

where $T$ = torque at motor shaft, in.-lb
$p$ = input pressure, psi
$A$ = exposed tooth area, sq in.
$r$ = radius from center of tooth area to center of shaft, in.

Usable pressure across the motor is the differential between input and discharge pressures. Also, mechanical efficiency of the motor determines the amount of torque available for useful work at the shaft. Inserting these two factors into our equation, we have

$$T = (p_i - p_0)Ar$$

where $p_i$ = input pressure and $p_0$ = discharge pressure.

### Vane Motors

Figure 8-3 shows an enlarged view of the rotor-vane assembly across the port area. Oil under high pressure enters the motor (Fig. 8-3) at *A* and moves toward discharge at *B*. Whereas the vane at *A* is balanced by pressure applying equally to both sides, pressure on the vane at *B* is not balanced. Torque is developed by this unbalanced pressure force, which is applied against the extended vane area at the major axis of the cam ring. The radius through which this force acts is the distance from the center of the extended vane area to the center of the rotor.

The movement of oil from the point of high pressure *A* to low pressure *B* carries each successive vane with it, thereby rotating the shaft. Since a vane motor is a constant-volume unit, shaft speed is directly proportional to the volume of oil pumped through the motor.

The balanced vane motor is the only type in commercial use. Like its pump counterpart (Sec. 7), the balanced motor has two pressure areas, which are directly opposite each other across the rotor. Equal pressure forces apply simultaneously in each pressure area. This not only balances the loads on the rotor but doubles the torque applied to

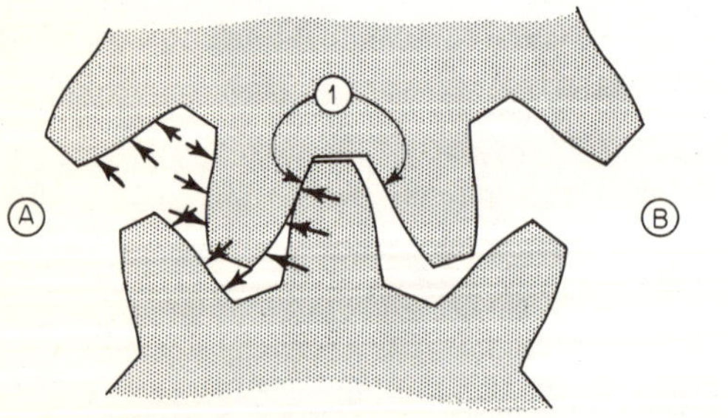


FIG. 8-2 *Forces on gear teeth.*

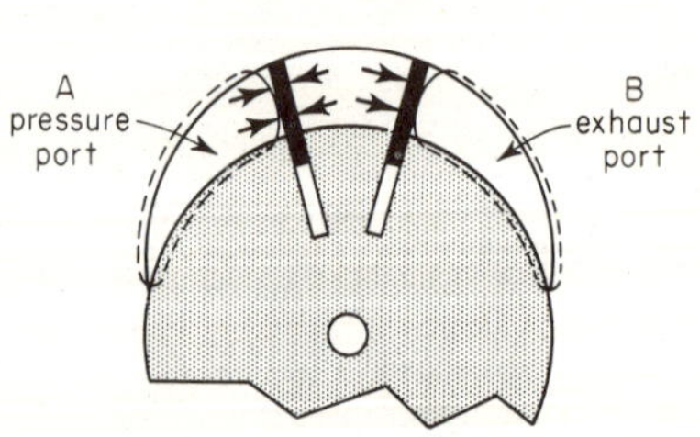


FIG. 8-3 *Pressures on vanes.*

 

the shaft. The structure of vane motors is such that the exposed vane surfaces are relatively small; the force couple produced by the double-pressure force application permits the generation of a total torque equivalent to other motors of similar pressure-volume specifications. The following equation is almost identical to the torque equation for gear motors. This is logical, since the mechanisms function in very similar manners.

$$T = 2Ar(p - p_0)$$

where $2A$ = areas of two extended vanes
$r$ = distance from center of vane area to center of rotor
$(p - p_0)$ = pressure differential across motor

**Piston Motors**

The operating principle of a piston motor is basically the same as that of a lift cylinder. Oil under pressure is applied to a piston and forces the piston to move linearly in its barrel. As it moves, the piston moves whatever load is attached to it through the piston rod.

In a piston motor, oil under pressure enters the piston chamber through intake port openings. Each piston in its turn is forced down against the inclined cam. Since the piston moves in a straight line, its contact with the cam moves the cam surface away from the piston, imparting a turning moment to the cam surface. This turning moment develops torque on the shaft. As in other motors, the pressure differential across the ports determines the magnitude of torque.

Piston motors are available in axial and radial styles with fixed and variable displacement. The cam surface mentioned above consists of a swash plate in axial designs and a cam rotor, or ring, in radial styles. Shaft speed is a function of oil volume delivered to the motor. In a fixed-displacement motor, shaft speed will vary directly with volume. Since torque is a function of cam eccentricity, it will remain constant regardless of shaft speed. Therefore, fixed-displacement motors are variable-horsepower units.

Variable-displacement motors are usually arranged to maintain constant shaft speed. Under this condition, torque varies inversely with input volume and horsepower remains constant.

Since calculation of torque involves rather complicated geometric determinations, it will not be included in this text.

*Efficiencies*

The volumetric efficiency of a hydraulic motor is the ratio between the geometric displacement at a given shaft speed and the actual measured input volume at the same shaft speed under various pressures. That is,

$$e_v = \frac{Q_g}{Q_a} \times 100 \qquad \text{(percent)}$$

where $Q_g$ = geometric displacement and $Q_a$ = actual displacement.

For gear motors this efficiency will be approximately 85 to 90 percent; for vane motors, approximately 90 percent; for piston motors, 90 to 95 percent.

Mechanical efficiency is the ratio between actual output power (determined from measurement of torque and shaft speed) and theoretical horsepower (calculated from the classic torque equation at the same shaft speed). That is,

$$e_m = \frac{hp_a}{hp_t} \times 100 \qquad \text{(percent)}$$

For gear motors, this efficiency is approximately 80 to 85 percent; for vane motors, approximately 85 to 90 percent; for piston motors, approximately 85 to 90 percent.

Total efficiency (overall efficiency, or $e_0$) is the product of volumetric and mechanical efficiencies:

$$e_t = e_v e_m \times 100 \qquad \text{(percent)}$$

For gear motors this efficiency will be approximately 70 to 75 percent; for vane motors, 75 to 80 percent; for piston motors, 80 to 85 percent.

*Power Calculations*

Horsepower output from a motor can be calculated from the equation:

$$hp_o = \frac{T2\pi n}{33{,}000}$$

where $T$ is in foot-pounds.

Torque can be measured by a Prony brake and revolutions can be measured with a tachometer.

Horsepower input to a motor can be determined from measurement of input volume and pressure differential across the motor.

$$hp_i = \frac{Q(p - p_o)}{33{,}000 \times 12}$$

where $Q$ = volume in cubic inches per minute and $(p - p_o)$ = pressure differential in pounds per square inch. This same equation can be written:

$$hp = \frac{G(p - p_o)}{1714}$$

where $G$ = volume in gallons per minute and $(p - p_o)$ = pressure differential in pounds per square inch.

Overall efficiency of the motor is the ratio between horsepower output and horsepower input.

$$e_t = \frac{hp_o}{hp_i}$$

This calculation should compare favorably with determinations made from the earlier equation for $e_t$.

*Performance Characteristics*

The primary purpose of a hydraulic motor is to develop torque. This it does very effectively. A hydraulic motor will deliver more horsepower from less space and weight than any other known power source. It can provide the widest range of shaft speeds of any power source. The torque output of a hydraulic motor is very nearly constant throughout the entire speed range. This means that the starting torque is almost as high as the running torque.

The availability of constant torque, variable horsepower, variable speed, and constant horsepower motors, coupled with constant- or variable-volume pumps, provides the opportunity for power transmissions and hydrostatic drive systems of almost unlimited performance combinations.

Hydraulic motors remain unaffected by most working atmospheric conditions. The only other type of motor that can be installed as easily is the electric motor. Control of hydraulic motors is extremely flexible and can be extremely delicate. They can be instantly reversed with relatively little strain on moving parts.

The disadvantages of hydraulic motors are that high losses occur when they are used with long transmission lines and that they have a relatively low efficiency as compared with straight-through mechanical drives. This usually precludes the use of hydraulic motors that are located at a substantial distance from the pump. For simple drive systems not requiring extreme performance variations, mechanical power trains are usually used.

## CYLINDERS

Motors provide the torque force in fluid power. Hydraulic cylinders produce the lineal force. These are the two principal methods used to produce useful work. Limited rotary actuators are available for special applications.

Cylinders are constructed in a wide variety of configurations to meet the needs of various fluid power systems. Special sizes and designs are manufactured to meet specific requirements of a machine or a circuit. Factors considered in design are as follows:

1. Force required (in both directions of travel)
2. Speed of rod travel required (in both directions of travel)
3. Operating and maximum pressures required
4. Weight and size limitations
5. Type of fluid used in the system

 

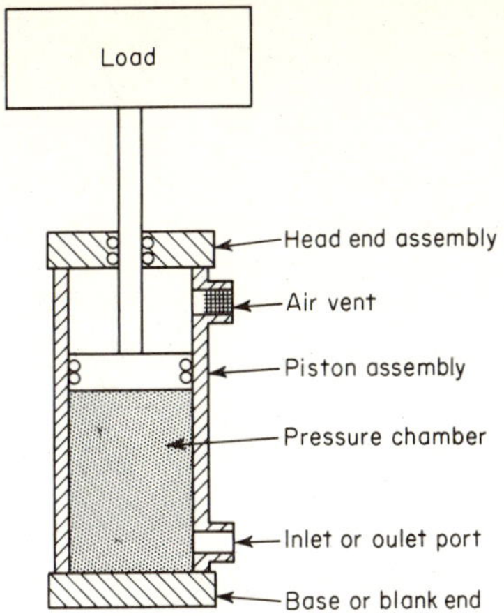


FIG. 8-4 *Single-acting hydraulic cylinder.*

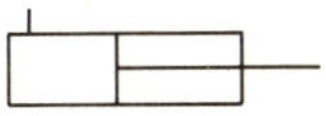

FIG. 8-5 *Symbol for single-acting hydraulic cylinder.*

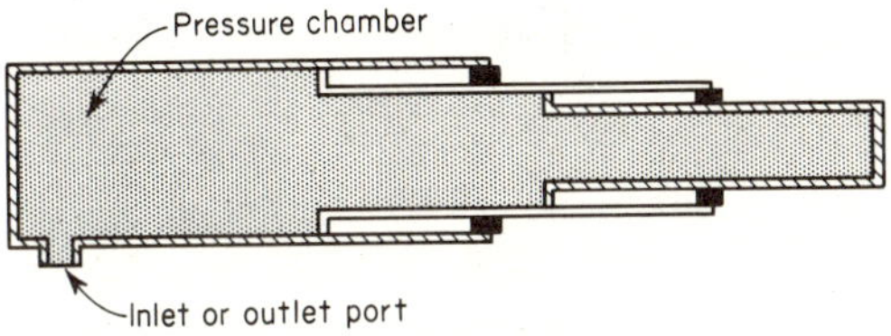


FIG. 8-6 *Telescopic cylinder.*

### Types

Basic cylinder design consists of a piston and rod, which are forced by fluid to travel in the cylinder. Four basic types are available: single acting; telescoping; double acting; and double rod end.

In *single-acting cylinders* the load moves the cylinder in one direction while the fluid from the pump is used to actuate its move-

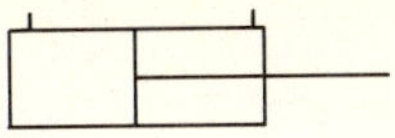

FIG. 8-8 *Symbol for double-acting cylinder.*

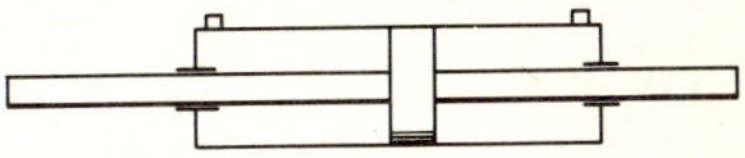

FIG. 8-9 *Double-rod-end cylinder.*

ment in the other. Figure 8-4 shows this cylinder and its component parts. The symbol for a single-acting hydraulic cylinder is shown in Fig. 8-5. These cylinders may be operated horizontally if a spring is installed either in the rod end of the cylinder or externally.

*Telescoping cylinders* are used where initial length is limited and long rod travel is required. Figure 8-6 shows a single-acting telescoping cylinder.

*Double-acting cylinders* are the type (Fig. 8-7) most commonly used. The piston and rod are moved by the fluid, which is directed either to the blank or to the rod end. During one direction of travel, the fluid from the opposite end is returned to the tank. Figure 8-8 shows the symbol for a double-acting cylinder.

*Double-rod-end cylinders* are used when work must be performed at both ends of the cylinder. They have the same features as do double-acting cylinders except that two rod ends are used. These cylinders have the same speed of piston travel in both directions with a given volume of flow. Also, their maximum-force capacity is the same in both directions. A double-rod-end cylinder is shown in Fig. 8-9.

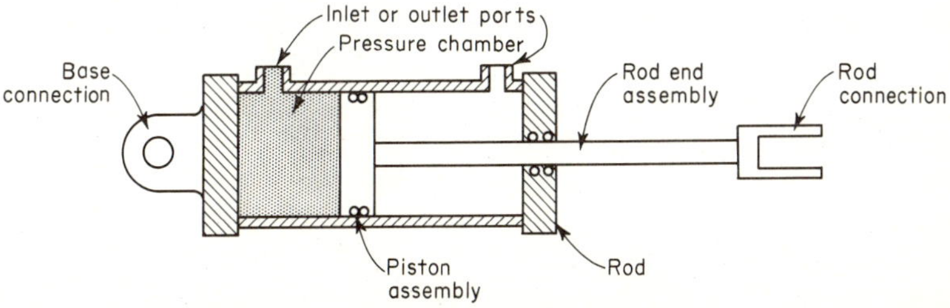


FIG. 8-7 *Double-acting cylinder.*

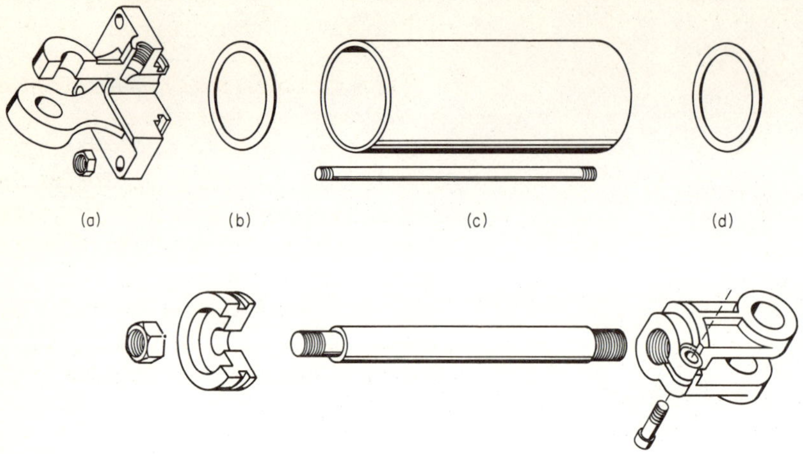


FIG. 8-10 *(a) Blank end. (b) Seal. (c) Cylinder and tie rod. (d) Seal. (e) Head end. (f) Piston. (g) Rod. (h) Clevis rod connection.*

**Component Parts**

Figure 8-10 shows the component parts of a double-acting hydraulic cylinder. Packing and seals are further discussed in Section 9.

**Cylinder Force**

The force or thrust available from a hydraulic cylinder (Fig. 8-11) is dependent on two factors: (1) the area of the piston; and (2) the pressure exerted by the fluid. The maximum force is the area multiplied by the maximum pressure of the system. The maximum force that a cylinder is capable of producing will vary depending on its direction of travel unless it is a double-rod cylinder, in which case its maximum thrust is the same in both directions. The rod reduces the available area on a piston that the fluid may act on; thus, a lower maximum force is available while a cylinder is retracting. (It must be remembered, however, that all systems have a maximum pressure available, which is set by the relief valve.)

*Example:* Given the following data for a cylinder, determine the pressure that the cylinder must generate to accomplish its task: Cylinder I.D. = 6 in.; rod diameter = 4 in.; rate of flow = 1 gpm (231 cu in./min); maximum pressure = 1000 psi (relief valve setting); maximum force while extending = area × maximum pressure ($F = Ap$); area = $(3.14 \times 6^2)/4 = 28.26$ sq in.; maximum force = $28.26 \times 1000 = 28{,}260$ lb; maximum thrust while retracting = effective area × maximum pressures; effective area = piston area minus rod area = $(3.14 \times 6^2)/4 - (3.14 \times 4^2)/4 \times 1000$ lb; area = $(28.26 - 12.56) = 15.75$ sq in.; maximum force = 15.75 sq in. × 1000 lb = 15,700 lb.

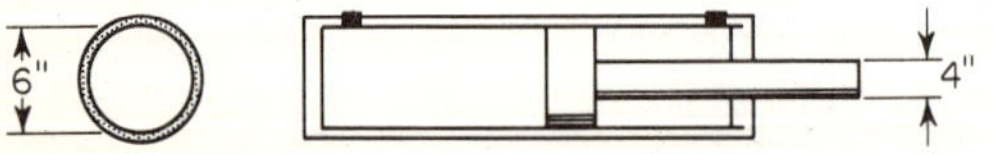


FIG. 8-11 *Six-in. cylinder with four-in. rod.*

*Solution:* The pressure in a cylinder will always be that which is required to do the work. This is also true of a motor. A pump does not produce pressure by itself; it is the work or the resistance to flow that causes pressure in a system. In the above example, if the load on the cylinder is 15,700 lb (note that this is maximum load that the cylinder can move using the maximum pressure while retracting), the pressure required while retracting is the maximum, or 1000 psi.

While this cylinder is extending, the pressure required to move the same load is less. Since the force of a cylinder is the area times the pressure, it follows that the pressure required is the load or force divided by the area:

$$p = \frac{F}{A}$$

Since here force is 15,700 lb and the area is 28.26 sq in., the pressure required is 15,700/28.26, or 555 psi.

These results are theoretical and do not account for friction, which would slightly increase the pressure required. Note also that a cylinder with the dimensions given above is capable of producing almost twice the force while it is extending that it can produce when it is retracting.

### Cylinder Travel Speed

The speed of travel is dependent on the rate of flow. It will vary directly as the flow increases or decreases. The pressure at which a cylinder operates has no effect on its travel speed unless there is leakage from one side of the piston to the other. This leakage can result from defective seals or cylinder warpage and can reduce the cylinder's efficiency.

The speed of travel is the volume of flow divided by the area of the cylinder. For example, if a cylinder had an area of 1 sq in., it would move 1 in. for each 1 cu in. of fluid that entered its chamber. The formula is:

$$\frac{\text{Flow}}{\text{Area}} = \text{rate of speed}$$

where flow is in cubic inches per minute, area is in square inches, and rate of speed is in inches per minute. If flow is given in gpm:

$$\frac{\text{Flow} \times 231}{\text{Area}} = \text{rate of speed}$$

where flow is in gallons per minute, area is in square inches, and rate of speed is in inches per minute (1 gal is equal to 231 cu in.).

For the cylinder discussed:

$$\text{Extending speed} = \frac{231 \text{ cu in./min}}{28.26 \text{ sq in.}}$$

$$= 8\tfrac{1}{4} \text{ in./min}$$

$$\text{Retracting speed} = \frac{231}{15.7} = 14\tfrac{3}{4} \text{ in./min}$$

It can then be concluded that the pressure in a cylinder is determined by the force which it must exert; the maximum force a cylinder can provide is dependent on the maximum pressure available. The maximum force will be greater when the cylinder is extending than it will when the cylinder is retracting. Travel speed is dependent on rate of flow; at a given rate it will retract at a faster rate than it will while extending.

## PROBLEMS

**1.** If a motor has an inlet pressure of 1000 psi and a discharge pressure of 100 psi, what pressure is available to produce torque?

________

**2.** If the above motor, equipped with a 2-in.-dia. gear, has 1 sq in. of effective exposed tooth area on each gear, what torque is delivered at 500 psi? ________; 1000 psi? ________; 2000 psi? ________

**3.** Define volumetric efficiency.

________________________________

________________________________

________________________________

________________________________

**4.** Define mechanical efficiency.

________________________________

________________________________

________________________________

________________________________

**5.** What is the formula for overall efficiency?

________________________________

**6.** What is meant by balance design in a vane pump?

________________________________

________________________________

________________________________

________________________________

**7.** Draw a sketch of an unbalanced vane pump.

**8.** Why are unbalanced vane pumps used?

**9.** What are the two principal types of piston pumps?

**10.** What are (*a*) the advantages of a piston motor?

(*b*) The disadvantages?

**11.** Label the basic component parts on Fig. 8-12:

1. Shaft
2. Blank end assembly
3. Piston
4. Piston seals
5. Tie rods
6. Cylinder
7. Shaft seal
8. Wiper
9. Rod end assembly

**12.** Using Appendix Table D-1, complete the missing data; assume that the cylinder is extending.

| *Cylinder area, sq in.* | *Pump delivery, gpm* | *Piston velocity, ft/min* |
|---|---|---|
| 100 | | 2 |
| 10 | 4 | |
| | 30 | 30 |

**13.** Using Table D-2, complete the missing data.

| *Cylinder dipl., cu. in.* | *Time to complete stroke, sec* | *Pump delivery, gpm* |
|---|---|---|
| 100 | | 9 |
| | 9 | 30 |
| 30 | 1 | |

**14.** Using Table D-3, complete the missing data; assume that the cylinder is retracting.

| *Force, lb* | *Cyl. or piston area, sq in.* | *Line pres., psi* | *Piston dia.* |
|---|---|---|---|
| 5,000 | | 500 | |
| 100,000 | 50 | | |
| | 40 | 5000 | |

**15.** Using Table D-4, complete the missing data.

| *Bore dia., in.* | *Shaft size dia., in.* | *Pres., psi* | *Push, lb* | *Pull, lb* |
|---|---|---|---|---|
| 6 | 3 | 3000 | | |
| 2 | 1$\frac{1}{16}$ | 2000 | | |
| 3½ | 2 | 500 | | |

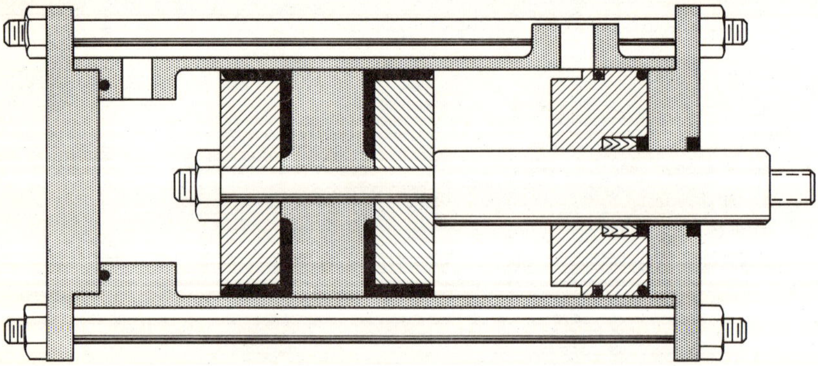

FIG. 8-12 *Hydraulic cylinder and its component parts.*

 

**16.** (*a*) Using Table D-5, give the cylinder displacement (extending) if the cylinder has a diameter of 10 in. and a 10-in. stroke.

_______ gal _______ cu in.

(*b*) What is the retracting displacement if the rod diameter is 5 in.?

_______ gal _______ cu in.

**17.** Using the left-hand column in Table D-1, give the area of a circle whose diameter is: (*a*) 6 in. _______ ; (*b*) 12 in. _______ ; (*c*) 25 in. _______ . Could the answers to this question be found in the center column of Table D-2? _______________

**18.** In what other table, in addition to Tables D-1 and D-3, could the answers for Prob. 17 be found? _______

**19.** Using the tables, what is the effective piston area of (*a*) a 9-in.-dia. cylinder with a 3-in. shaft? _______ ; (*b*) a 30-in.-dia. cylinder with a 15-in. shaft? _______

**20.** Using the right-hand column in Table D-1, find the following:

14 ft = _______ in. 650 in. = _______ ft
90 ft = _______ in. 65 in. = _______ ft
50 ft = _______ in. 150 in. = _______ ft

**21.** Using the right-hand column in Table A-2, find the following:

2 gal = _______ cu in.
30 gal = _______ cu in.
100 gal = _______ cu in.
30,000 cu in. = _______ gal
4000 cu in. = _______ gal
400 cu in. = _______ gal

**22.** Assume that, in a given cylinder, flow = 6 gpm, cylinder = 4 in., rod diameter = 2 in., stroke = 24 in., and maximum pressure = 1000 psi. Find:

(*a*) Speed extending _______
retracting _______
(*b*) Maximum force extending _______
retracting _______
(*c*) Displacement extending _______
retracting _______
(*d*) Time extending _______
retracting _______

# SEALING DEVICES

# 9

Seals and packings play a very important role in fluid power systems. It was not until effective packings were developed for hydraulic pistons in cylinders that the application of hydraulic power to industry was made possible. Before that time, losses of liquid due to inadequate sealing caused inefficiencies of operation which limited the number of applications in which fluid power could be used.

Today, with the advances made by industry in sealing technology, pumps, cylinders, and valves have been developed with many times the pressure ratings that existed 30 years ago. Wherever there is a pressure seal to be made, either in a rotary, reciprocating, or static application, hydraulic packings or seals are used.

The type of seal to be used in a particular application depends upon four primary factors: (1) temperature; (2) type of motion (dynamic-reciprocating, rotary, or static); (3) pressure; and (4) the fluid medium which is being used. For example, a fluid in a valve may best be sealed by rubber or a rubber compound; however, if the temperature is extremely high, rubber will cook and become brittle and cracked, which causes leaks. Therefore, a metal or teflon plastic seal must be used.

### TYPES OF PACKINGS AND SEALS

Seven general types of packings and seals are most often used in hydraulic systems today: O-ring, quad-ring, V-packing, cup-packing, U-packing, flange, and metallic.

*O-ring Seal*

This is one of the most popular seals used in the hydraulics industry today (see Figs. 9-1 and 9-2). It is used where there is either reciprocating, rotary, or static motion. On the rotary shafts, O-ring seals are limited to applications where pressures are low (500 psi) and speeds are low (approximately 200 or less surface feet per minute). With back-up rings they may be used with pressures up to 5000 psi. On static seals, where there is no motion, they have been successfully used for pressures as high as 15,000 psi. Figure 9-3 shows the effects of pressure on O rings.

O rings are made of many different combinations of rubber or plastic to handle the many types of fluids used in hydraulics.

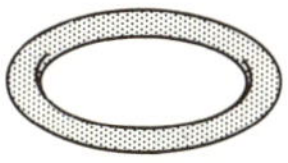

FIG. 9-1 *O ring.*

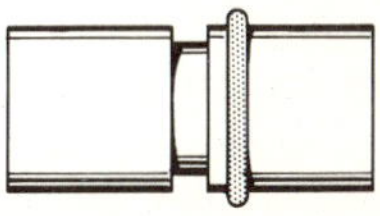

FIG. 9-2 *O ring and groove.*

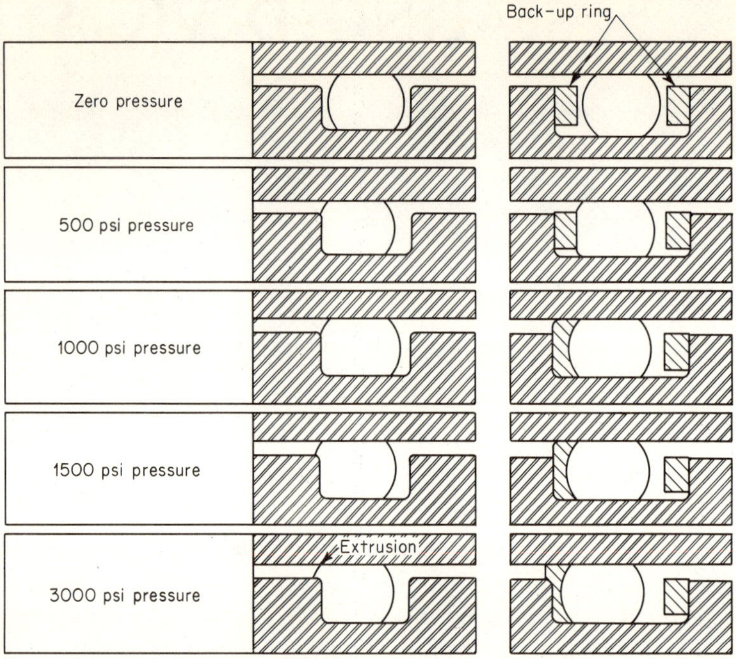


FIG. 9-3 *Seal configuration.*

Some of the more interesting types of rubber used and their applications are as follows:

1. *Buna-N* is the most common type of rubber used; it will handle most hydraulic and petroleum oils and greases, automotive gasoline, alcohols, glycols, propanes, and butanes. It will not handle phosphate ester based nonflammable hydraulic oils. It has a temperature range of −40 to +300°F.
2. *Butyl* is for use where phosphate esters are being used as the base in the fluid being handled.
3. *Silicone* is a high-temperature compound (−100 to +500°F); it is for use in air or high-aniline-percent oils (not hydraulic oils).
4. *Viton* is also a high-temperature compound (−20 to +500°F), and is resistant to a large variety of chemicals and gases.
5. *Teflon* is a thermoplastic material which has very high temperature resistance (up to +500°F), is unaffected by almost all chemicals, and has a low coefficient of friction.

Another factor which affects O-ring seals is the durometer (hardness or resiliency) of the rubber compound used. The durometer range is from 10 (which is a light sponge) to 90 (which is extremely hard). The two durometer specifications commonly used in O rings are 70, for pressures under 1000 psi, and 90, for pressures over 1000 psi. The 90-durometer rubber is used for these high pressures because it is more resistant to extrusion and distortion.

### *Quad-ring Seal*

The quad-ring seal (Fig. 9-4) is similar to the O-ring seal in that it is composed of either synthetic rubber or thermoplastic com-

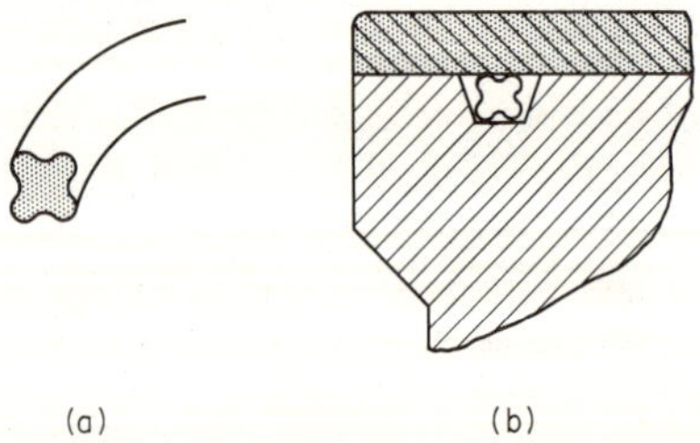


FIG. 9-4 *(a) Cross-section. (b) Quad-ring seal.*

 

pounds; however, it has a modified square-type cross-section. It may be applied over a wide range of pressures. It is especially useful where lubrication is difficult, yet where a minimum friction is required.

*V-packing Seal*

V-packing seals (see Fig. 9-5) are commonly found on hydraulic cylinders (reciprocating seals) and generally are furnished in sets. Usually they are made of leather; however, other materials, such as rubber, or fabric combinations, can be used.

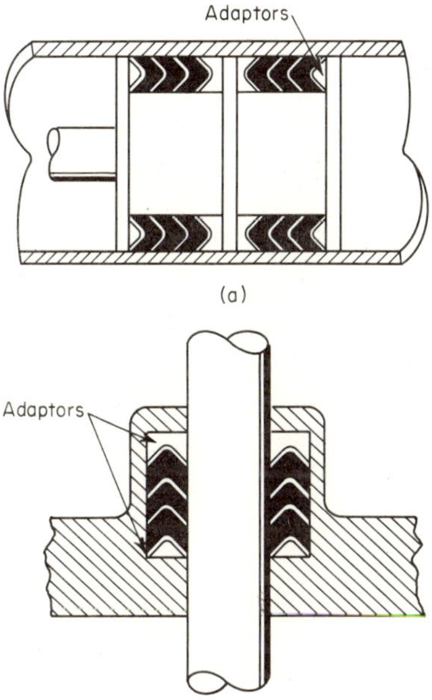


FIG. 9-5 *V-packing seal.*

*Cup-packing Seal*

Cup-packing seals (Fig. 9-6) are used in many areas of high-pressure service. They are generally used on the ends of piston rods.

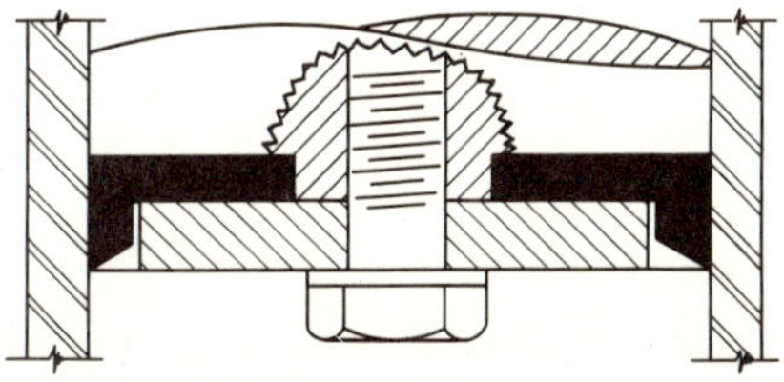

FIG. 9-6 *Cup seal.*

*U-packing Seal*

U-packing seals (Fig. 9-7) are used as single-unit seals, whereas V-packing seals are used in sets.

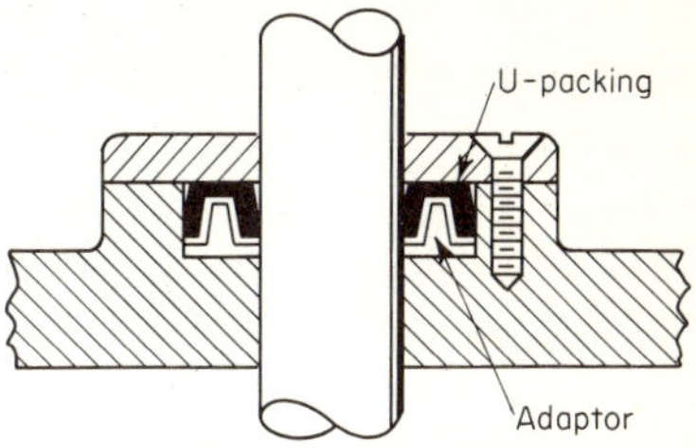


FIG. 9-7 *U-packing seal.*

*Flange Seal*

Flange seals (Fig. 9-8) are single-molded types of seals; they are used where the configuration of the existing space limits the type of seal to be used.

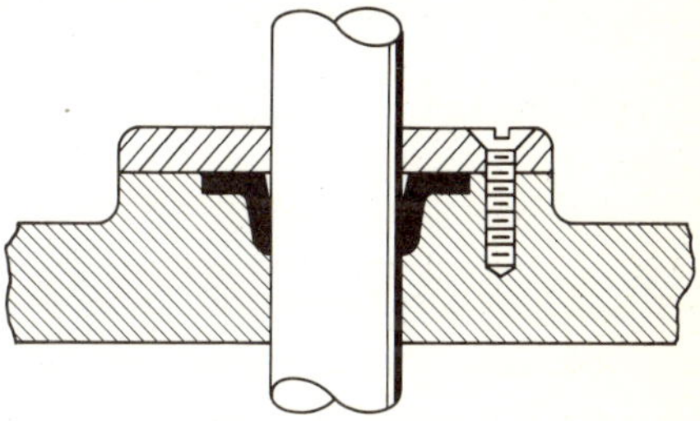

FIG. 9-8 *Flange seal.*

*Metallic Seal*

The most common example of a metallic seal is the piston ring in the internal-combustion gasoline engine. The metallic seal is particularly useful where the seal is subjected to an extreme heat that would cause even the most exotic rubber or thermoplastic compounds to fail.

## INSTALLATION AND LIFE OF SEALS

Where sealing rings are installed either in a valve or in the grooves on a shaft, the groove must be clean and free of dirt. Care in installation is imperative. Leakage to ensure lubrication of the seal is provided by the construction of the component. In the case of O rings, for example, lubrication must be provided, for without it the seal will twist and abrade (pull apart). Lack of lubrication produces excessive friction, and an O ring subjected to such friction would fail in a relatively short period of time.

In most cases, the fluid being sealed will provide sufficient lubrication for the ring. However, pneumatic seal installations must be treated with special care, for dependable performance is due in a large part to proper lubrication. Tests have indicated that grease is a better lubricant than oil for both reciprocating and static pneumatic seals; however, to be a suitable lubricant, grease must be compatible with the O-ring material in order to avoid swelling or shrinkage.

In the case of O rings, it is a good rule of thumb to change the seal when the unit has been taken apart for any reason, primarily because dirt may have gotten into the seal causing it to become cut or scratched.

 

# CIRCUIT COMPONENTS

# 10

## RESERVOIRS

Hydraulic reservoirs serve the following purposes:

1. They serve as supply sources and storage places for the hydraulic fluid.
2. They help dissipate heat and air from the fluid.
3. They provide a location for removing fluid contaminants.

### Component Parts

A drain plug is provided to facilitate removal of the fluid from the tank. Removal is necessary for purposes of cleaning the tank or replacing the fluid. Access for cleaning the tank is provided by a large removable cover or a large removable plate bolted to the tank. Frequency of cleaning is determined by the filtration system and the conditions under which the system operates.

A sight gauge or a dipstick is provided for quick oil-level determination.

Tanks must be provided with excess capacity, since the amount of fluid in the tank will vary. This is particularly true in circuits in which hydraulic cylinders are used. Expansion resulting from heat generated also requires additional capacity. The size of the tank and amount of fluid in a system are determined by the ambient temperature (that temperature immediately around the system under which the system operates) and heat-dissipation requirements. If heat exchangers are used, the tank does not need so much excess capacity. As a general rule, the tank capacity is usually two to three times the pump output.

The filler cap usually has a screen to remove particles as the tank is being filled and to act as a breather, or vent, for the tank as the volume in the tank changes. This breathing action, under dusty conditions, will contaminate the fluid; therefore, these caps should have a filter which can be cleaned to remove these contaminants. The necessity for using clean fluid cannot be overemphasized.

The positions of the pump inlet and the return line are important; they must be located such that repeated use and recirculation of part of the oil are prevented. This is accomplished by spacing the location of the return line on the tank. Baffles are installed to prevent rapid recirculation. These are installed in a manner such that the return oil is trapped on one side and is forced to flow to the inlet side through an indirect route. The return line is installed so that oil as it enters the reservoir is below the oil level. This will eliminate foaming and aeration.

If it is decided to use an internal strainer

(also known as a sump-type strainer), provisions must be made for its easy removal. Internal strainers must be cleaned periodically, and installation should be such that oil need not be drained to accomplish this task. External strainers are more accessible and can be equipped with a gauge that will indicate their condition.

In some instances magnetic plugs are installed in the tank. The purpose of these plugs is to gather metal particles as they accumulate in the tank and to prevent their recirculation. These plugs are mounted, so that they can be periodically removed, cleaned, and replaced.

The features discussed above can readily be constructed in the separate-type reservoir that is generally used in industrial power units. In this type of installation the pump and motor can be mounted on top of the reservoir as a matter of convenience. They are often elevated to conserve space and assist the gravity flow in a prefill system.

### Types of Reservoirs

The *integral system* is often used in mobile equipment. The machine's tubular members, part of its frame, and any available area are closed and used as a reservoir. The advantages to such a system are saving of space, cost reduction, minimized plumbing, and additional area for cooling. Further, a neater appearance often results. On the other hand, with integral systems, localized heating may occur, oil-level determination may be difficult, and draining of the oil and cleaning of the reservoir are sometimes a problem.

*Dual-purpose reservoirs* are used in some instances. In this system the hydraulic fluid acts as both the pump supply and the lubrication oil for the machine's transmission or for other areas on the machine. The difficulty in this system lies in the fact that the oil serves a dual purpose; its composition may serve one of its purposes but be lacking in necessary qualities for the second. Further, temperature control may be difficult without a heat exchanger. Its advantages are saving of space and the elimination of the necessity of having two types of oil available for a machine.

The configurations used in reservoir construction are many and varied; they depend on available space, cost, and performance requirements of the circuit.

## STRAINERS

Strainers are used to remove particles from the fluid. They are installed in the pump inlet line either internally or externally. Internal strainers are mounted inside the reservoir; their general construction is shown in Fig. 10-1. The straining element consists of a wire mesh. Frequently this is a *100 mesh,* which means there are 100 meshes per linear inch. These trap relatively large particles and should be distinguished from filters, which are used to remove fine particles. Filters offer greater resistance to flow and for this reason must be installed on the return line. A filter would cause a vacuum and cavitation of the pump if used on the inlet line. A vacuum greater than 3 or 4 in. Hg is not desirable in the intake line.

External strainers and their construction are shown in Fig. 10-2. The mesh is usually 100, as in the internal strainer. They may have magnets installed that assist in the removal of metallic particles.

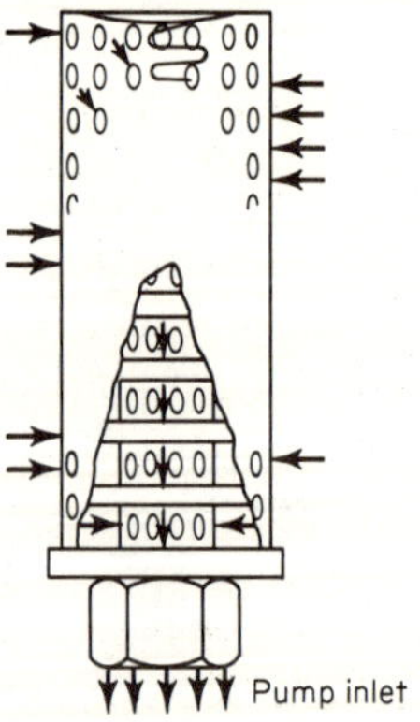


FIG. 10-1 *Strainer.*

 

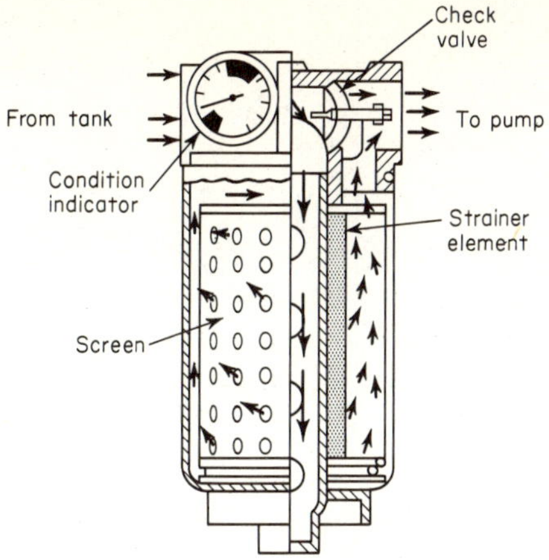


FIG. 10-2 *External strainer.*

In both types of strainers a by-pass is provided so that if the screen becomes clogged, the check valve will open and permit flow directly to the pump. If this feature were not provided, and the strainer were not cleaned, cavitation at the pump would occur, and damage to the pump would ensue. It is particularly important to clean the filters and strainers on a new machine. Contaminants are left in the system when it is being assembled, and these must be removed.

On the external strainer, a vacuum gauge is often installed to indicate the condition of the element. A red line often appears on the dial at a vacuum of 10 in. Hg. Strainers often operate at a vacuum of approximately 2 to 6 in. Hg. As the strainer becomes clogged, the vacuum will increase, and the indicator will show that cleaning is required. If this were neglected, additional restraint of flow would cause the poppet-type check valve to open, which would permit flow directly to the pump.

Many designs, configurations, and features are available. Which strainer is best suited for a system depends on the conditions under which the system performs, and it must be remembered that strainers are ineffective if not cleaned. Self-cleaning filters have not been developed. It cannot be overemphasized that hydraulic fluid must be kept clean for a dependable fluid power system.

## FILTERS

Return-line filters are installed for the purpose of removing fine contaminants, that is, those which the strainer cannot remove, and are usually made with a disposable element constructed from paper, felt, wood pulp, glass fiber, or woven wire mesh. The degree of filtration achieved depends on the material of the filter that is installed. The materials are rated in microns ($\mu$). (A micron is 1/1000 of a millimeter; 25,400 $\mu$ = 1 in.)

Table 10-1 shows values of common objects and their size in microns.

The materials used in the filters are usually rated from 25 to 100 $\mu$. Filtration as fine as 10 $\mu$ is frequently achieved. A 25-$\mu$ filter will pass material which has a diameter of 1/1000 in. or smaller. A 100-$\mu$ filter will pass material which has a diameter of 1/250 in. or smaller. These filters provide for two methods of contaminant removal: (1) screening; and (2) absorbing of the contaminants.

It can be visualized that with such small openings pressure is required to force the oil through these filters. For this reason they cannot be used as intake filters, since they would cause a vacuum and cavitation at the pump.

Filters should be equipped with a pressure gauge which will show the condition of the filter. As the filter becomes contaminated, higher pressure will be required to force the oil through the element. These gauges are red lined at the maximum pressure at which the filter will operate efficiently, thereby affording an easy check of the filter's condi-

TABLE 10-1

| *Material* | *Size, $\mu$* |
|---|---|
| Table salt | 100 |
| Human hair | 50–75 |
| Barely visible to naked eye | 40–50 |
| Talcum powder | 10 |
| Bacteria | 2 |
| 100-mesh screen | 150 |

tion. The pressure drop across a filter will vary depending on the design, the element used, and the volume of the flow.

A filter is equipped with a check valve, which can open and permit flow directly to the tank. This check valve will open if the filter becomes contaminated or if the filter capacity is insufficient for surge flows. In this respect, it acts as a relief valve. Cold oil will often require a pressure greater than that preset on the valve. Generally the pressure drop through a filter is 5 to 15 lb. It will vary depending on design and the material used. Oil temperature and its viscosity will also vary the pressure drop.

Full-flow in-line filters, which are capable of sustaining the full system pressures, are available. They are installed after the pump, but before the relief valve or any other component, and have the advantage that they filter the fluid before it passes through any of the circuit components.

Filters and strainers should be cleaned and replaced regularly for the following reasons:

1. Neglect will result in by-passing of the oil; a filter's value when fluid is being by-passed is zero.
2. The life of the components and the life of the system depend on the condition of the oil.
3. If contaminants lodge in valve spools, the system will not work properly.
4. Soot and oil form a lapping compound which causes rapid and excessive wear on expensive components.

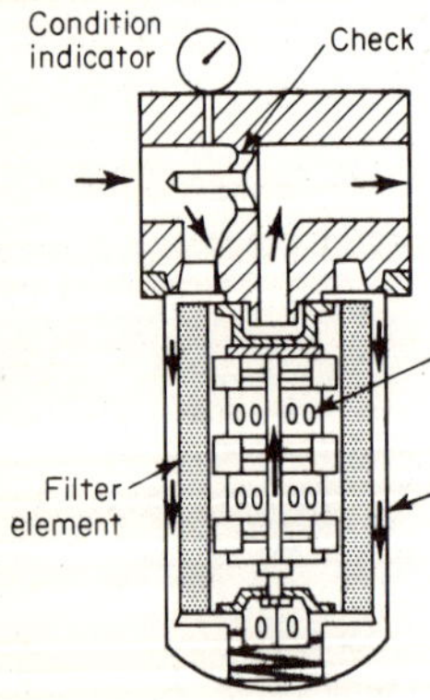


FIG. 10-3 *Return-line filter.*

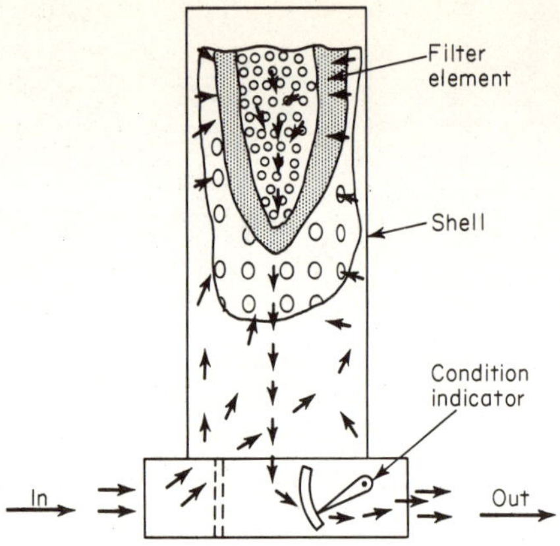


FIG. 10-4 *In-line filter.*

5. If additives formed in oil or added with the oil are not removed, they will affect rubber hose and seals; generally they will have a detrimental effect on the system.
6. The use of clean oil is the best "maintenance program" that can be instituted. Consult the manufacturer for recommended filter materials which will have properties required to assure effective filtration in a particular system.

### Maintenance Suggestions

1. Keep oil at proper level. If sight gauge is provided, use it. If sight gauge is not provided, determine a method of checking the oil level, and make provision to keep the oil at its proper level.
2. Clean strainer and replace filter elements when required (as indicated by the condition gauge). If a gauge is not provided, set up a time schedule for its cleaning and keep a record. Follow the schedule. It will vary with the conditions under which the system operates.
3. Check oil for aeration. If aeration occurs, find the cause of trouble and correct it.
4. Use the manufacturer's recommended fluid and filter elements.

## HEAT EXCHANGERS

Fluid used in a hydraulic system is continually being heated. Pressure constantly drops as the fluid flows through the lines and across all components; and since energy cannot be created or destroyed, this loss in potential (pressure) energy and kinetic (velocity) energy is converted into heat energy. Only the actuators can convert kinetic or potential energy into useful force—either linear or torque—and even the actuators convert part of the energy into heat. The fluid is constantly heated as it flows through the tubing—when making turns, or when passing through valves, pumps, motors, or any other component.

Some heat is naturally dissipated in the hydraulic lines, and proper design can hold excess fluid heating to a minimum. For example, one of the functions of the fluid reservoir is to exchange heat. If sufficient heat is not being lost, and the oil is consequently not at proper operating temperatures, the size of the reservoir can be increased. This allows more time for the fluid to cool and reduces the number of circulations the fluid makes in a given amount of time. Unfortunately, this is often impractical because of problems of bulk and weight. Where the system cannot dissipate, by radiation and convection, sufficient heat to maintain proper temperature, a heat exchanger can be installed.

Two basic types of heat exchangers are used—one type uses air to transfer the heat, and the other uses water. Water is generally used for industrial installations, whereas in mobile equipment, for obvious reasons, air is most often used.

Heat exchangers on mobile equipment act as do engine radiators; they have a fan to carry away the heat. In other installations the oil flows around a series of tubes in which the cooling water is circulated. The heat from the oil is dissipated to the water and into the atmosphere by the engine radiator. The engines in these installations have radiators large enough to give them sufficient capacity to dissipate the heat from both the engine and the hydraulic circuit.

The oil-to-air system is not limited to mobile equipment. Industrial plants use this type for two purposes: (1) hydraulic oil coolers, and (2) heat for buildings. Since excess heat from a hydraulic system must be removed in any case, if this heat is utilized to heat the building, appreciable savings can be realized. This type of heat exchanging has recently been perfected; in industrial plants, and also in office buildings, the excess heat from machinery, lights, people, and other sources is exchanged (air to water) during the day. During the night this heat is again exchanged (water to air) and utilized. The cost of heating the building is substantially reduced in many instances.

Industrial-type heat exchangers utilizing the shell-and-tube principle consist of a shell which encloses a series of tubes. The shell is baffled to induce turbulent flow of the hydraulic fluid (Fig. 10-5). As a result of this baffling, the flow of the fluid will, for the most part, flow perpendicularly to the tubes. This increases the efficiency of the heat transfer. Since water will transmit heat to metal approximately four times faster than oils, exchangers are often constructed so that water will make from two to four passes within the unit. These additional passes increase the velocity of the water; this causes turbulent flow, which increases the water's cooling efficiency.

Manufacturers have made technical data about their heat exchangers available. In selecting size and models, the manufacturer's charts, information, and recommendations should be utilized.

## PRESSURE INTENSIFIER

Pressure intensifiers or boosters are used to increase pressure in a system. These components utilize a high-volume flow at relatively low pressures and convert a portion of the flow to a high pressure; the balance is returned to the tank. They use the energy in that portion returned to the tank to increase the pressure in the balance of the flow.

Figure 10-6 shows a type of pressure

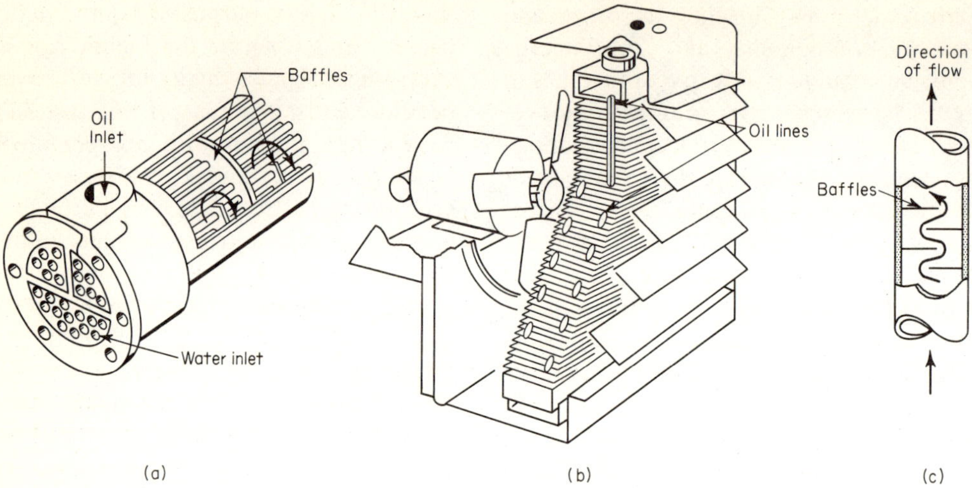


FIG. 10-5 *(a) Oil-to-water heat exchanger. (b) Oil-to-air heat exchanger. (c) Tube construction.*

booster which uses an oscillating piston to increase the pressure. The ratio of intensification may be anywhere from 1:2 or 1 to almost any desired number; it is determined by the ratio of the small piston to the large piston.

Figure 10-6 also shows a large piston reciprocating in a cylinder. It has two small ends which reciprocate with the large piston and are an integral part of it. The piston is oscillated by a four-way two-position pilot-operated valve which directs flow to either

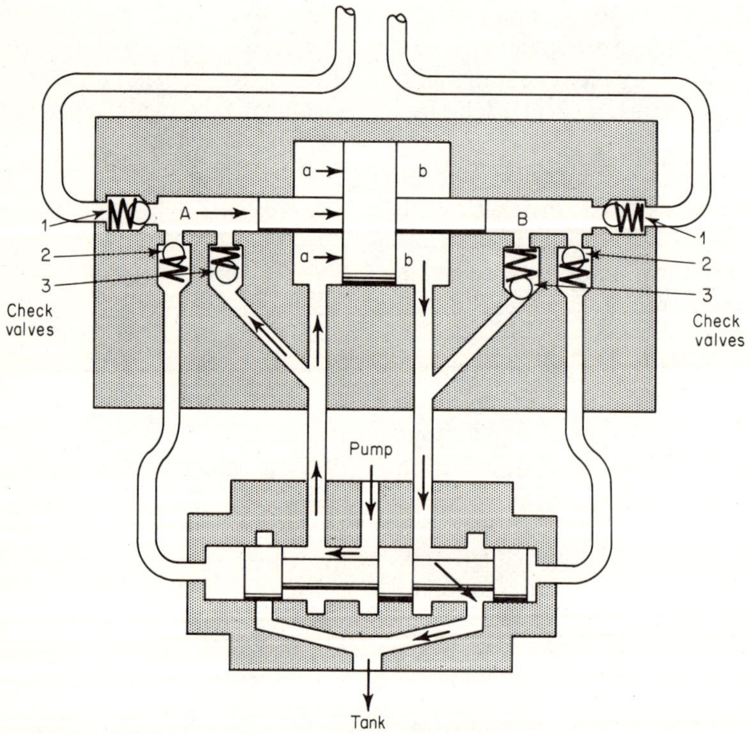


FIG. 10-6 *Pressure intensifier.*

 

side of the large piston while discharging the fluid from its opposite end. One small chamber is filled while the opposite end forces the fluid to the high-pressure system.

Check valve 1 prevents high-pressure fluid from reentering chamber *A* or *B* after it has been intensified. Check valve 2 is preset to open at the high pressure desired. When fluid in chamber *A* or *B* attains this pressure, it opens and reverses the spool setting in the four-way valve. This reverses the direction of travel of the piston. Check valve 3 permits low pressure to assist piston movement against the opposite end; at the same time oil which has just been forced into the high-pressure line fills the chamber. The four-way valve will, in this manner, move the piston in a reciprocating manner, thus intensifying pressure to the high-pressure line, which utilizes the energy in the oil returned to the tank. The pilot chamber is drained through a vent line in the valve.

*Example:* Let us assume that area *A* or *B* is 5 sq in.; area *a* or *b* is 15 sq in.; and the low pressure is 500 psi. Low-pressure fluid acts on the piston, which has 20 sq in. (*A* plus *a*) acting against 5 sq in. (*B*). Area *b* is open to the tank; 20 sq in. × 500 lb = a 10,000-lb force. This 10,000-lb force is available to increase pressure in *A* or *B* depending on the direction of the piston movement. The piston area given for *A* or *B* is 5 sq in.; thus the maximum pressure in this area is 10,000 ÷ 5, or 2000 psi. We now have the maximum pressure of 2000 psi leaving chamber *A* or *B* as it is forced into the high-pressure line. The ratio of this unit would be 1 to 4. If the pump were delivering 20 gpm, the ratio would remain the same: 15 gpm would be used to reciprocate the piston and 5 gpm would be delivered to the high-pressure system.

# FLOW CONTROL COMPONENTS

# 11

## CHECK VALVES

Valves of this kind permit flow in only one direction. They close against reverse flow, or if spring loaded, or when flow ceases. The force of the liquid in motion opens the valve, and it is closed by back flow. It may be closed by the action of a spring, or by gravity. If no spring is installed, this valve should be installed so that gravity assists closure.

Figure 11-1 shows a simple gravity non-spring-loaded swing-type check valve. Another type, the vertical-flow check valve, is shown in Fig. 11-2. This type may be modified to require the oil to change direction, as shown in Fig. 11-3.

The disk in many check valves is left free to rotate (Fig. 11-2) in place so as to distribute wear around its circumference. Swing-type check valves offer less resistance to flow than lift types, but the lift type will close more tightly on low-pressure lines, espe-

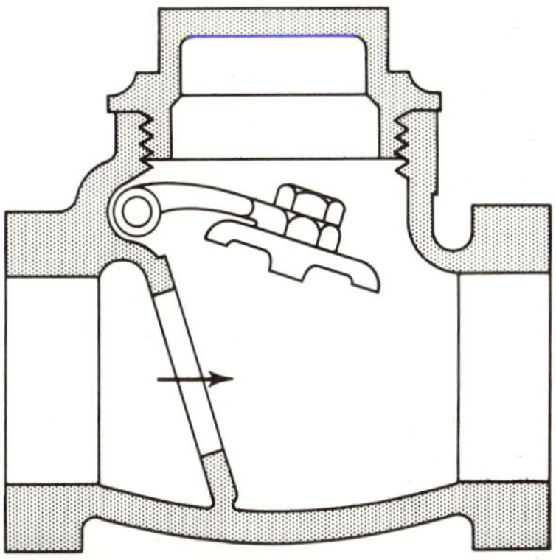

FIG. 11-1 *Check valve.*

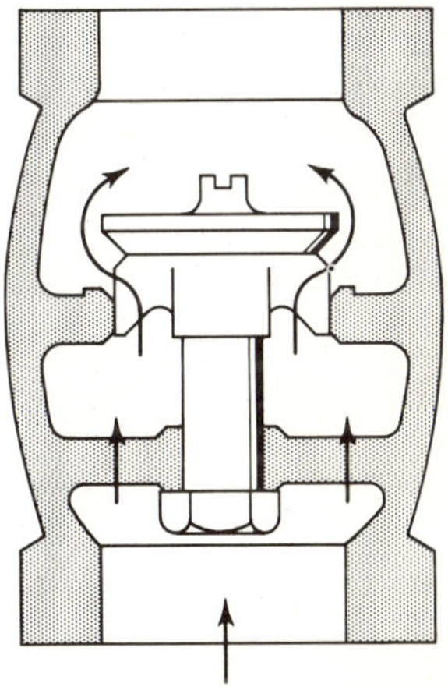

FIG. 11-2 *Check valve.*

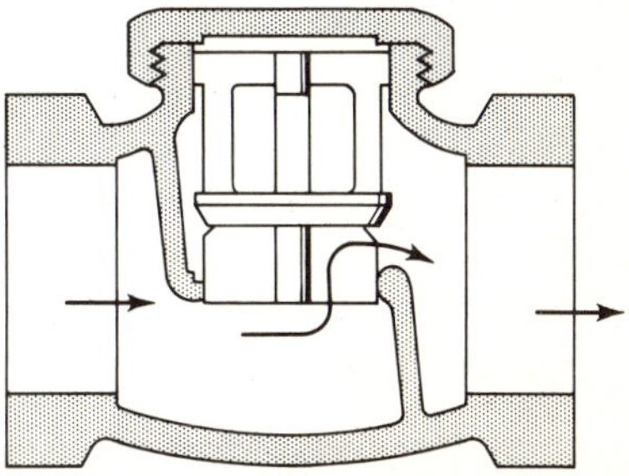

FIG. 11-3 *Check valve.*

cially if a spring is added. Conversely, however, it will be more difficult to open. Lift valves are less likely to slam shut than swing valves. As pressure and velocity increase, however, lift valves are more liable to chatter and pound—a fault to which all kinds of check valves are liable.

Check valves are used in hydraulic systems for several purposes, but they are used chiefly to prevent reverse flow. The type used depends on the particular application. They are often used by themselves in a system. They are, however, used in many components as an integral part of the device. The check valve with other mechanical features make up components capable of doing precise fluid power functions. Figure 11-4 shows the symbol for a check valve.

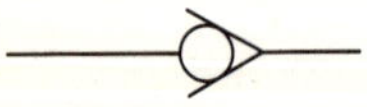

FIG. 11-4 *Symbol for a check valve.*

#### RELIEF VALVES (MAXIMUM-PRESSURE VALVES)

Relief valves are used to make certain that system pressure does not rise above a certain level. Check valves such as those shown in Fig. 11-5 could be used to do this if they were equipped with a spring adjusted to compress when the pressure reached a definite level. Liquid is then by-passed out of the system or back to the reservoir. The valve is controlled by means of an adjustment screw against which the top of the spring is seated (Fig. 11-6). An inward adjustment of this screw increases the compression of the spring, making it able to resist a higher pressure.

Relief valves can be considered as safety valves, and must be large enough to handle the entire flow from the pump in case of a sudden pressure demand that is greater than that for which the system is designed. They must be designed to be capable of handling the entire pump output at the maximum pressure. For this reason they are often called maximum-pressure valves. They also act as a shock damper, by opening momentarily, to prevent high-pressure shock waves. For example, when a closed center valve is moved to the closed position, blocking flow from the pump, the velocity (kinetic) energy of the moving fluid is changed to potential (pressure) energy. This causes a surge in pressure, and a small amount of fluid is by-passed, which prevents excessive pressure peaks. Figure 11-7 shows the symbol for a relief valve.

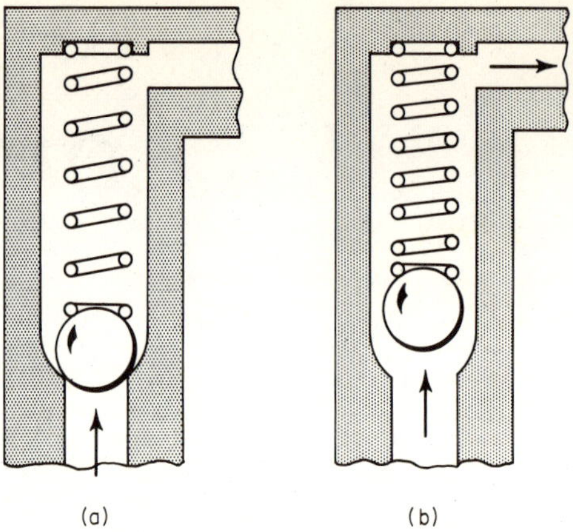



FIG. 11-5 *Spring-loaded check valve.*

##### Installation and Maintenance of Valves

The kind of metal to be used in a valve—brass, iron, steel, or special alloys—will de-

FIG. 11-6 *Adjustable spring-loaded check (or relief) valve.*

 

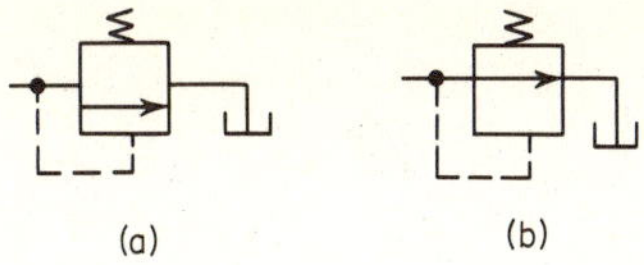

FIG. 11-7 *(a) Relief-valve symbol (closed). (b) Relief-valve symbol (actuated).*

pend on the liquid employed and the working temperatures and pressures of the system.

Manufacturers of valves specify the range of safe usage for each of their valves, and these specifications should be carefully heeded.

Valves should be handled carefully. If they are dropped, a part vital to dependable operation may break or become deformed. Valves should not be exposed to damage from blows, and must not carry the weight, sag, or expansion of a line.

Gate or globe valves should be in closed position during installation to avoid twisting of operating parts and to protect areas of closure. Overlength threading should be avoided on the joining pipe so that the end of the pipe will not project into the valve and injure the valve seat or other machined parts. Sealing compound or tape should be applied to the pipe threads, with the exception of the first two or three threads, and care taken that sealing material does not get into the valve. The pipe should be cleaned before a valve is installed, and blown out again with compressed air after installation. In attaching a valve, the wrench should be used on the hex nut nearest the pipe to which the valve is being connected. The packing or seal in a new valve should be tightened after it has been in use for a short time, but it should not be allowed to become tightly compressed.

Open and close valves *slowly*. Do not force the adjusting screw against the seat. If a valve refuses to seat properly, open it slightly so that sediment on the seat will be flushed away. If it still fails to seat, the system must be shut down and the valve disassembled to locate and correct the trouble. Valves that have been shut when hot should be tested for closure after the system has cooled.

Inspect all valves regularly, and test them at least four times a year. Keep them locked if there is any chance of unauthorized manipulation. Do not turn a valve stem with a wrench. Except in special cases, do not remove the turning wheels from valves.

In disassembling, do not use a long-handled wrench to release the valve. A short wrench and a few light hammer taps over the threads are more effective, and will not twist the valve.

Use only the correct tools in reassembling. Acquaint yourself thoroughly with the construction of the valve before putting it together. Repair parts should be ordered by giving the name of the *manufacturer*, the *serial number* on the valve, its *size* and *type*, and the *name* (and if possible the number) of the valve part.

The successful operation of a hydraulic system depends upon the proper operation of the valves. The operator should know which valves must be open or closed. Check on the open or closed position of hand-operated valves before starting the system. If the system has been operated previously, the automatic valves, such as directional, relief, and safety valves, should not be disturbed until found faulty or until time for periodic checkup.

Usually the presence of foreign matter in the valve seat, scoring or grooving of mating parts, and the plugging of openings cause improper action of a hydraulic system. The valve may stick, or may fail to close completely, so that the system registers low or fluctuating pressures, or even no pressure at all.

The usual remedy for such conditions (the remedy for practically all valve troubles) is to dismantle the valve, thoroughly clean all parts, replace those that are damaged, and reassemble. Any valve can easily be taken apart if careful notice is taken of just how the different parts are related. Sometimes foreign matter can be blown out of a valve, especially a relief valve, by momentarily increasing the pressure in the system so that the particles will be forced out of the way.

Leakage may be due to foreign matter, to scoring, or to distortion of the valve seat because the valve is not adequate in design or is inappropriate for the system. A scored valve can be reground, or a new seat provided. A distorted valve part should probably be replaced. At the time of installation, care must be exercised to avoid distortion of the valve body caused by overtightening the inlet and outlet pipes.

### Operation of Simple Relief Valves

As we already know, even when a hydraulic system is operating normally, it may develop temporary dangerous excess pressures (surge), as for example, when an unusually strong work resistance is encountered. Relief valves are used to control this excess pressure. Their purpose is not to maintain a given flow or pressure but to keep pressure from rising above a definite level, and they establish the maximum system pressure.

It is because simple relief valves sometimes are unsatisfactory in operation that compound relief valves were developed. A simple relief valve with a suitable spring adjustment can be set so that it will open when the system pressure rises to, for example, 500 psi, which is the valve *cracking pressure.* When it does open, the volume of flow will cause a rise in pressure above the cracking pressure. This pressure is the *full-flow* pressure value, which, in this instance, could be as high as 600 to 650 psi. A simple relief valve, because of the large difference between the cracking pressure and the full-flow pressure, is at a disadvantage when compared with a compound relief valve, which has a much smaller difference between cracking and full-flow pressure; this valve will be discussed later.

In addition, when the system pressure is reduced, the simple relief valve will close at a pressure *below* the preset cracking pressure. As indicated in Fig. 11-8, the seating surface of the valve must be wider than the

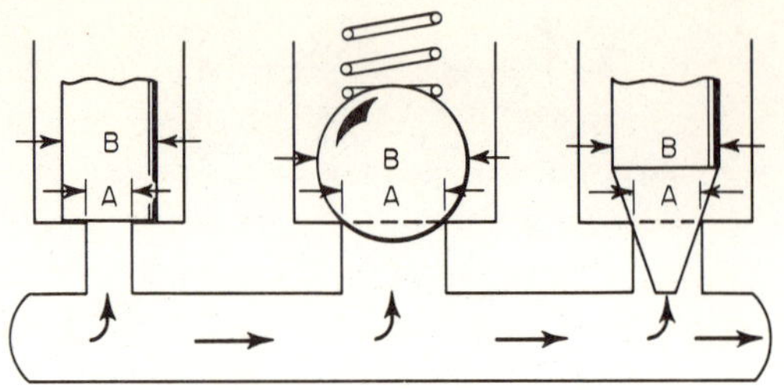


FIG. 11-8 *Forces on a check valve.*

pressure opening if the valve is to seat satisfactorily. The pressure in the system acts on the area of the valve open to it. In each case in Fig. 11-8 the force exerted *directly upward* by this pressure when the valve is closed depends on the horizontal area across the valve at *A*. The moment the valve opens, however, the upward force exerted by the pressure will depend on the horizontal area of the valve at *B*, which is necessarily greater than the area at *A*. This leads to an immediate upward jump in the action of the valve. It will, at this point, have a greater force resisting the closing (spring action) of the valve as a result of the increased area. The valve will close at a lower pressure than the pressure at which it opened. The same pressure acting over different areas produces forces proportional to the areas. The pressure at which a valve will close is known as the *reseat pressure,* which may be as low as 400 to 450 psi.

For example, the valve may be set to open at 500 psi. If the upward area at *A* exposed to this pressure when the valve is closed is 0.5 sq in., an upward force of 250 lb ($500 \times 0.5$) will be exerted on the valve at the moment of opening. With the valve open, however, the pressure acts on the area at *B* which might be 1 sq in. The upward force would then be 500 lb, or double the force at which the valve actually opened. For the valve to close, pressure in the system would have to fall well below the figure at which the valve opened. The exact pressure at which the valve would close would depend upon certain geometrical relations of the particular shapes involved which will not be further discussed here.

 

It follows that simple relief valves are liable to open and close rapidly as they *hunt* above and below the set pressure, causing pressure pulsations and undesirable vibrations in the system and producing a noisy chatter. Compound relief valves use the principle of action of simple relief valves for one phase of their action—that of the pilot valve—but provision is made to limit the amount of liquid that the pilot valve must handle, and the weaknesses of simple relief valves are thus avoided.

### Structure and Operation of a Compound Relief Valve

These relief valves consist of two moving parts, the pilot valve and the main valve. The purpose of the pilot valve is to sense the maximum pressure setting and then to start the operation of the main valve, which will then return the oil to the reservoir. While it is permitting the oil to return to the tank, the pressure to the work will be held nearly to the maximum setting of the relief valve.

Note the following:

1. The pilot valve is located inside the main valve.
2. The main valve has a line through its body so that the pressure is equalized with that to the work area, and it also acts on the bottom of the pilot spool, tending to lift it.
3. The pilot spring holds down the pilot valve against this pressure.
4. The main valve is held down by a light spring, since it does not work against system pressure.
5. The pilot spring chamber is vented to the tank (Fig. 11-9).

When the relief valve is closed, there is equal pressure on the top and bottom of the main valve, the mainspring holds down the main valve (hydraulically balanced due to

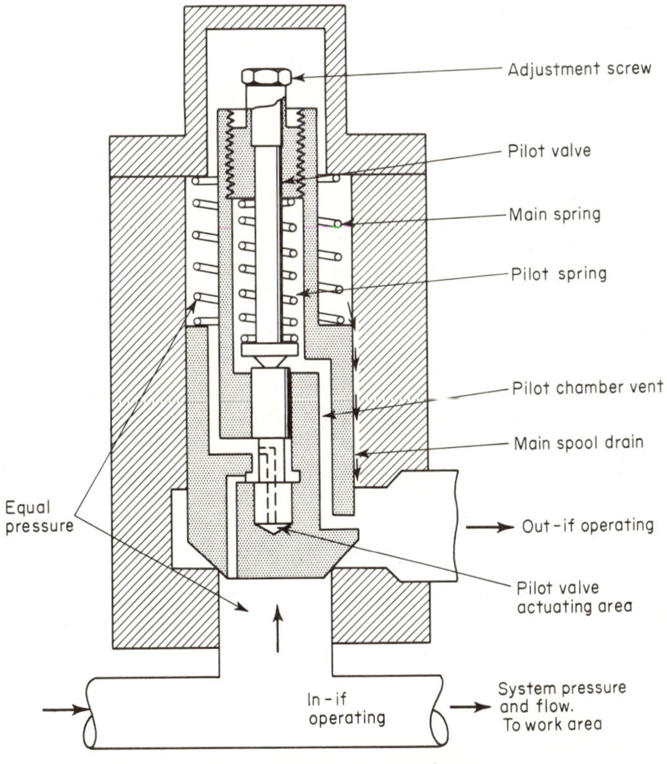


FIG. 11-9 *Compound relief valve.*

equal pressure top and bottom), and the pilot spring holds down the pilot valve.

As the load of the work increases the pressure to the maximum setting on the pilot spring, we have the following sequence taking place. The pilot valve lifts and cuts the flow from the main system to the top of the main valve; thus we no longer have equal pressure on the bottom and the top of the main spool. The top drains to the discharge port. With the greater pressure on the bottom, the main spool will lift, permitting oil to pass through the valve and go to the reservoir. As the pressure drops in the system, the pilot valve reseats again, permitting the work pressure to enter the chamber on top of the main valve and causing equal pressure on both sides. This allows the main valve spring to reseat the main valve. The oil that is trapped in the upper chamber above the main spool is discharged to the tank along the wall of the spool and that of the housing. This slows down the opening and prevents the chattering which would occur if the action were not dampened.

Under normal conditions, liquid flows past the valve and no valve action takes place. If pressure in the system rises above the setting of the pilot valve, however, the pilot valve is forced open, and this leads to the opening of the main valve. Figure 11-10 shows a relief valve in its seated and operating conditions. In Fig. 11-10*a* the system pressure acts on the base and on the top (outer chamber) of the main spool in chamber *A*. The pressure is equalized through passage *A*. The inner chamber *B* is always drained to the outlet side of the valve. The pilot spring holds down the pilot spool until pressure at area *B* increases and overcomes the pilot spring and moves the pilot spool up. A small movement up will block off passage *A*, and consequently the pressure in chamber *A* will drop, permitting the main spool to rise. The pilot spool will keep passage *A* blocked off as long as the system's pressure is greater than that set on the pilot spool. Consequently, the main spool will remain open and permit passage of the fluid back to the tank (Fig. 11-10*b*). As the pressure drops, the pilot spool will seat and open the passage, equalizing pressures at the top of the main spool, and the spring will force it down to its seated or closed condition.

In this manner the valve will divert either a portion (or total) of the flow in the pressure line to the reservoir, and maintain the set pressure in the system. These valves are always installed on the high-pressure side of the pump, as close to it as is convenient.

### Malfunction of Relief Valves

The most frequent cause of malfunctioning of relief valves is foreign matter lodging in

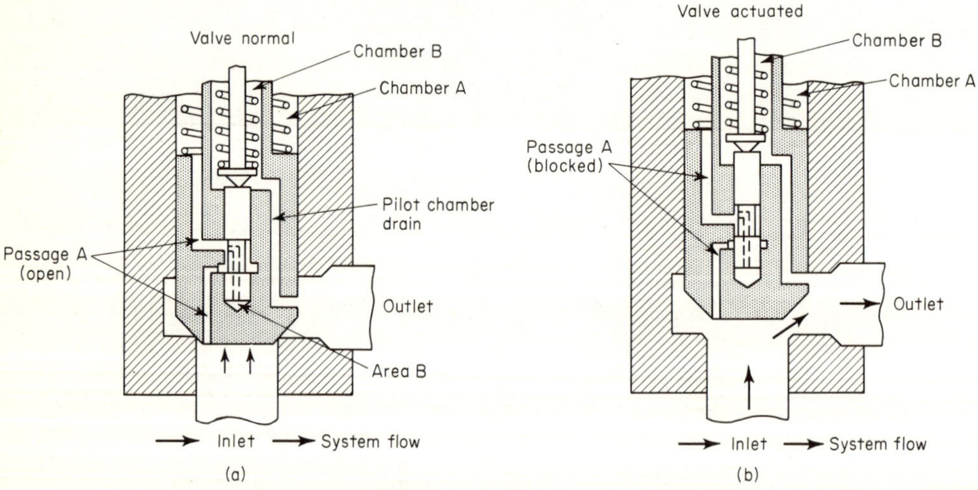


FIG. 11-10 *(a) Valve normal. (b) Valve actuated.*

 

the openings of the seats, particularly the pilot seat. (This has the same effect as excessive pressure, since they will both permit a decreased pressure on top of the main valve and cause the flow to discharge back to the tank.) The foreign material will cause the flow to be discharged regardless of system pressure and thus causes a malfunctioning system. Cleaning the pilot valve of this material will correct the problem. Contamination that cannot be cleaned out during operation will necessitate disassembly and cleaning of the valve.

When mounting a relief valve, care should be taken not to warp the casting, since this causes the spools to stick in the open or closed position. This is true of any component. *Care in circuit assembly* will eliminate many problems in the operation of a hydraulic system.

### PRESSURE-REDUCING VALVES

The purpose of a pressure-reducing valve is to provide for a low-pressure line from a high-pressure main system. Often a lower pressure is required in a secondary line, and a pressure-reducing valve will accomplish this in the following manner.

The valve is in a tee to the high-pressure line as shown in Fig. 11-11. The branch-line outlet is at *B*. The maximum pressure at this outlet is set by the adjusting screw against a spring (as shown). Both the primary and branch lines will have the same pressure up to the maximum set in the valve. As the pressure is increased in the main system to a point where it is higher than the setting on the spring, the following sequence will take place. The pressure in line *B* (or the branch-line pressure) will act on the bottom of the spool (area *B*). When the pressure in the branch line reaches a magnitude where the force at the bottom of the spool (area *B*) is greater than the spring setting, it will move the spool up and restrict the flow, as shown in Fig. 11-11*b*. As the flow is restricted, the pressure drops at line *B* and also at area *B*, stopping the spool movement. Note that the pressure in the main stream will act on the spool at *a* and *b*. Since they have equal areas, the mainline pressure will have no effect on the spool movement (hydraulically balanced). The only effective force on the spool is the pressure in line *B* (branch line) at area *B* and the spring. Thus, with this valve the maximum pressure desired in line *B* can be set by the adjusting screw, and it will not exceed the setting in the branch line. A drain line is provided to permit escape of oil in the upper portion of the

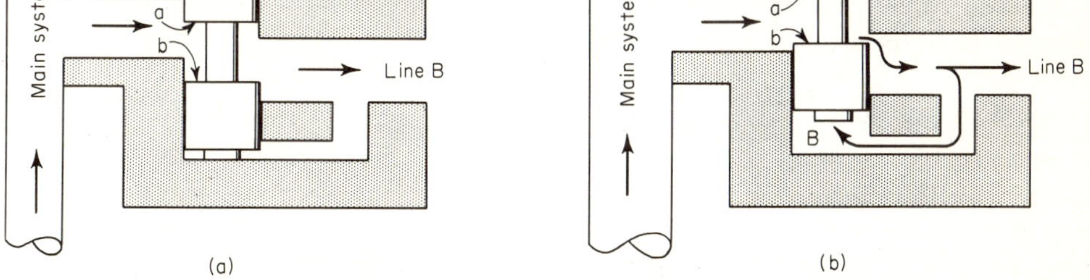


FIG. 11-11 *(a) Mainline and branch line same pressure. (b) Mainline higher pressure than branch line.*

spool so that a *hydraulic block* may be precluded. If this area were not drained, it could become filled with oil—by leakage past the spool—and any movement of the spool would be prevented. The symbol for a pressure-reducing valve is shown in Fig. 11-12.

## DIRECTIONAL VALVES

Directional valves are designed for the specific purpose of directing the flow of liquids in hydraulic systems. It may be desired, for example, to perform a work operation by driving a piston back and forth in its cylinder. A directional valve, whose movable parts change position so as to alternately introduce and drain off liquid from each end of the piston cylinder, would make this possible.

Directional valves may operate hydraulically (by differences of force set up on opposite sides of their movable parts—pilot operated), manually (by hand), or by mechanical means or electric power (solenoid). Sometimes two or more methods of operating the same valve will be utilized.

Four-way valves are directional valves with four pipe connections: one from the pressure source; a second to the reservoir; and the other two connected to the work hydraulic cylinder or motor. This type of valve is frequently used.

Directional valves are of two designs—spool and rotary. In the spool design a specially shaped sliding piston, as it moves back and forth in its housing, opens and closes passages through the valve. In the rotary design a round block or core is rotated inside a sleeve. Recesses and passages in the block make the desired connections between passages in the sleeve, which lead to different parts of the system. Valves of each of these designs will now be discussed in more detail. The symbols for two- and three-position directional valves are shown in Figs. 11-13 and 11-14. In the three-position directional valve the center block denotes the condition of flow in the valve in its neutral position, which will be discussed in detail later.

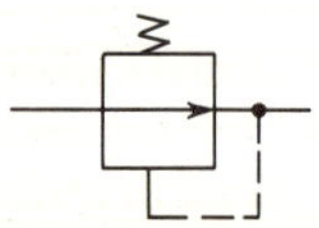

FIG. 11-12 *Symbol for pressure-reducing valve.*

### Spool Directional Valves

A two-way sliding two-position two-way valve is shown in open and closed positions in Fig. 11-15. As the piston is moved back and forth, it either allows liquid to pass through the valve or prevents flow.

The piston of the valve illustrated in Fig. 11-15 cannot move back and forth as a result of utilizing differences in hydraulic pressure set up within its cylinder, since the forces there are in balance. When hydraulic forces are equal, the spool in a valve is known to be hydraulically balanced. As indicated by the arrows against the piston heads, the same pressure acts on equal areas on their inside faces. When the input passage is blocked, small grooves around the spool permit the pressure to act equally around the piston. This assists in lubrication and balances the spool radially to permit free movement.

A number of features common to most spool-type valves are shown in Fig. 11-15. The drain openings at either end of the cylinder are needed so that back pressure that would hinder movement of the spool will not be built up in the cylinder. If this area were to become filled, no movement would be possible. This is known as a *hydraulic block*. These areas are internally drained and require no external drain lines, as the drawings indicate. When spool valves be-

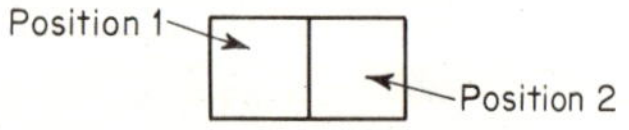


FIG. 11-13 *Two-position valve basic symbol.*

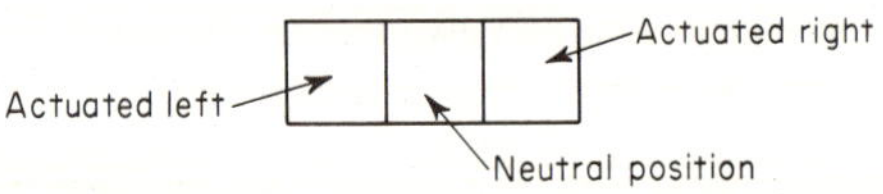


FIG. 11-14 *Three-position valve basic symbol.*

 

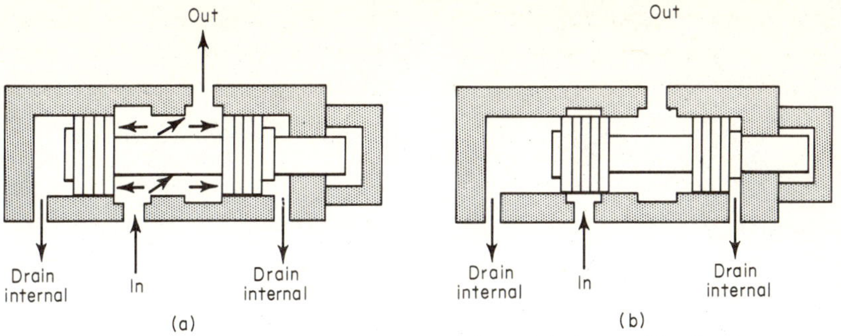

FIG. 11-15 *Two-position two-way valve. (a) Open. (b) Closed.*

come worn, they may lose balance because of greater leakage on one side of a spool than on the other. In that event the piston tends to stick while moving back and forth. The grooves machined around the sliding surfaces of the spools allow leaking liquids to encircle the spools and keep them lubricated and centered.

The valves in Fig. 11-15 have two positions and two ports. In the open position the flow through the valve is permitted from one port to another, and in the closed position the ports are blocked off, and no flow through the valve is possible. These valves may be spring loaded so that they will automatically take one of the two positions. If the spring moves the spool to a position where flow is permitted, it is called "normally open," and if the opposite is true, it is called "normally closed." See Fig. 11-16 for symbols.

### Four-way Spool Valves

Valves of this kind connect with four separate lines. In a typical installation, such as the one shown in Fig. 11-17, one line connects the valve to the pump. Another connects it with the reservoir (tank) by means of two exits from the inside of the valve. The other

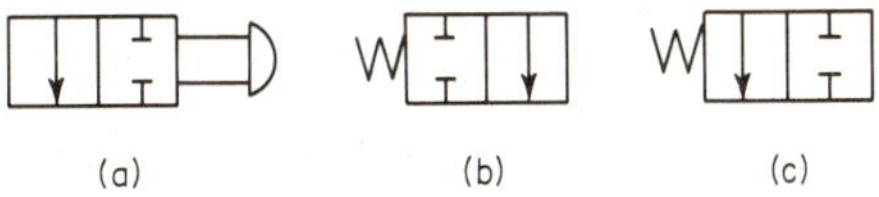

FIG. 11-16 *(a) Two-position two-way manually operated valve. (b) Spring-loaded two-position two-way valve, normally closed. (c) Spring-loaded two-position two-way valve, normally open.*

two run to either end of the hydraulic cylinder.

As the spool of the four-way valve is moved back and forth, each end of the work cylinder is connected in turn with the pressure line from the pump, while the other end of the work cylinder is connected with the tank. This enables the work piston to be driven back and forth without meeting hydraulic resistance. The valve in Fig. 11-17 is manually operated.

Figure 11-18, taken in conjunction with Fig. 11-17, shows how a four-way valve works. The position of the piston in Fig. 11-18*a* corresponds to the situation shown in Fig. 11-17. With the piston of the valve to the far right in its cylinder, liquid from the pump flows to port 1 and the rod end of the cylinder while liquid from the blank end of the cylinder is being returned to the reservoir. In Fig. 11-18*b*, with the spool to the far left in its cylinder, relations are reversed, since liquid from the pump now flows to the blank end of the cylinder while liquid from the rod end is being returned to the reservoir. In Fig. 11-18*c*, where the piston is in an intermediate position, flow through the valve from the pump is shut off, and both ends of the work cylinder can drain to the reservoir unless other valves are used to control flow from the cylinder.

The spool in the valve can be positioned by the following methods: (1) manually; (2) mechanically, through the action of a cam brought into position at the proper moment by the work operation or by means of gears

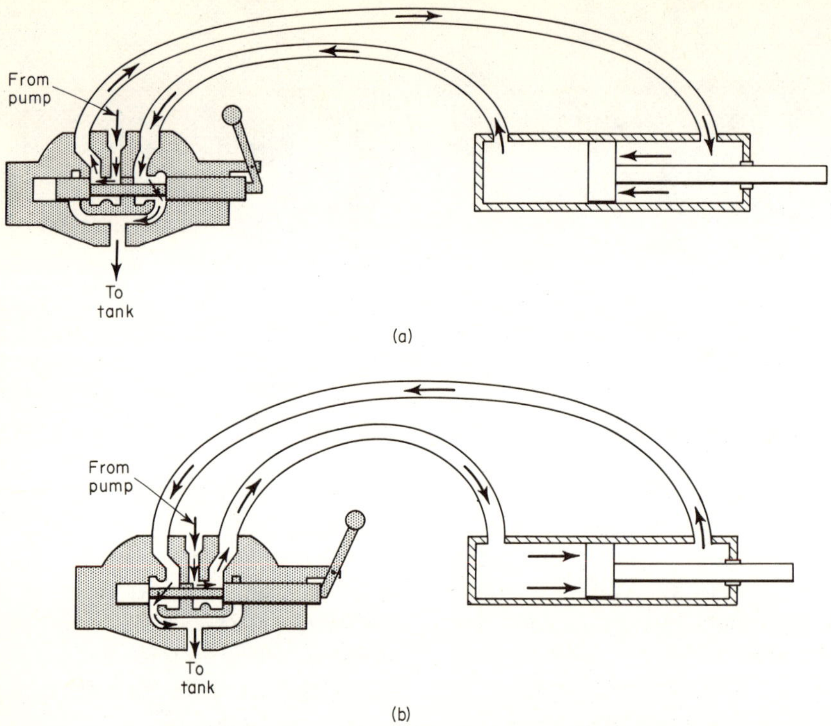


FIG. 11-17 *Four-way valve, double-acting cylinder circuit.*

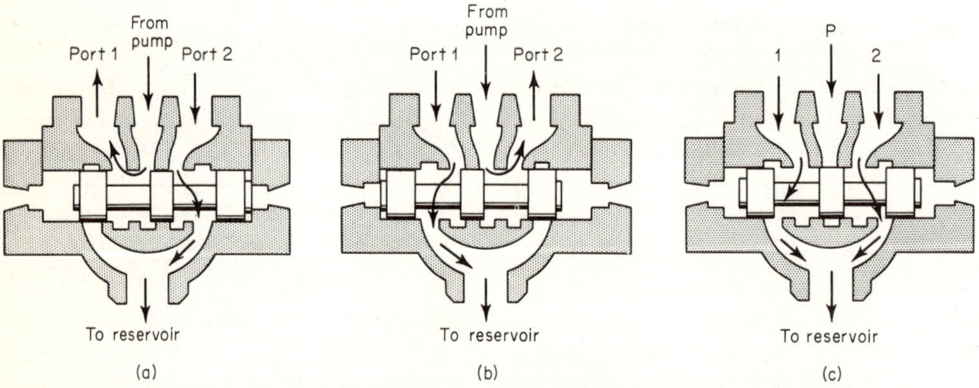


FIG. 11-18 *Cross-section of three-position four-way valve.*

and shafting; (3) electrically, by a solenoid; (4) hydraulically, through the action of a pilot valve. Large valves are frequently controlled by a small solenoid-operated valve which will direct flow of oil to either side of the spool to move it to the desired position. This combination of valves is called "solenoid controlled pilot operated."

Figure 11-18 illustrates a four-way valve designed to be operated by a pilot valve. Hydraulic pressure and a small flow to either side will move the spool. The pilot valve alternately delivers pressure to one end and then to the other of the cylinder of the four-way valve, in exactly the same manner as the four-way valve might be used to operate a hydraulic cylinder (see Fig. 11-19). In Fig. 11-18*a*, for example, the pilot valve would stop delivering pressure to the left-hand end of the spool and would begin delivering

 

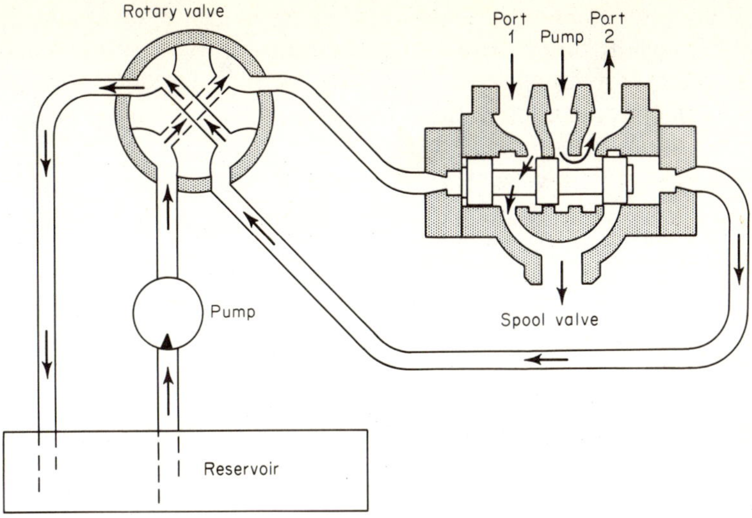


FIG. 11-19 *Pilot operation of a four-way valve.*

pressure to the right-hand end, thus reversing the position of the valve spool (Fig. 11-18*b*).

In the three-position four-way valve (Fig. 11-18) arrows show direction of flow: In one position (*a*) flow is from the pump to port 1 and from port 2 to the reservoir. In the second position (*b*) flow is from the pump to port 2, while the return flow comes from port 1 to the reservoir. In its center position (*c*) the flow from the pump is blocked. The three-position four-way valve symbol is shown in Fig. 11-20. The type of spool used is indicated in the center block, and will be discussed later. *Note:* Arrows indicate direction of flow in each of the two end positions.

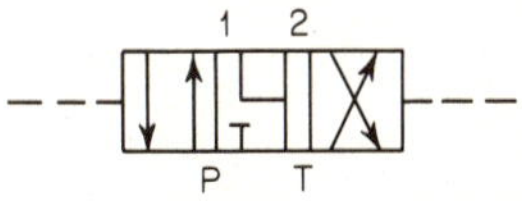


FIG. 11-20 *Symbol for three-position four-way pilot-operated valve.*

A four-way valve may be constructed to have two positions. In certain operations it may be an advantage not to have a center position. The construction of these valves is shown in Fig. 11-21.

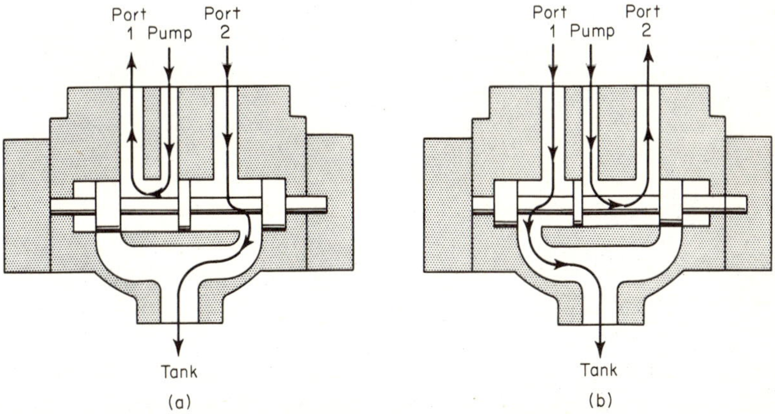


FIG. 11-21 *Two-position four-way valve.*

Note that in this valve the flow is always directed to one port or the other. It would not be possible to stop a hydraulic cylinder in any mid-position. It would move in either one direction or the other at all times. Also, a motor could be operated in either direction but not stopped. The symbol for a two-position four-way valve is shown in Fig. 11-22.

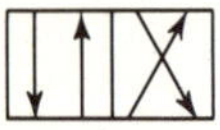

FIG. 11-22 *Symbol for two-position four-way valve.*

**Closed- and Open-center Spool Valves**

In closed-center valves the spool and housing are so constructed that all passages through the valve are blocked when the spool is centered in its housing. Figure 11-23 shows the flow through the valve in its other two positions.

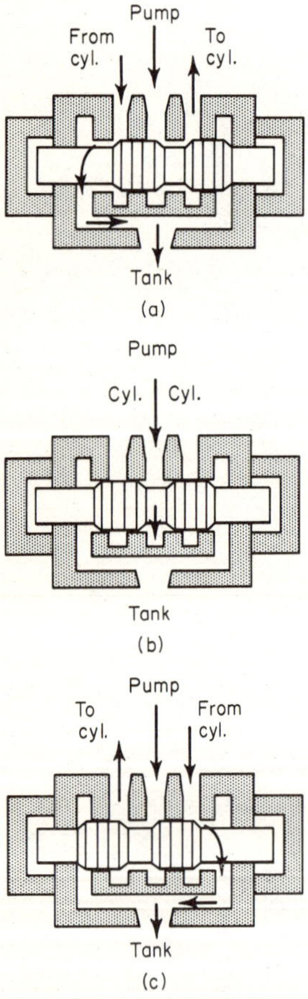


FIG. 11-23 *Closed-center valve.*

In open-center valves the spool and the housing are so constructed that all passages are open to each other when the spool is in its neutral or centered position (Fig. 11-24*b*). Parts (*a*) and (*c*) of Fig. 11-24 again show flow through the valve in the other two positions.

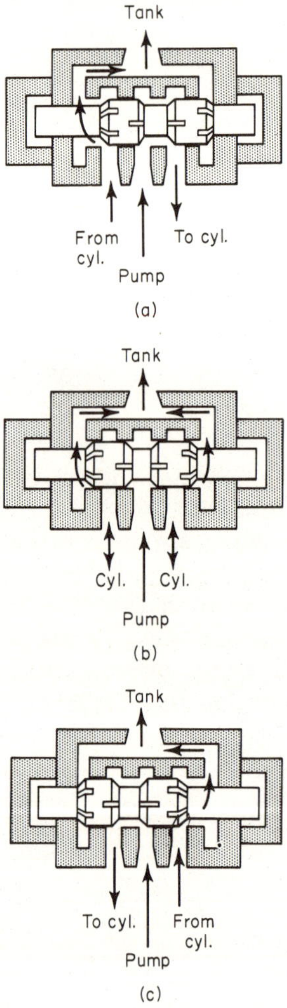


FIG. 11-24 *Open-center valve.*

In some open-center valves, passages to the work cylinder are blocked when the valve is centered, while liquid from the pump is carried in bared passages through the spool and out to the reservoir (Fig. 11-25*b*).

In valves of this kind liquid must be carried to both ends of the directional valve to keep it balanced, that is, to *A* and to *B*

 

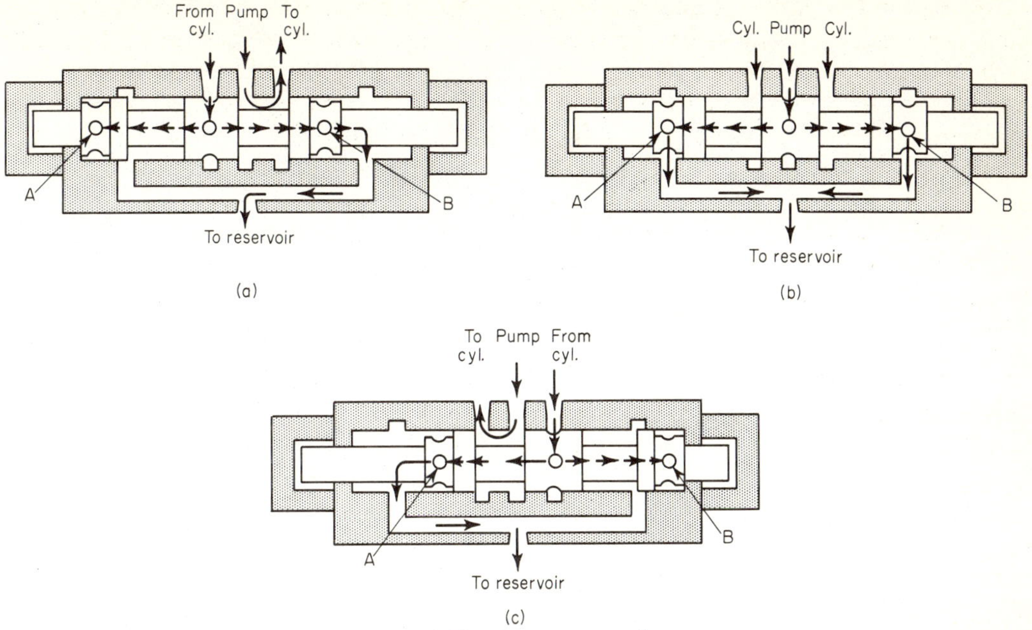


FIG. 11-25 *Open-center valve.*

in Fig. 11-25. Instead of discharging into the reservoir when the valve is centered, liquid may be directed to other valves so that a set of operations is performed in sequence.

The condition of the center depends on the configuration of the spool and housing. The condition of the valve, in the center position, is shown in the center block of its symbol (Fig. 11-26). $P$ = pump flow, $T$ = tank (reservoir), and 1 and 2 are control ports.

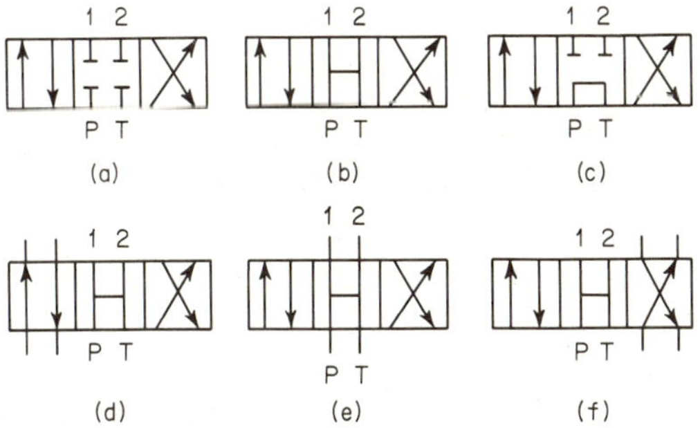


FIG. 11-26 *(a) All ports blocked. (b) All ports open. (c) Pump and tank open, cylinder ports blocked. (d) Open center, actuated left. (e) Open center, center position. (f) Open center, actuated right.*

Several other common spool configurations are used to serve a specific purpose. Symbols are shown in Fig. 11-27.

**Methods of Actuating Valves**

Symbols are shown in Fig. 11-28. Open-center valves are used when the work cyl-

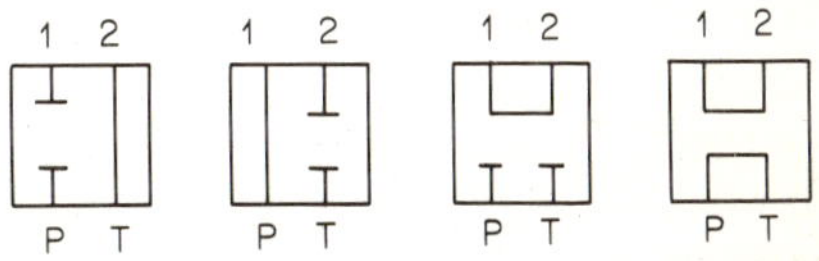


FIG. 11-27 *(a) Pump blocked, one port blocked, tank and second port common. (b) Pump and port 1 common; tank and port 2 blocked. (c) Pump and tank blocked; ports 1 and 2 common. (d) Pump and tank common; ports 1 and 2 common.*

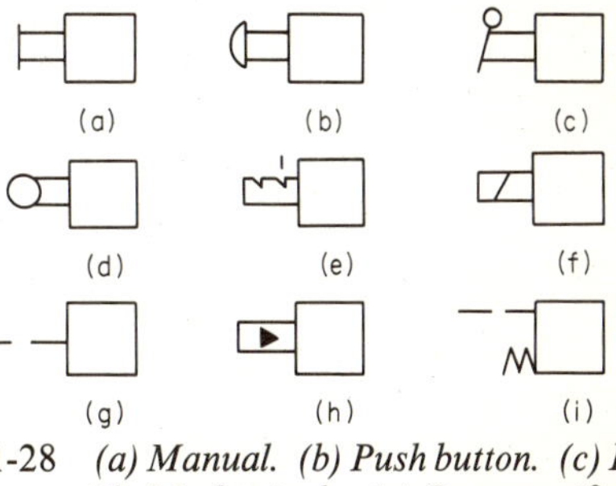


FIG. 11-28 *(a) Manual. (b) Push button. (c) Push-pull lever. (d) Mechanical. (e) Detent. (f) Solenoid. (g) Pilot. (h) Pilot. (i) Pilot, spring return.*

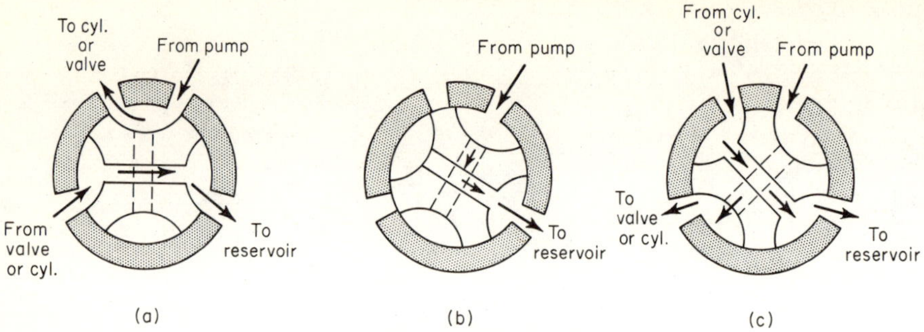


FIG. 11-29 *Rotary-type valve.*

inder does not have to be held in position by pressure and where the power is used to perform a single operation. They also avoid shock to the system when the valve spool is moved from one position to another, since in the intermediate position pressure is temporarily relieved by the passage of liquid from the pump directly to the reservoir. Reversal of action is therefore smooth.

Closed-center valves are used when a single pump or accumulator performs more than one operation, and where there must be no pressure loss in shifting the direction of stroke at the point of work.

Spool valves may be spring centered or spring offset, according to the position in which the spool of the valve stands when it is in neutral. A spring may return the spool to the center position, or it may be returned to an off-center position whenever the force controlling it is released.

## ROTARY DIRECTIONAL VALVES

Valves of this kind are most frequently used as pilot valves to direct flow to other valves. A schematic view of a rotary directional valve used to control flow to a four-way spool valve is given in Fig. 11-19. Liquid is passed from the pump through the rotary valve on to the spool valve, where it positions the spool valve to direct flow from another pump to one side of a work piston (which is not shown in Fig. 11-19).

Through a return line, liquid is carried from the other end of the spool valve through the rotary valve to the reservoir.

Figure 11-29 shows how a rotary-type valve works. Figure 11-29 views (*a*) and (*c*) show it set to deliver liquid through a spool valve to different sides of a work piston or a pilot operated valve, while (*b*) shows the valve in neutral position with all passages through the valve blocked. The valve shown is an open-center model, since it contains two cross passages (one indicated by a dotted line in the figure). Other rotary valves are made with closed centers.

Rotary valves may also be operated manually, mechanically, hydraulically, or electrically.

## METERING VALVE

Figure 11-30 is a partial cutaway, showing that a rotating core can be used to meter the flow of a liquid.

As the position of the core is changed, a greater or smaller quantity of liquid is allowed to pass through the valve. In this way the speed with which an operating part moves

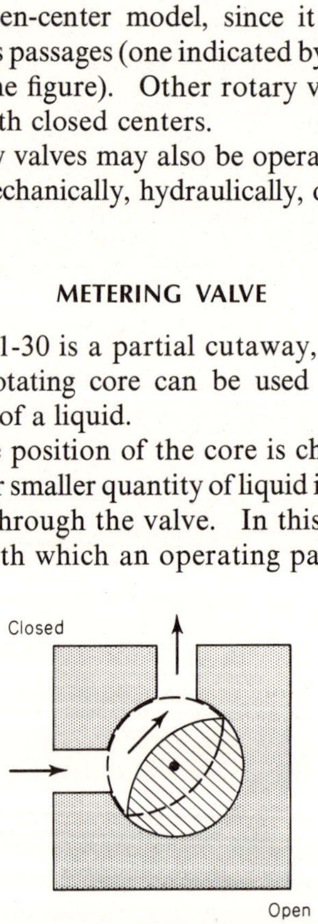


FIG. 11-30 *Two-way rotary metering valve.*

 

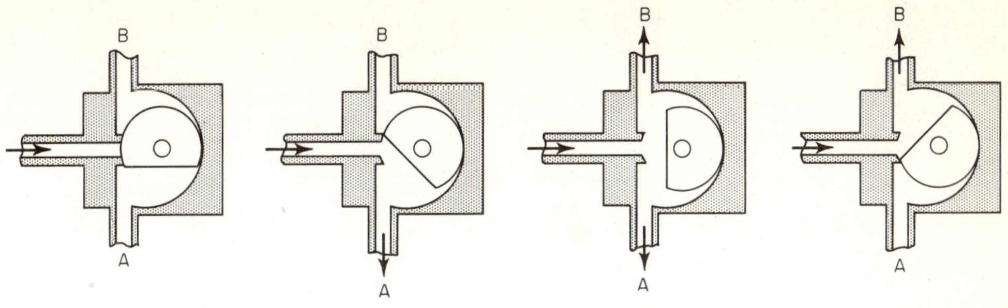


FIG. 11-31 *Three-way rotary valve.*

can be controlled, the action of a valve delayed, or spools in cylinders cushioned at the end of a work stroke. The valve could be used to prevent a tool from breaking through too rapidly at the end of its stroke without straining the system.

Using the same principle, two ports can be metered with one rotating core. Figure 11-31*a* shows both ports blocked; Fig. 11-31*b* shows full open to port *A*; Fig. 11-31*c* shows both ports open; and Fig. 11-31*d* shows meter or full flow to port *B*.

This metering device provides for an adjustable orifice in two directions. It is the basic metering device for a pressure-compensated flow divider if a spool is added to keep the pressure the same at *A* and *B*. Consequently, the pressure drop will be the same, since the pressure at the inlet is common. If orifices were set to be equal in size, with the same pressures at both sides, equal flow would result. This will be discussed in a later section.

## NEEDLE VALVE

A simple device for controlling flow is a needle valve (Fig. 11-32). The opening is adjusted by the screw, and flow can be restricted and thus controlled.

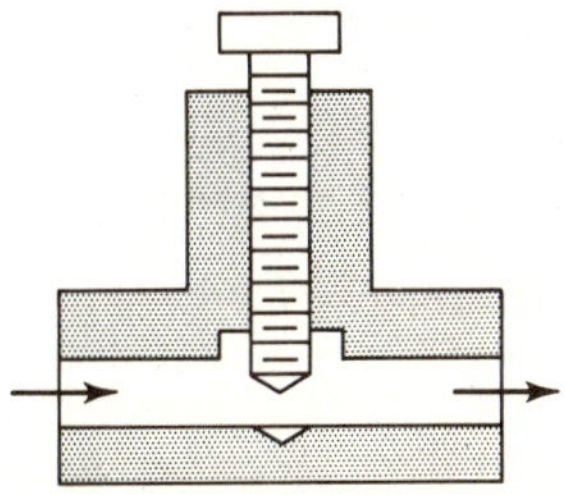

FIG. 11-32 *Needle valve.*

## SHUTTLE VALVE

Another simple but useful valve is a shuttle valve (Fig. 11-33). Flow can be directed to port *C* from port *A* or *B*. It would receive flow from the port with the highest pressure. At the same time it would block flow from *A* to *B*. It is often used on a pilot-operated valve where it is required to receive a signal from one of two sources at different times.

## DELAY VALVE

Figure 11-34 diagrams the action of a delay valve. Liquid moving from the pump through a control valve (not shown) at different times takes three paths: passage 1 goes from the pump *around* the time-delay valve past check valve *A* on to the point where work is to be done; passage 2 is *through* the time-delay valve (when spool is seated); and passage 3 goes *through* check

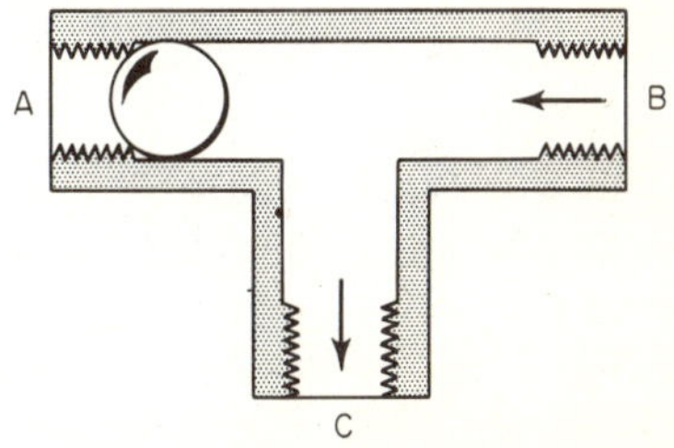


FIG. 11-33 *Shuttle valve.*

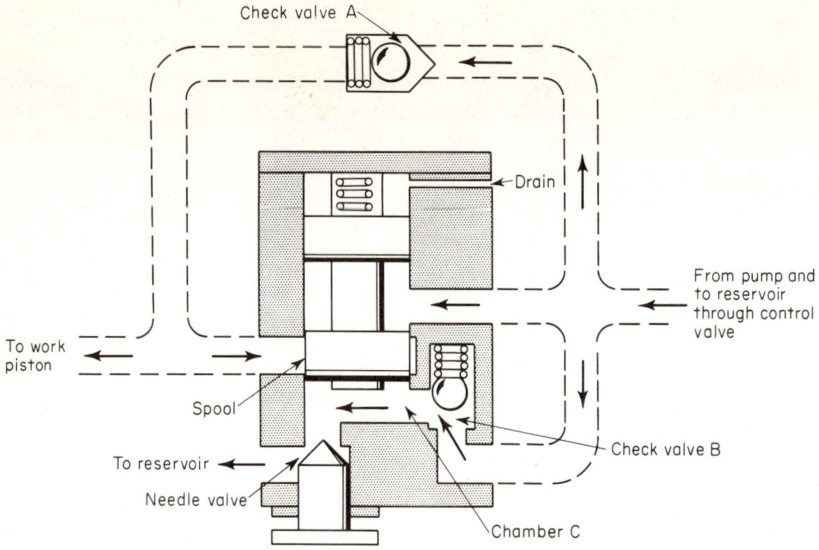


FIG. 11-34 *Time-delay valve.*

valve *B* to produce a force on the bottom of a piston which is held in place by a spring in the time-delay valve. As an increase in hydraulic pressure compresses the spring on the spool, it rises to shut off flow through the delay valve. Figure 11-34 shows the valve in position 3.

When the position of the valve controlling flow from the pump is changed, liquid returns from the point of work, but is blocked from passage 1 around the delay valve by the upper check valve, and blocked from passage 2 through the delay valve by the lower piston head of the valve. But a needle valve under the piston slowly drains liquid from chamber *C*, so that the piston can slowly descend, allowing liquid to pass over it and on to the reservoir by the same path that liquid from the pump originally entered the valve. This is made possible by a different positioning of a control valve. The adjustment of the needle valve will determine the length of time that return flow will be delayed.

### SEQUENCE VALVE

The flow in this valve (Fig. 11-35) runs in one passage until a preset pressure is reached. It is then also carried along a second pas-

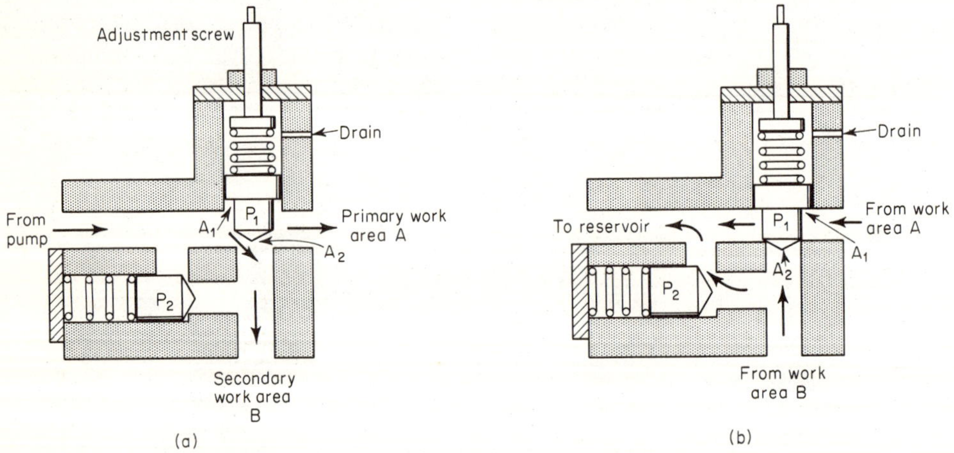


FIG. 11-35 *Sequence valve.*

 

sage to another work operation. Liquid flows around $P_1$ against the resistance of its spring. Then liquid flows through the second passage of the valve to do other work in the secondary work area. $P_1$ is held open by a smaller pressure than is needed to open it because of the greater area against which this pressure can then act. While flow is to work area $A$, the pressure acts on the spool at $A_1$. When the pressure is reached that will overcome the spring, the spool moves up and now both areas $A_1$ and $A_2$ are available to hold the spool up, permitting flow to work area $B$. When the position of the control valve governing the (sequence) valve is changed, liquid returns through the valve from work area $A$. Flow from work area $B$ depresses the check valve $P_2$ and flows past the base of this spool (Fig. 11-35).

## FOOT VALVE

Here free flow is possible in one direction, while flow in the other direction is impossible until predetermined pressure is reached. A nonadjustable foot valve is shown in Fig. 11-36. The plunger is hollow and has an inner passage with a check valve at its base and an opening at the top, so that liquid can flow upward through the valve but is prevented from returning until pressure on top of the spool increases sufficiently to unseat the valve by compressing the spring at its base. Then the whole plunger descends, permitting downward flow until pressure above the valve is lowered. At this lower pressure the spring reseats the plunger and shuts off the flow.

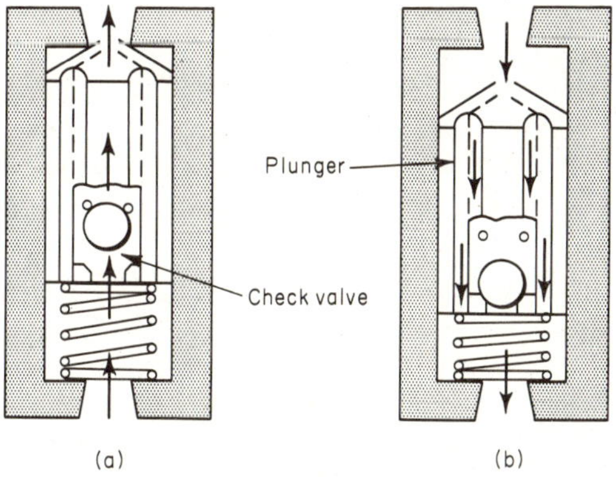


FIG. 11-36 *Foot valve.*

## ADJUSTABLE FOOT VALVE

The construction of an adjustable foot valve is shown in Fig. 11-37. Liquid moving from the pump to the point of work passes $P_2$ by compressing a light spring (Fig. 11-37*a*), while Fig. 11-37*b* shows liquid returning to the reservoir past $P_1$ against the resistance of a spring set to compress at a definite pressure.

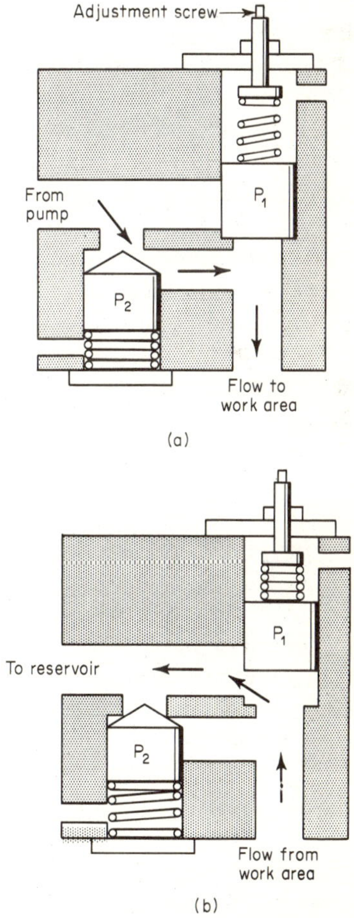


FIG. 11-37 *Adjustable foot valve.*

## PREFILL CHECK VALVE

This valve permits liquid from a storage reservoir to fill the work cylinder of a hydraulic system during the first phase of the work action when the work plunger is being moved rapidly to the point of work by gravity or other means (Fig. 11-38). In this way a small-volume pump can be used to do the work, instead of the large pump that would be needed if all the liquid to fill the work cylinder had to be furnished by the pump.

In the surge type illustrated, when a partial vacuum is set up inside the work cylinder as the piston in the hydraulic cylinder descends, the check valve is opened by atmospheric pressure acting on the free surface of the liquid in the reservoir. The check valve is closed by its spring when the movement of the piston comes to rest or when pressure in the work cylinder has risen sufficiently. From this point on, further movement of the work piston is accomplished by pump action.

The check valve might be opened by liquid under pressure directed to it by a pilot valve, and it might also be closed by that valve instead of by a spring.

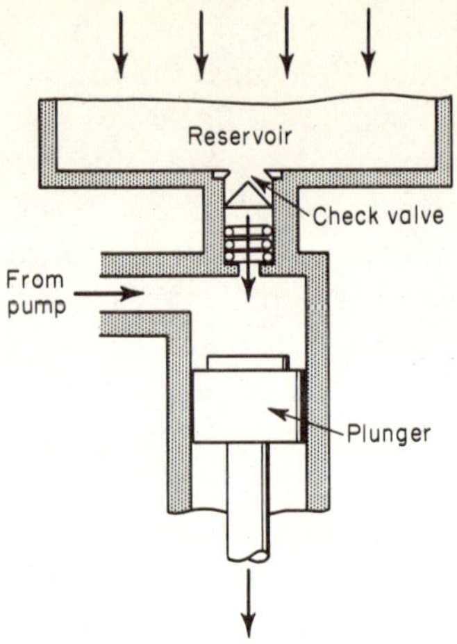


FIG. 11-38 *Prefill check valve.*

## UNLOADING VALVE

When a system calls for a valve that will permit the return of oil back to the tank at zero pressure, an unloading, or dump, valve is used. Pressure from another source is used to hold this valve in its unloading position. It is generally used where a high-low volume dual pump is used—in each case two independent supplies of fluid are available.

A high-low dual pump is used where speed is first required in the cylinder or motor and then a high force or torque is needed with slow speeds. By using two pumps at a low pressure, speed is gained until the point at which force, and not speed, is required. When a preset pressure is reached, the high-volume pump unloads, and the low-volume pump completes the work. The dual pump with its (high-low) flow relationship permits high speed initially, and, after dumping, a high force is obtained without requiring additional horsepower from the prime mover. For example, if a 20-gpm pump and a 4-gpm pump were used, speed could be first obtained by using a 12-hp motor and a 750-psi system. At this point we would be utilizing the total hp available. When speed is no longer required, but force is, the 20-gpm pump could be returned to the tank, and the 4-gpm pump could deliver pressure up to 4500 psi without additional horsepower. The total force available would be dependent on the cylinder size.

This valve can be set by the adjusting screw to unload at a preset pressure. It is pilot operated, so system pressure will actuate the unloading operation with a signal from the low-volume pump system. The pressure in the high-volume system will drop to almost zero after the valve is actuated. Figure 11-39 shows this valve in its operating and unloading condition. A check valve must be used to prevent back flow. Without a check valve, flow from the high-pressure system could enter port *B* and return to the tank. Either a check valve must be installed or an internal check valve provided.

The valve is balanced with the spring holding the spool down (Fig. 11-39*a*). When

 

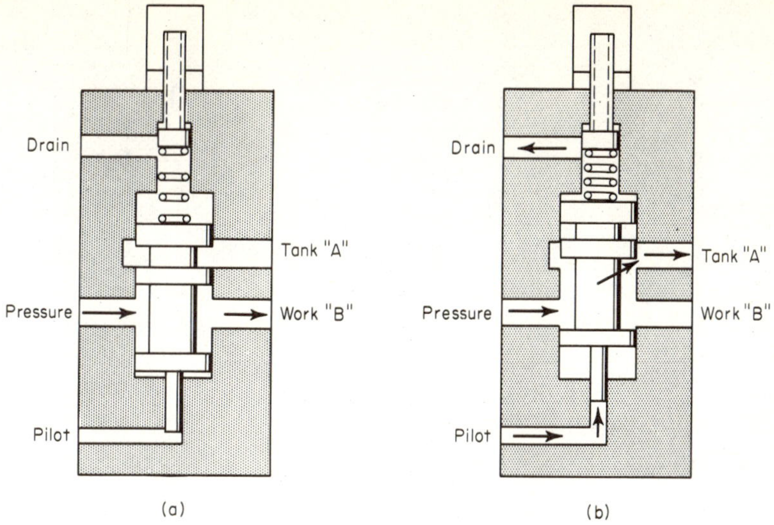


FIG. 11-39 *Unloading valve.*

the pressure from the remote source becomes great enough to overcome the spring, the spool will move into the second position (Fig. 11-39*b*) and permit the oil to return to tank *A* at the discharge outlet under no pressure. Thus the second system is left to do the work. Note that the spool is balanced so that pressure in the valve will have no effect on spool movement. Figure 11-40 shows a system in which the hydraulic cylinder can move up or down at a fast rate. Both pump outlets are directed to the cylinder ports. As the rod hits the work, the pressure will rise, unloading the high flow outlet from the pump and leaving the low flow to complete the work at the pressure that is required. Note that the cylinder will always move down at a rapid rate (away from the work). If the pilot line were also connected to the rod end of the cylinder line, it would then unload in this direction if the preset pressure on the valve were reached.

## COUNTERBALANCE VALVE

These valves are used for maintaining back pressure to counteract a hydraulically sustained load. They are used as a load-lowering valve. Figure 11-41*b* shows the valve in its blocked position. No flow is permitted through the valve, and the work is thus sustained regardless of the pressure created by this load at the pressure port. When the

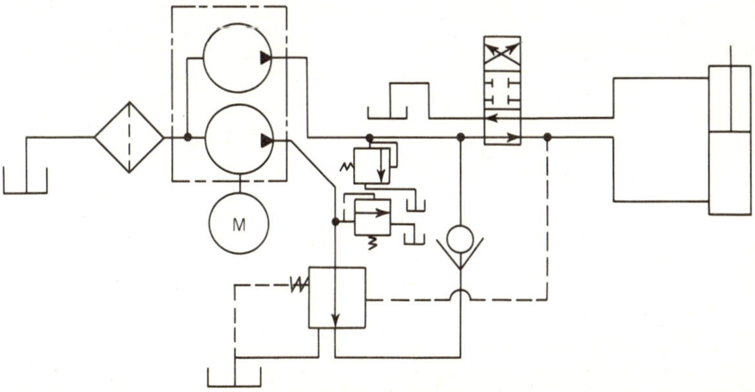


FIG. 11-40 *Unloading circuit.*

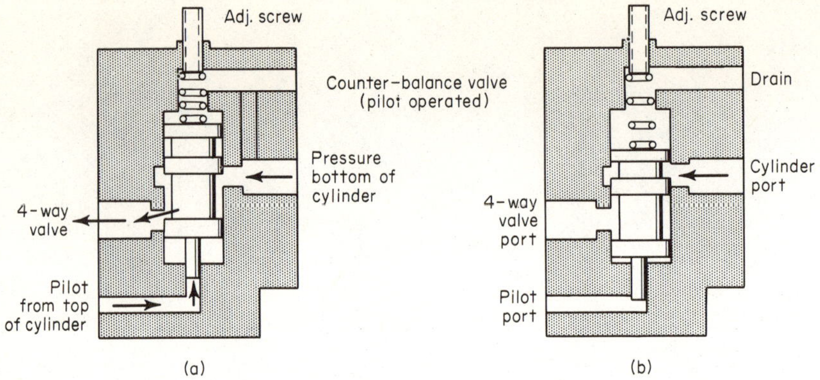


FIG. 11-41 *Counterbalance valve (pilot).*

pump induces additional flow to the top of the cylinder, pressure is increased, and this higher system pressure is sensed at the pilot port. When it becomes sufficient to overcome the spring force, the spool will shift, as shown in Fig. 11-41*a*. In this position it will permit flow and, consequently, movement of the work. A reduction of pressure will permit the spring to reseat the spool, thus again blocking flow.

These valves may be constructed to be pilot operated, as shown in Fig. 11-41, or directly operated, as shown on Fig. 11-42. In this configuration oil is blocked until the pressure again overcomes the spring and moves the spool to the position shown on Fig. 11-42*a*. To raise the load in a directly operated valve, a check valve must be provided to permit load raising, as shown in Fig. 11-42*b*.

## PROBLEMS

**1.** What is the purpose of a relief valve in a system?

______________________________

______________________________

______________________________

______________________________

**2.** Figure 11-8 is a sketch of a simple relief valve. What are the advantages of this valve as compared with the pilot-operated type?

______________________________

______________________________

______________________________

______________________________

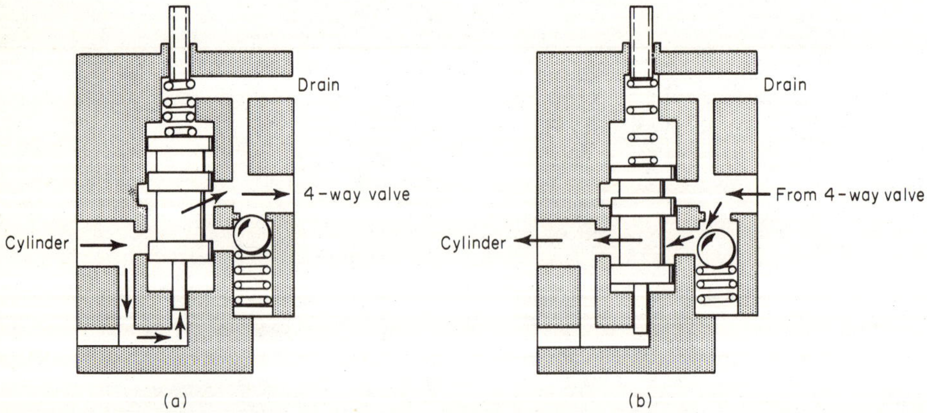


FIG. 11-42 *Counterbalance valve (direct).*

 

**3.** What are the advantages of a pilot-operated relief valve?

______________________

______________________

______________________

______________________

**4.** What are two principal causes of relief-valve malfunction?

______________________

______________________

______________________

______________________

**5.** What is the symbol for a relief valve?

**6.** What is the purpose of a pressure-reducing valve?

______________________

______________________

______________________

______________________

**7.** What is the pressure balance in a hydraulic component?

______________________

______________________

______________________

______________________

______________________

______________________

**8.** What is the symbol for a pressure reducing valve?

**9.** Figure 11-43 shows a three-position four-way valve. Complete the information requested below:

(*a*) Flow from pump to port number

______________________

Flow from port number ______ to

______________________

(*b*) Center open or closed? __________

Port blocked or open? __________

(*c*) Flow from pump to port number ___

______________________

Flow from port number ______ to

______________________

(*d*) Flow from pump to port number ___

______________________

Flow from port number ______ to

______________________

(*e*) Center open or closed? __________

Port blocked or open? __________

(*f*) Flow from pump to port number ___

______________________

Flow from port number ______ to

______________________

**10.** What are three principal methods of actuating a spool valve?

______________________

______________________

______________________

**11.** Describe the different conditions of the three-position valve (Fig. 11-44) in its center position.

(*a*) ______________________

(*b*) ______________________

(*c*) ______________________

______________________

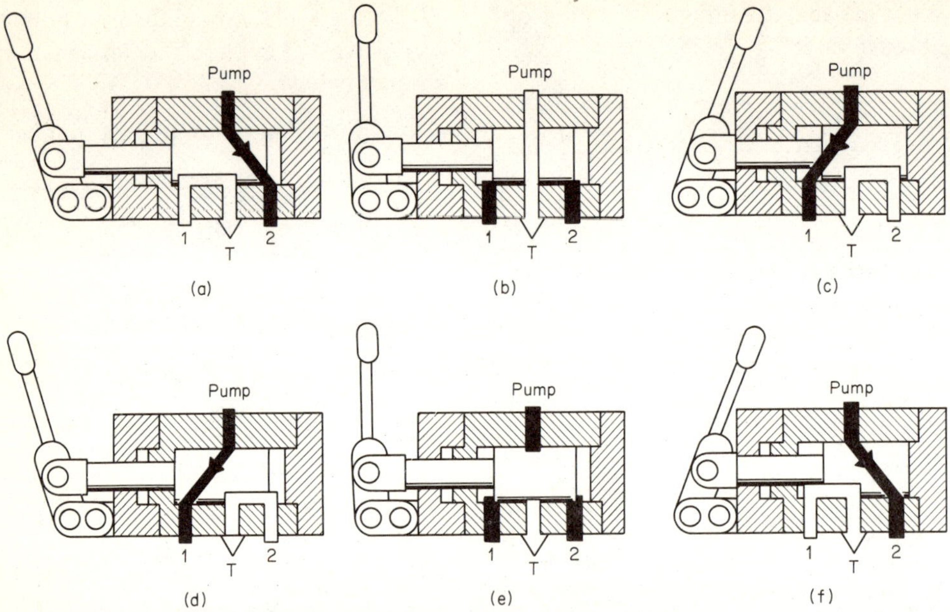


FIG. 11-43 *Three-position four-way valve.*

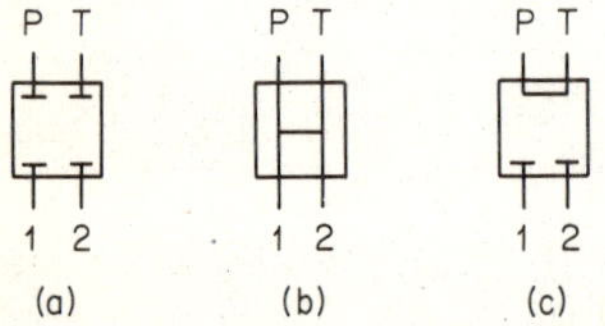


FIG. 11-44 *Center-position three-position valve.*

**12.** How would a sequence valve be used in a circuit?

______________________________

______________________________

**13.** What is the symbol for a sequence valve?

**14.** In what type of circuit would a prefill check valve be used?

______________________________

______________________________

**15.** Explain why an unloading valve is necessary and in what type of circuit it is used.

______________________________

______________________________

______________________________

______________________________

______________________________

______________________________

 

# PRESSURE-COMPENSATED FLOW CONTROL

# 12

Any flow control, with a pressure-compensation feature, will automatically maintain a rate of flow regardless of pressure changes that are caused by the work. If a cylinder is moving 500 lb at the rate of 1 ft/min, and suddenly it is required to move 1000 lb, it will move the heavier load at the original speed. This means that the volume of oil going to the cylinder must remain the same. The pressure, of course, must be doubled in order to move twice the load. Flow control valves are designed and used to accomplish this feature.

Before we analyze the mechanism of these valves, we must remember a fluid flow law which states that a fluid rate of flow will be constant through an orifice of a given size as long as the pressure drop remains constant. This means that the pressure can vary as long as the change is the same.

*Example:* Let us say that $P_1 = 1000$ psi, and $P_2 = 900$ psi. The pressure differential is now 100 psi. Assume the flow to be 10 gpm. If the pressure of $P_1$ is increased to 1200 psi and $P_2$ is increased to 1100 psi, the flow is still 10 gpm because the pressure differential is the same (or still 100 psi).

This characteristic of fluid flow permits a wide variety of automatic flow control valves, a few of which we will study and analyze.

Figure 12-1 shows a pressure-compensated flow control valve which will provide a constant flow at the outlet port regardless of the pressure required to do the work.

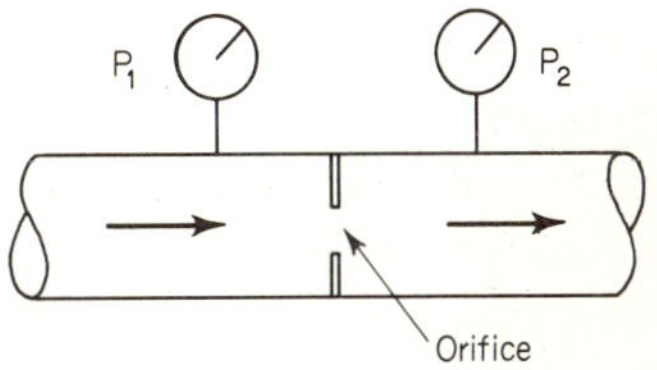

FIG. 12-1 *Orifice in a pipe.*

The orifice is adjustable, so a selection of any gpm may be obtained. The volume is limited only to the amount available at the inlet, and the size of design of the unit itself.

This constant pressure differential is maintained as follows. As the pressure (Fig. 12-2) increases at C (which is the pressure required to do the work), the pressure at $C_1$ will also increase, forcing orifice 1 open. This reduces the pressure drop across this orifice and increases the pressure in area B, again equalizing the difference between B and C. If the pressure at C decreases, so will the pressure at $C_1$. This means the pressure at point B (which is now greater) will force the spool against the spring, thus closing orifice 1 and reducing the pressure at B and at $B_1$. It is again in equilibrium. In this manner this spool will be actuated by a difference in pressure at area B and area C and will always take a setting that

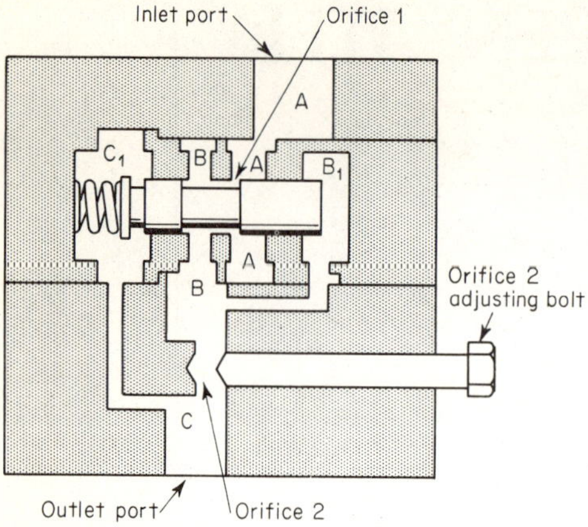


FIG. 12-2 *Two-port flow control valve (pressure compensated).*

will keep the pressure differential the same between B and C, thus providing the outlet with a constant volume of flow.

The excess oil in this valve, provided by the pump, will pass over a relief valve and back to the tank, or may be used in another part of the circuit. If a secondary circuit is used, the pressure to operate it must be greater than that required for the primary circuit. If it were lower, all the flow would go to the secondary circuit, and zero flow would be available through this valve.

*Example:* If area A has 1000 psi available, and area C requires 100 psi to do the work (and if it requires 100 psi pressure drop across orifice 2 for 10 gpm of flow), then area B must have a pressure of 200 psi. It follows that to maintain 10 gpm, area B must always have a pressure of 100 psi greater than area C. If orifice 2 were reduced, the pressure differential in C and B would increase, and the flow would be reduced depending on the reduction of orifice size. Therefore, a rate of flow can be established and maintained depending on the size of orifice 2.

*Example:* If the work suddenly required 500 psi (at area C), the following would take place. The pressure at C and $C_1$ would rise to 500 psi. At $C_1$ this increased pressure would overcome the pressure at $B_1$ (with the help of the spring), thus opening orifice 1 and increasing the pressure in area B until pressure at B and $B_1$ would stop further opening of orifice 1 so that pressure at B would be 600 psi. Again, we have a 100-psi differential, and the flow continues at the original 10 gpm. Because oil is essentially incompressible, these operations occur almost simultaneously.

The speed of the work can now be controlled by the manual control of orifice 2. It will maintain the speed because of the constant flow of oil to the work regardless of the pressure required at the work end of the circuit. The maximum pressure at the work is, of course, limited by the setting of the maximum-pressure (relief) valve.

#### Pressure-compensated Flow Divider

Another type of flow control commonly used is the flow divider. This has advantages over the control valve previously discussed. The advantage is that excess flow is diverted out of a second port and may be returned to the tank, or may be used as part of the overall circuit. It can be used and controlled to do work. These valves can be adjusted to divide flow in any proportion, such as 50–50, 90–10, or 100 percent out of either port. If the volume available at the inlet port changes, the flow at the outlet ports will vary proportionately.

On Fig. 12-3 we can follow the flow and see how this principle is used to produce a constant flow from the outlet ports regardless of the pressure that is required to do the work at the actuator.

There are three pressure areas: A, B, and C, as shown in Fig. 12-3. There are two adjustable metering orifices, 1 and 2. Orifices 1*a* and 2*a* are opened and closed by the action of the spool as it moves. The spool will close orifice 2*a* and open 1*a* as it moves to the right. This is a result of the hydraulic pressure in areas $B_1$ and $C_1$. It will close orifice 1*a* when the pressure in area $C_1$ increases and forces the spool to the left. Note that areas at points $B_1$ and $C_1$ are equal. The

 

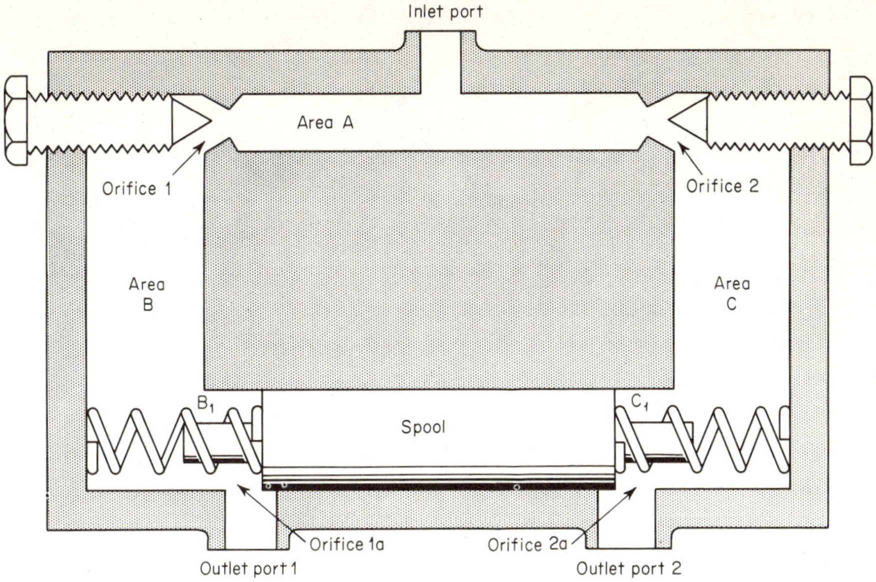


FIG. 12-3 *Flow divider, pressure compensated.*

spool is balanced, and only a differential pressure at the end of the spool will effect a movement.

As orifice 1 or 2 is opened manually by the adjustment screw, the flow is increased. To maintain a constant rate of flow, it is necessary to maintain a constant pressure differential between areas A and B and areas B and C. Thus the pressure at B and C must remain the same. The spool, by metering flow at orifices 1*a* and 2*a*, maintains equal pressure at B and C. For example, pressure in area A is always that pressure which is required by the heaviest load at port 1 or port 2 plus the pressure loss in the valve. Pressure in areas B and C must always be the same so as to maintain a constant flow across orifices 1 and 2. The spool keeps the pressure in these areas the same, as described below. Orifices 1 and 2 are adjustable, so they can be set for the desired volume of flow at each outlet port. As we know, the flow remains constant if the pressure drop remains constant. Therefore, we can set both orifices 1 and 2 to the desired flow, and it will be maintained as follows.

If the pressure is zero at ports 1 and 2, the pressure at C and B will be zero (except for pressure loss from B or C to the outlet), and the flow will be governed by the orifice sizes at 1 and 2.

If the pressure required at 2 increases, it increases the pressure on the end of the spool at $C_1$. This moves it to the left and restricts flow at orifice 1*a*, thus increasing the pressure in B until it is the same as it is in C. So the same pressure drop is maintained at orifices 1 and 2, and again the same flow results at each of the orifices and at each of the ports.

Conversely, the same thing happens if the pressure increases at port 1. Again the spool will move (because of the higher pressure at $B_1$ and B) until the pressures at C and B are equal. Again there is constant flow across orifices 1 and 2 and ports 1 and 2.

If the load or pressure required at ports 1 and 2 were the same, the spool would center, since there would be no differential pressure at the end of the spool. The pressure in B and C would be the same, since orifices 1*a* and 2*a* would be the same. If adjusting knobs were set with the same size of orifice at 1 and 2, the valve would provide a 50–50 split in the flow. And, as explained previously, it would maintain that proportion regardless of the pressures required at outlets 1 and 2.

# HYDRAULIC FLUID

# 13

Hydraulic systems transmit power by means of a liquid regardless of size or the apparent simplicity or complexity of the installation. The fundamental principle of power transmission is the same in any hydraulic system. A hydraulic fluid must:

1. Transmit power
2. Provide adequate seal and lubrication
3. Protect system from rust and corrosion
4. Minimize wear
5. Permit settling and separation of dirt and contaminates that enter the system.

*Water* was the first hydraulic fluid used and is still used in some types of systems, particularly in those that require large volumes of fluid. Water, however, meets only the first requirement of a hydraulic fluid—transmittal of power. Being too light, it does not provide adequate seal within the system, and it certainly will not protect from rust and corrosion. Water cannot provide proper lubrication, so it cannot minimize wear. In addition, water freezes solid at a moderate temperature and boils away to vapor at operating temperatures of some hydraulic systems. For these reasons, water is a poor hydraulic fluid for nearly all systems.

*Mineral oil* of the proper characteristics is the most satisfactory hydraulic fluid because it fulfills all the essential requirements listed above.

## SELECTION OF PROPER FLUID

The basic factors which should be considered in selecting an oil should include the following: viscosity; viscosity index (V.I.); stability; foaming; film strength; and demulsibility.

*Viscosity* is the most important property of a satisfactory hydraulic oil. Viscosity refers to the "body" which the fluid possesses. Another way of putting it is that viscosity is the measure of "flowability" at a definite temperature.

By definition, viscosity of an oil is the duration of time in seconds (Saybolt Seconds Universal, or SSU) that is required to measure 60 milliliters (ml) through a standard orifice at a definite temperature (usually 100 and 210°F).

For example, an SAE 10 oil has a viscosity of 160 to 170 SSU at 100°F and 44 SSU at 210°F. An SAE 30 oil has a viscosity of 570 SSU at 100°F and 67 SSU at 210°F.

From the standpoint of power consumption alone, the lowest viscosity which can be used is the most satisfactory. Operating temperatures will have a bearing on the selection of an oil. Those systems that operate at very low temperatures naturally must have an oil with sufficient fluidity to be handled properly. Likewise, high operating temperatures require heavier oil.

A gear pump will handle a relatively heavy oil; excessive slippage may be encountered with too light an oil. On the other hand, a vane-type pump requires a lighter oil.

Low-viscosity oils will leak more readily past pistons, packing glands, and other parts of the system than will oils with a higher viscosity. No attempt, however, should be made to overcome mechanical deficiencies by the use of heavy oil.

Proper viscosity is a prime consideration where power and friction loss demand the use of the lightest possible fluid consistent with sealing requirements of the piston and packing glands and the lubrication of the bearings and mechanical parts within the system.

*Viscosity index* (V.I.) is the rate of viscosity change with respect to change in temperature. Some oils thin out more rapidly than others as temperatures increase and thicken more readily as temperatures decrease. Such oils are said to have a low V.I. and are not satisfactory as hydraulic oils. A good hydraulic oil should remain relatively fluid at the lowest expected temperature and remain sufficiently heavy at the highest operating temperature to provide good lubrication and minimum leakage. This necessitates the use of an oil with a high V.I.

| GENERAL RECOMMENDATIONS—VISCOSITY SSU AT 100°F | | |
|---|---|---|
| Vane-type pumps | up to 1000 psi | 150–225 or SAE 10 |
| | above 1000 psi | 275–315 or SAE 20 |
| Piston type | | 275–315 or SAE 20 |
| Gear type | | 150–315 or SAE 10 |

Stability refers to an oil's ability to resist oxidation. The pressures and various restrictions through which an oil passes produce high oil temperatures. Since it is impossible to exclude air from the system, the combination of high temperature and oxygen causes severe oxidation conditions.

Oil oxidation thickens the oil and is most undesirable because oxidation products are deposited on the closely fitting surfaces as varnish. This interferes with proper action of pistons and control valves. For these reasons a satisfactory hydraulic oil must have a chemical stability sufficient to resist oxidation under the most adverse conditions.

*Foaming* is caused by air entrained in the oil. In many cases it is caused by air leaks in the suction side of the system; however, some systems will always have some tend-

TABLE 13-1 TROUBLESHOOTING WITH RESPECT TO FLUIDS

| *Trouble indication* | *Cause of problem* | *Correction* |
|---|---|---|
| Loss of power | 1. Oil too light | Check manufacturer's recommendations and consult manufacturer |
| | 2. Low oil supply | Add fluid |
| | 3. Aeration of oil | Check inlet lines |
| | 4. Excessive temperature | (1) Check oil level. (2) Is there a restriction in the line? (3) Check with manufacturer |
| Spongy movements | 1. Aeration of oil | Check oil level and inlet lines |
| | 2. Excessive temperature | (1) Check with manufacturer's recommendations. (2) Find cause of oil heating, such as restriction and faulty components |
| Sticky controls | Contaminated oil | Change oil filter and clean strainer elements |
| Malfunction | Contaminated oil | Change oil, clean filter and strainer elements |

 

ency to foam because of the rapid movement of the oil through the system and the high-velocity discharge into the reservoir. The use of an oil containing an antifoam agent is a must in all hydraulic systems.

The *film strength* of hydraulic oil must, of course, be adequate to prevent metallic contact of the moving surfaces within the systems. This requirement can be met with film strength and/or extreme pressure additives.

*Demulsibility* refers to an oil's ability to separate from water. The oil in some hydraulic systems suffers considerable water contamination through leakage. Under these conditions, an oil of high demulsibility is required. An oil containing an antirust agent should be used in a system where rusting is a problem.

Although we are primarily concerned with petroleum-base hydraulic fluids, there is another type that should be mentioned. This type, known as fire-resistant fluid, is usually one of the following: (1) water in oil; (2) glycol and water; and (3) synthetics.

These fire-resistant fluids have their applications where fire hazards are high, for example, in molten metal pots in steel mills. When a fire-resistant fluid is required because of environment, the primary consideration is, of course, effective fire resistance. Other factors, such as pump life and cost, become secondary.

## HYDRAULIC FLUID MAINTENANCE

The following rules, together with Table 13-1, can be used as a guide to proper fluid maintenance.

1. Avoid the addition of water to the oil even in small quantities.
2. Do *not* mix oils. Use one type as instructed or recommended by manufacturer.
3. Do *not* contaminate oil. Store in a clean container and keep covers tight.

# FLUID POWER DEMONSTRATIONS

# 14

## 1 • VANE OR PISTON-PUMP CHARACTERISTICS

PURPOSE: To demonstrate vane or piston-pump characteristics through their range of operating pressures.

COMPONENTS: Needle valve.

CIRCUIT: Connect pressure supply 1 to the metered return line with the needle valve in the line (Fig. 14-1).

PROCEDURE: (Do NOT exceed 800 psi for this demonstration. Pressures greater than 800 psi will be analyzed in the next demonstration.)

**1.** Set the relief valve at maximum pressure (full in).

**2.** Open the needle valve.

**3.** Move the pump flow control counterclockwise to full flow.

**4.** Start the unit. Set the panel flow control to maximum or full flow to pressure supply 1.

**5.** On Fig. 14-6 record the supply pressure and flow: _______ psi at _______ gpm.

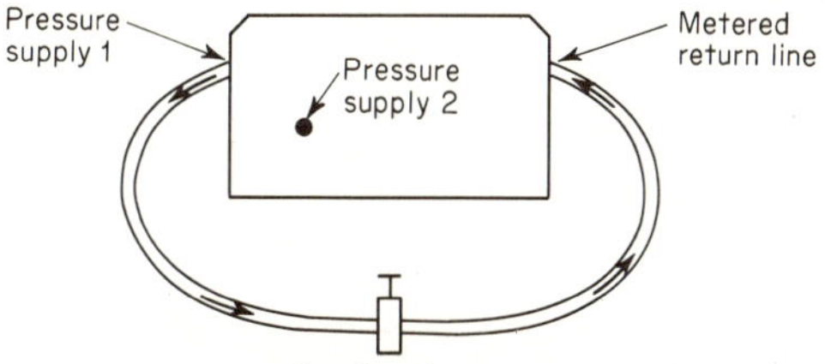


FIG. 14-1 *Circuit.*

FIG. 14-2 *Symbol for a line restriction (fixed).*

FIG. 14-3 *Graphic symbol for needle valve. Line restriction.*

**6.** Increase the supply pressure using the needle valve to 200 psi and record the flow _______ gpm.

**7.** Increase the pressure to 400 psi and record _______ gpm.

**8.** Increase the pressure to 600 psi and record _______ gpm.

**9.** Increase the pressure to 700 psi and record _______ gpm.

**10.** Increase the pressure to 800 psi and record _______ gpm.

Using the pump flow control, set the flow to 4 gpm. Repeat the exercise and plot the data on the same graph. If pressure is 0, record 4 gpm for either pump.

**11.** 200 psi, record _______ gpm.

**12.** 400 psi, record _______ gpm.

**13.** 600 psi, record _______ gpm.

**14.** 800 psi, record _______ gpm.

FIG. 14-4 *Symbol for flexible line.*

FIG. 14-5 *Symbol for reservoir.*

RESULTS: As the pressure increases, the flow that is delivered by the pump decreases. Complete the graph (Fig. 14-6).

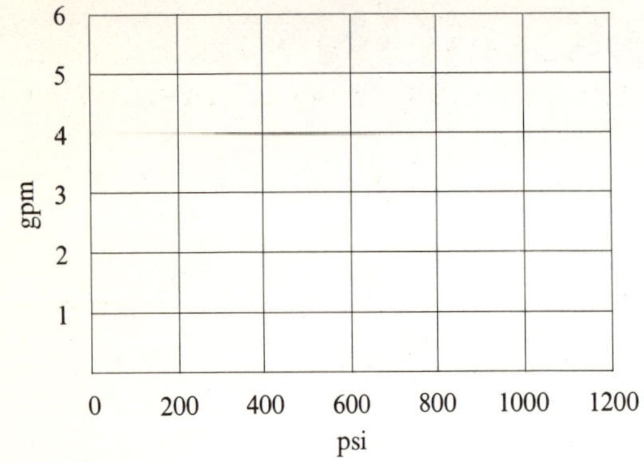


FIG. 14-6 *Demonstration results.*

PROBLEMS

**15.** Is this a typical curve for (*a*) a vane pump? ______ (*b*) a piston pump? ______

**16.** What causes the drop-off or loss of flow as the pressure is increased?

______________________________

______________________________

______________________________

______________________________

**17.** Would a vane pump of the same capacity or rating, made by another manufacturer, have an identical curve to this pump?

______

**18.** Would all vane pumps have an identical curve if they were the same model and manufactured by the same company? ______ Why?

______________________________

______________________________

______________________________

______________________________

**19.** Would a gear pump have a similar curve?

______

**20.** Would a piston pump have a flatter curve? ______ Why?

______________________________

______________________________

**21.** Would a piston pump cost more? ______ Why?

______________________________

______________________________

**22.** What are the principal differences in construction between pumps rated at 1000 psi and 2000 psi if they have the same pumping principle (i.e., vane, gear, or piston)?

______________________________

______________________________

______________________________

______________________________

______________________________

______________________________

**23.** If a pump is rated at 2000 psi maximum (and 6 gpm at 0 psi), draw its approximate pressure flow curve (Fig. 14-7).

**24.** From the manufacturer's data sheet (regarding the pump on the power unit), answer the following questions:

(*a*) What is the make and model number?

______________________________

(*b*) What is the weight of the pump?

______________________________

(*c*) What type of mounting is used?

______________________________

(*d*) What is the recommended operating oil temperature?

______________________________

CONCLUSION: Pumps do not pump pressure but produce pressure when it is required by the actuator or caused by restrictions. The amount of pressure is determined by the actuators, components, pressure losses, and restrictions in the circuit.

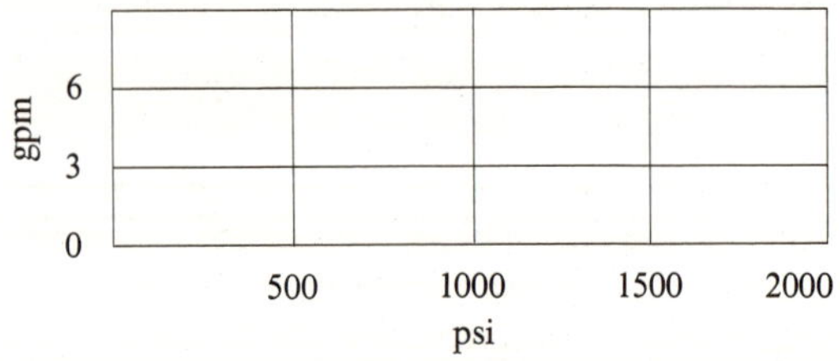


FIG. 14-7 *Demonstration results.*

 

A pump will provide a constant flow at a given pressure setting. If the pressure required increases, the pump will provide the additional pressure but will produce slightly less flow. The higher the pressure, the smaller the volume of flow. As the upper limit of the pump's pressure capacity is reached, the flow will drop more rapidly. This loss is caused by oil slippage and warpage in the pump. (Efficiency of the pump decreases as the pressure increases.)

Pumps should always be operated within the manufacturer's rating with respect to pressures and rpm so as to assure reasonable efficiency and long life.

## 2 • PRESSURE-COMPENSATING FEATURE OF A VANE PUMP

PURPOSE: To show the operation of the pressure-compensating feature of a vane pump.

COMPONENTS: Needle valve.

CIRCUIT: Training unit circuit with supply pressure 1 connected to the metered return line. A needle valve is installed in this line (Fig. 14-8).

PROCEDURE

1. Open the needle valve.
2. Set the pump flow control to maximum flow, counterclockwise, until no further tension is apparent on the control knob. The cam ring is now resting against the housing.
3. Set the relief valve at maximum pressure.
4. Start the unit and set the panel flow control to maximum flow or 100 percent of pump delivery to pressure supply 1.
5. Close the needle valve until zero flow is recorded (maximum pressure).

At this point, the pump has zero net displacement, since it is not delivering any fluid to the system. It is rotating in a position so that all oil that is picked up is brought to the pressure setting of the pump and then rotates with the rotor. The cam ring and rotor are now symmetrical, as shown in Fig. 14-11.

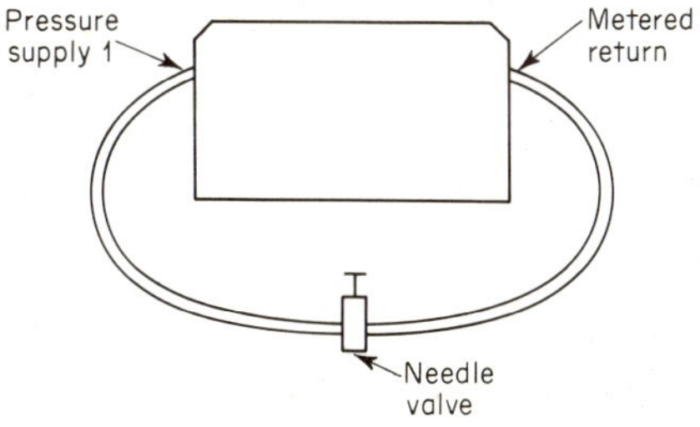


FIG. 14-8 *Circuit.*

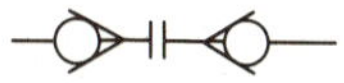

FIG. 14-9 *Graphic symbol for double-ball-check quick-disconnect (disconnected).*

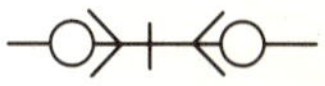

FIG. 14-10 *Graphic symbol for double-ball-check quick-disconnect (connected).*

Figure 14-11 shows this "no-flow" position, and Fig. 14-12 shows the "full-flow" position. When an intermediate volume of flow is produced by the pump, the cam ring will be in an intermediate position from the two positions shown.

It can be seen that as the pressure increases, it will move the ring against the spring to the no-flow position, as shown

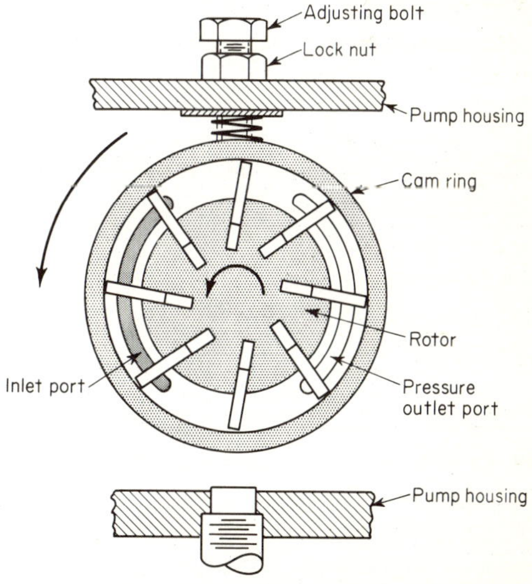


FIG. 14-11 *No-flow condition.*

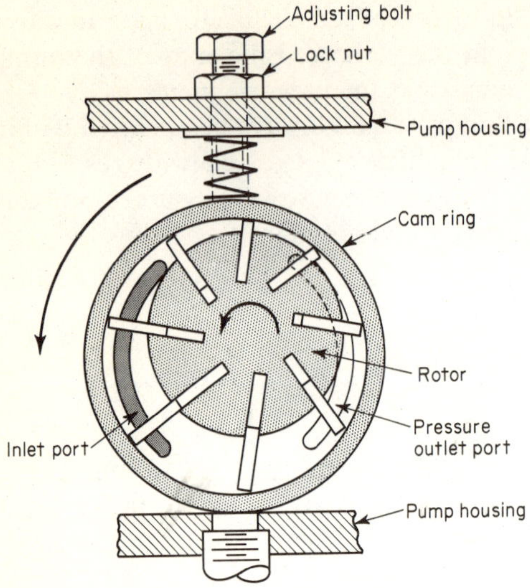


FIG. 14-12 *Full-flow condition.*

in Fig. 14-11. As the pressure builds up in the discharge side and is unable to leave through the port, it will increase until it moves the cam ring to the position shown. It can be seen that a displacement exists in Fig. 14-12, but zero net displacement is in effect as shown when no-flow conditions exist. The spring, as the pressure is reduced, will force the cam ring down to an intermediate or full condition as shown in Fig. 14-12.

**6.** The pressure recorded at a no-flow condition is that which is set on the individual pump. It can be adjusted by increasing or decreasing the tension on the spring using the adjusting bolt in the back of the pump. "In" or "clockwise" rotation of this adjusting bolt increases the maximum pressure, and the reverse decreases it.

**7.** Slowly open the needle valve to 2, 3, 4, 5, and 6 gpm, and record the pressure on the graph (Fig. 14-13).

(*a*) 0 flow, _______ psi.
(*b*) 2 gpm, _______ psi.
(*c*) 3 gpm, _______ psi.
(*d*) 4 gpm, _______ psi.
(*e*) 5 gpm, _______ psi.
(*f*) 6 gpm, _______ psi.
(*g*) Max. gpm ____ psi.

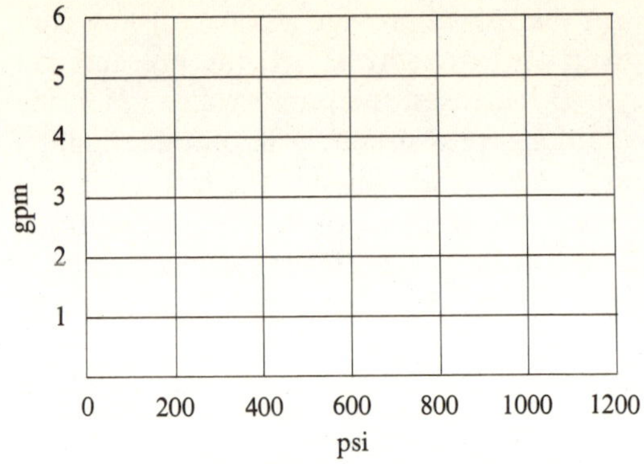


FIG. 14-13 *Demonstration results.*

SUPPLEMENTAL PROBLEM: Close the needle valve to the no-flow condition (maximum pressure). Turn the pressure-control-adjusting bolt counterclockwise until the pressure gauge indicates 600 psi. (Note lock nut on this bolt.) Repeat the above exercise and plot the pressures on the same graph.

To reset the pump to 1000 psi, close the needle valve to zero flow and turn the pressure-adjusting bolt clockwise until the pressure indicates 1050 psi. Set the lock nut.

CONCLUSION: A pressure-compensated pump will protect itself against shock and overloads. It acts as a relief valve and will produce no flow at its set maximum pressure. This pressure can be adjusted by the adjusting bolt. Upon approaching the set pressure, it will start producing less flow until zero flow pressure is reached. A pressure compensator can be built into a vane or piston pump, but not into a gear pump. Gear pumps are always fixed-volume pumps and require a relief valve for their protection.

PROBLEMS

**8.** What pressure is produced at zero flow? _______ psi.

**9.** What pressure is produced at 4 gpm? _______ psi.

**10.** What is the maximum flow? _______ gpm.

**11.** What pressure is produced at the maximum flow? _______ psi.

 

NAME .............................................. DATE .............. CLASS ...........

**12.** Would a relief valve be required if this pump were used in a system? ________ Why?

________________________________________

________________________________________

________________________________________

________________________________________

**13.** What is the range of pressure compensation of this pump (zero flow pressure minus the pressure at which flow starts to fall rapidly)? ________ psi to ________ psi.

**14.** From the manufacturer's data sheet determine (*a*) how many types of springs are available ________; and (*b*) what the maximum pressure is for each.

________________________________________

**15.** What are the minimum and maximum drive speeds recommended for the pump?

________________________________________

**16.** What is the graphic symbol for a needle valve?

**17.** What is the graphic symbol for a connected double-check quick disconnect?

**18.** What is the graphic symbol for a disconnected double-check quick disconnect?

## 3 • VARIABLE-VOLUME FEATURE OF A PUMP

PURPOSE: To show the operation of the variable-volume feature of a pump.

COMPONENTS: Needle valve.

CIRCUIT: Training unit and needle valve, as shown in Fig. 14-14.

PROCEDURE

**1.** Connect pressure supply 1 to the metered return line, with the needle valve installed.

**2.** Open the needle valve.

**3.** Set the pump flow control to zero flow (clockwise to the stop position).

**4.** Set the relief valve to maximum pressure (full-in position).

**5.** Start the unit, and set the panel flow control to 100-percent flow to pressure supply 1.

**6.** Open pump flow control until 1 gpm is recorded.

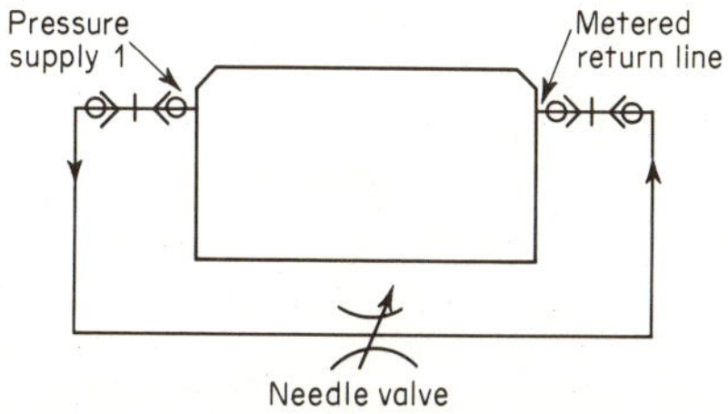


FIG. 14-14 *Circuit.*

FIG. 14-15 *Symbol for fixed displacement pump (unidirectional).*

FIG. 14-16 *Symbol for variable-volume pressure-compensated pump (unidirectional).*

**7.** Close the needle valve until the supply pressure is 500 psi. Record the gpm at this pressure. ________.

**8.** Using the pump flow control, set the flow at 2 gpm and record the pressure. ________ psi.

**9.** Set the flow at 4 gpm, using the pump flow control. Record pressure. ________ psi.

**10.** Increase flow to the maximum, and record the pressure. ________ psi. Figure 14-17 shows the no-flow condition and the full-flow condition. It also shows the adjusting bolt that manually moves the ring to an intermediate position. Intermediate cam ring position will vary the displacement of the pump and will produce flow in any increment from zero to maximum. Note that pump performance will vary with individual pumps depending on general tight-

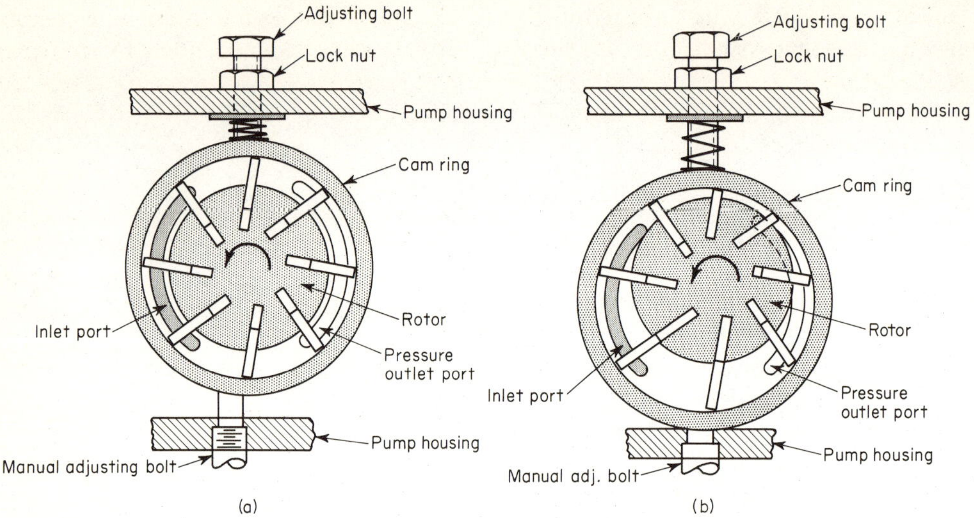


FIG. 14-17 *(a) No-flow condition. (b) Full-flow condition.*

ness and manufacturer's tolerances for the individual unit.

RESULTS: Flow volume can be set on a pump with the following feature. The adjusting bolt can be turned clockwise, which moves the cam ring so as to cause a zero displacement position; consequently, the no-flow position is reached.

Any intermediate position can be set which will result in the desired maximum volume of flow. The pump will have performance curves similar to those shown in the demonstration of vane or piston-pump characteristics (i.e., the flow will drop off as the pressure is increased).

Figures 14-18 and 14-19 show the adjusting bolt and its two extreme positions. Figure 14-18 shows the cam ring forced against the spring unit. Zero displacement is present. This is the same position that the cam ring will take when the pressure in the discharge side of the cam is great enough to force the cam ring against the spring to a no-flow position.

Figure 14-19 shows the full-flow position with the ring resting against the pump housing as a result of the spring action. It will stay in this position until it is moved manually or by pressure.

PROBLEMS

**11.** Draw force arrows on Fig. 14-17*b* showing unbalanced forces acting on the rotor.

**12.** Can any desired flow be set on a variable-volume pump? ______

**13.** How does the increased pressure affect the gpm delivered by this pump? __________
______________ Why? ______________
______________________________
______________________________
______________________________

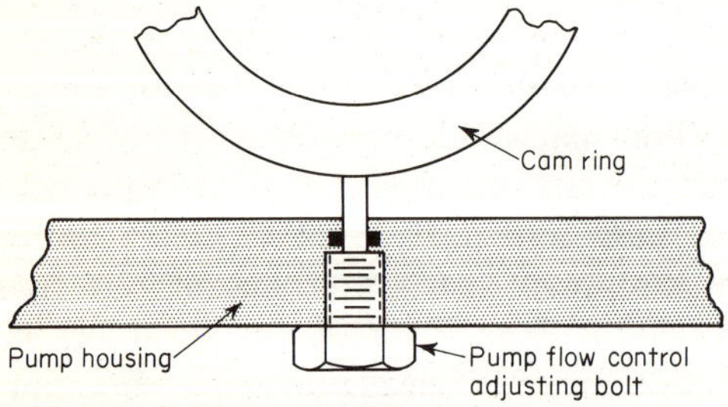


FIG. 14-18 *No-flow condition.*

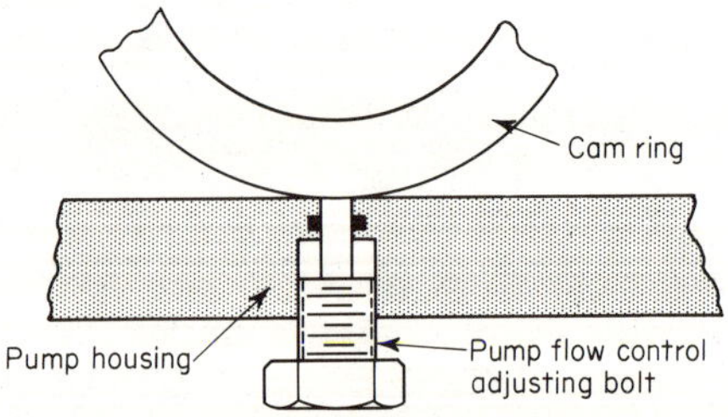


FIG. 14-19 *Full-flow condition.*

 

**14.** (*a*) Does this pump have characteristics similar to a fixed-displacement vane pump from 0 to 800 psi? ______

(*b*) Does it have similar characteristics from 800 to 1000 psi? ______ Why?

**15.** In a fixed-displacement pump, what must be installed to protect the pump?

**16.** How does a relief valve protect a pump?

**17.** If you could vary the rpm on the pump, and it were reduced to 900 rpm (or one-half), what would happen to the flow, assuming it was initially set at 6 gpm?

**18.** Would a variable-volume pump be best suited for a gas engine or a fixed rpm electric motor? ______ Why?

## 4 • EFFECTS OF OIL STARVATION

PURPOSE: To show the effect of oil starvation on a pump (cavitation).

COMPONENTS: Needle valve.

CIRCUIT: Connect pressure supply 1 to metered return line with a needle valve installed in this line, as shown in Fig. 14-20. If training unit has the cavitation unit installed on the panel, the dynamometer unit must be used for this demonstration.

PROCEDURE

**1.** Set the pump flow control to maximum flow.

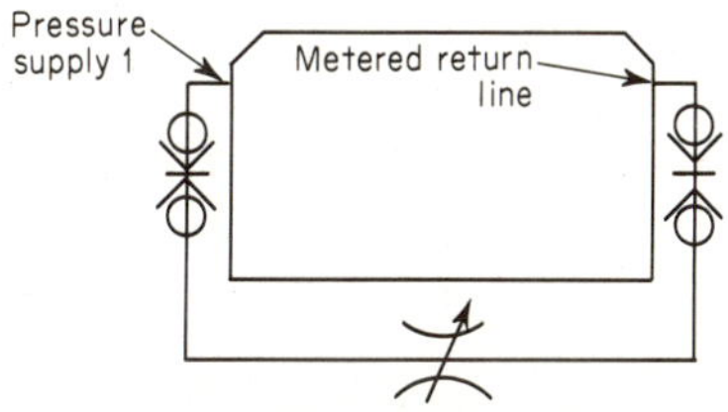


FIG. 14-20 *Circuit.*

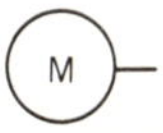


FIG. 14-21 *Symbol for electric motor.*

FIG. 14-22 *Drain line, working line, and pilot line.*

**2.** Set the panel flow control to maximum flow to pressure supply 1.

**3.** Set the relief valve to maximum pressure (full in).

**4.** Start the unit, and note the vacuum gauge, or the strainer condition indicator. Record the vacuum. ______ in. Hg.

**5.** Set the needle valve to induce 500 psi into the circuit. *Slowly* close the gate valve until infrequent or intermittent cavitation occurs. This will be apparent by vibration, sound, and rapid pressure-gauge fluctuations. Record the vacuum at this condition. ______ in. Hg.

**6.** Close the gate valve further until maximum sound and vibrations occur. Record the vacuum at this condition. _______ in. Hg.

**7.** Further close the gate valve, and note the sound and vibrations. Record the maximum vacuum obtained. _______ in. Hg.

**8.** Open the gate valve fully.

*Note:* To prevent unauthorized cavitation of the pump, it is recommended that the knob be removed and kept in the instructor's possession. It should be noted that the effects are pronounced and a prolonged cavitation will result in damage to the pump. It is particularly severe on a piston pump.

The cavitation was not complete since the gate valve has a drilled 3/16-in.-dia. hole in its gate to provide lubrication and a small flow during demonstrations.

RESULTS: Loss of an oil supply to the pump at approximately 10 in. Hg resulted in shock waves in the system. This is indicated by an unpleasant sound, pressure impulses, and a malfunctioning system.

CONCLUSION: Cavitation is oil starvation to a hydraulic pump. It can be caused by a plugged strainer, a restriction in the intake line, undersized inlet lines, excessively long inlet lines, or cold-weather operation with too heavy a fluid.

Cavitation is detrimental to a pump and to other components in a circuit. It causes shock loads resulting in damage to the pump's vanes, seals, and wear plates. These shock loads in a 1000-psi system can be as high as 8000 to 10,000 psi.

PROBLEMS

**9.** At what vacuum does the pump operate under normal condition with a clean strainer at 6 gpm and 500 psi? _______ in. Hg.

**10.** At what vacuum does the pump first indicate signs of cavitation? _______ in. Hg.

**11.** At what vacuum did the maximum cavitation occur? _______ in. Hg.

**12.** What was the maximum vacuum obtained? _______ in. Hg.

**13.** What are the principal causes of cavitation?

________________________________________

________________________________________

________________________________________

________________________________________

**14.** In a hydraulic pump what are the two factors that cause flow to the pump?

(*a*) __________________________________

(*b*) __________________________________

(*c*) Which of these is the principal factor?

________________________________________

**15.** What are five factors that should be considered in designing a system to avoid cavitation?

________________________________________

________________________________________

________________________________________

________________________________________

________________________________________

**16.** What is the graphic symbol for a fixed-displacement pump?

**17.** What is the graphic symbol for a variable-volume pump?

**18.** What is the graphic symbol for a variable-volume, pressure-compensated pump?

 

## 5 • EFFECTS OF AIR IN THE INTAKE LINE

PURPOSE: To show the effects of air in the intake line.

COMPONENTS: Needle valve.

PROCEDURE

**1.** Connect supply pressure 1 to the metered return line, as shown in Fig. 14-23.

**2.** Set the pump control and the panel flow control at maximum flow to pressure supply 1.

**3.** Set the pressure to 500 psi, using the needle valve.

**4.** Open the air valve, and note the pressure gauge, sound, and vibration of the unit. If air valve is not present, open union in the intake line.

RESULTS: As the air was introduced, pump impact, shock loads, and gauge fluctuation were apparent. Air and oil were mixed as pumping continued, and the oil supply was emulsified by the addition of air particles.

CONCLUSION: A hydraulic system with an introduction of air will result in a cavitation-like effect. It will result in emulsified fluid and an erratic and malfunctioning system. Excessive foaming and complete failure of the system would occur if a prolonged period of air introduction were present.

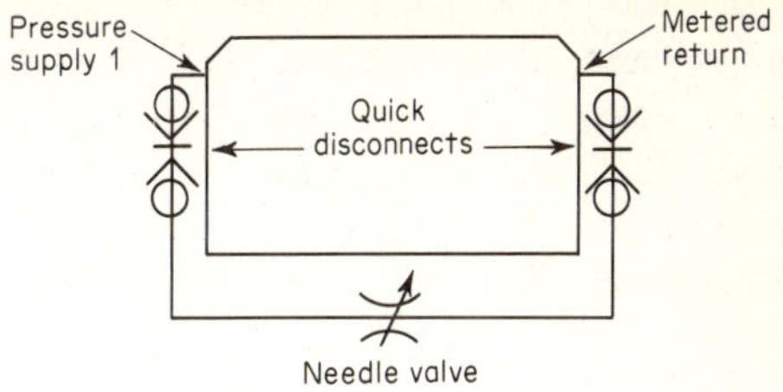


FIG. 14-23 *Circuit.*

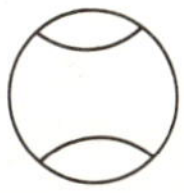

FIG. 14-24 *Symbol for flowmeter.*

FIG. 14-25 *Symbol for filter or strainer.*

## 6 • RELIEF VALVE

PURPOSE: To show the operation of the relief valve or the maximum-pressure valve.

COMPONENTS: Needle valve and maximum-pressure valve (relief valve).

PROCEDURE

**1.** Connect the needle valve in a line between pressure supply 1 and the metered return line (Fig. 14-26).

**2.** Start the unit and set the pressure with the needle valve to approximately 400 psi using full flow to pressure supply 1 on the panel flow control.

**3.** Turn the relief valve counterclockwise until fluid just starts to by-pass back to the tank, as indicated by a slight drop in the flow rate. Lock the relief valve in this position.

**4.** Increase the pressure by using the needle valve until full flow (FF) passes through the relief valve, and zero flow is indicated on the flowmeter. What is the pressure? _______ psi.

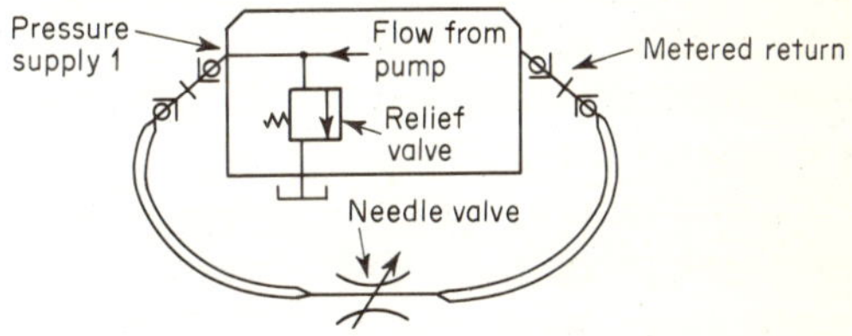


FIG. 14-26 *Circuit.*

**5.** Open the needle valve *slowly* until a slight flow is indicated on the flowmeter, and record this pressure. _______ psi.

**6.** Continue to open the needle valve until FF is indicated on the flowmeter, and record this pressure. _______ psi. This is the relief-valve reseat (RS) pressure.

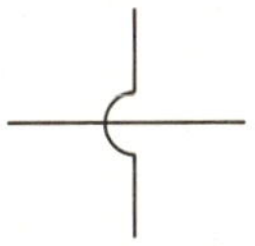

FIG. 14-27 *Symbol for lines crossing.*

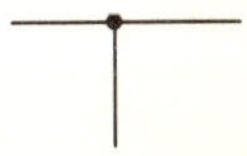

FIG. 14-28 *Symbol for lines joining.*

**7.** Increase the pressure until the flow starts to decrease. This is the relief-valve cracking pressure (CP). Record this pressure. _______ psi.

**8.** Further increase the pressure until zero flow is recorded on the flowmeter and note the pressure. This is the FF, or relief valve full-flow pressure.

**9.** With the data obtained, plot and label the full flow (FF), reseat (RS), and cracking-pressure (CP) results on the graph (Fig. 14-29).

Repeat the exercise by setting the full bypass flow at 600 psi by use of the needle valve and the relief valve: (1) set system pressure by needle valve to approximately 600 psi; (2) turn relief valve control counterclockwise.

Again determine the three items listed above: full-flow, reseat, and cracking pressures. Record these three points on the graph (Fig. 14-29).

RESULTS: A relief valve will determine the maximum pressure which can be used in the system. This maximum pressure can be determined by an adjusting knob which determines the tension on the spring which holds a steel ball in place. Pressure acts against this ball and at a set pressure will move it against the spring, creating an opening and permitting some oil to flow back to the tank. This is the cracking pressure. As the pressure increases, the opening will increase until full flow is returned to the tank. This is known as full-flow pressure. Upon reducing pressure, the spring will reseat the ball and cause the flow again to be directed through the system. The pressure at which the ball reseats is known as the reseat pressure.

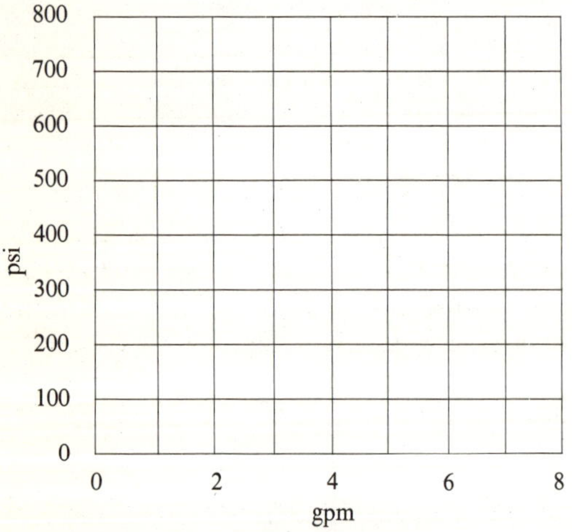


FIG. 14-29 *Demonstration results.*

CONCLUSION: A relief valve is used to (1) establish the maximum pressure in a system, and (2) to protect the pump and other components from shock and overload. These two features are the reason for the two names often used for this component: relief valve and maximum-pressure valve.

PROBLEMS

**10.** If full-flow pressure is 400 psi, what is the CP pressure? _______ RS? _______

**11.** If full-flow pressure is 600 psi, what is the CP pressure? _______ RS? _______

**12.** What is the difference between CP and RS pressure with a full-flow pressure of approximately 400 psi? _______ psi.

**13.** What is the difference between CP and RS pressure with a full-flow pressure of approximately 600 psi? _______ psi.

**14.** With an FF pressure of approximately 400 psi, what is the difference between RS and FF pressures? _______ psi.

**15.** With an FF pressure of approximately 600 psi, what is the difference between the RS and FF pressures?
_______ psi.

**16.** Why is there a difference between the reseat and cracking pressures? (Check Fig. 14-30 and explain.)

_______________________________________

_______________________________________

_______________________________________

_______________________________________

**17.** Why is a greater pressure required for FF than for the RS pressure?

_______________________________________

_______________________________________

_______________________________________

(a) (b)

FIG. 14-30 *Forces acting on a ball-type check valve. (a) Cracking pressure. (b) Reseat pressure.*

**18.** What is symbol for double-check quick disconnect when they are disconnected?

**19.** Show the symbol for two hydraulic lines when they cross but are not connected.

When two lines form a connected cross.

## 7 • OIL-TEMPERATURE THERMOSTATIC CONTROL

PURPOSE: To show operation of oil-temperature thermostatic control.

COMPONENTS: Needle valve.

CIRCUIT: Training unit and needle valve as shown in Fig. 14-31.

PROCEDURE

**1.** Set the thermostat at 140 and start the unit with full flow at pressure supply 1. Record the oil temperature. ______ °F.

**2.** Set the pressure by the needle valve to 800 psi, and operate for approximately 5 min.

**3.** The thermostat is adjusted by using a screwdriver, below the dial. Reduce the setting until the unit shuts off. (*a*) Record that temperature. ______ (*b*) Record the temperature on the oil gauge. ______

**4.** Touch the needle valve. Observe the increase in temperature after the unit has operated a short time at high pressures.

RESULTS: The thermostat acts as a safety device to shut off the unit if the oil temperature exceeds the maximum control setting.

CONCLUSION: All hydraulic actuators or components have a pressure drop across them. When an actuator converts pressure and velocity energy to either torque or linear force, it is combined with a rise in oil temperature. A pressure drop across the systems of other components (e.g., needle valve) will cause a loss of energy which is converted to heat.

PROBLEMS

**5.** What was the pressure at the upstream side of the needle valve (supply pressure)?

______

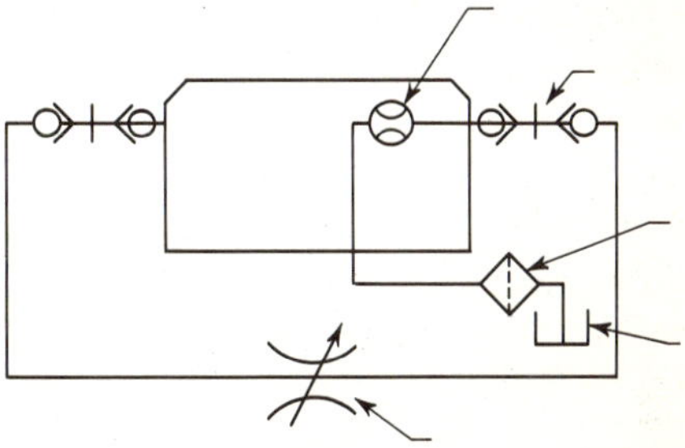

FIG. 14-31 *Circuit.*

**6.** What is the oil-velocity relationship in the upstream and downstream sides of the needle valve? Is downstream faster, slower, or same? ______

**7.** What was the pressure at the downstream side of the needle valve? ______

**8.** What was the oil temperature rise during the demonstration? ______

**9.** Describe the law of conservation of energy.

______

______

**10.** What happened to the energy from one side of the needle valve to the other?

______

**11.** Draw flow arrows in the partial circuit shown in Fig. 14-31. Write a description of the graphic symbols next to the symbols.

**12.** Draw the symbol for the electric motor.

**13.** What type of hydraulic line is indicated by a dotted line? ______

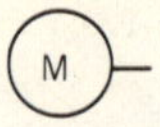


FIG. 14-32 *Symbol for electric motor.*

**14.** What type of hydraulic line is indicated by a dashed line? ____________

**15.** What type of hydraulic line is indicated by a solid line? ____________

## 8 • TWO-POSITION TWO-WAY VALVE

PURPOSE: To show the operation of a two-position two-way valve.

COMPONENTS: Two-position two-way valve.

CIRCUIT: Training unit and two-position two-way valve as shown in Fig. 14-33.

PROCEDURE

**1.** Connect pressure supply 1 to the inlet port of the valve and the outlet to the metered return line.

**2.** Start the unit with the valve open. If it is spring loaded, leave it in its normal position. Note the pressure and flow on the meter.

**3.** Change the valve to its 2 position, and note the pressure and flow.

RESULTS: Free flow from the pressure side to the discharge side is permitted in one position. The second position permits no flow.

CONCLUSION: This valve acts the same as a needle valve but requires less time to operate. It will act as a restriction in the line if the spool is carefully held in an intermediate position.

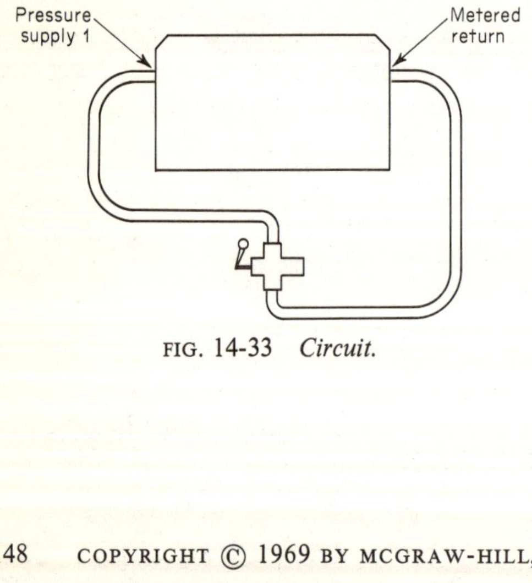


FIG. 14-33 *Circuit.*

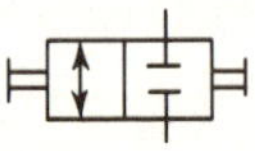

FIG. 14-34 *Symbol for two-way valve, manually operated, off.*

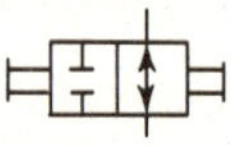

FIG. 14-35 *Symbol for two-way valve, manually operated, on.*

PROBLEMS

**4.** Draw on Fig. 14-36 the flow arrows for the two positions.

**5.** If the valve has a spring installed which returns the spool to position 1, the valve is called a spring-loaded two-way valve "normally open." If the spring held the spool in the opposite position, what would it be called?

____________

**6.** What is the advantage of this type of valve over a needle valve?

____________

____________

**7.** What are the three principal methods of actuating a spool valve?

(*a*) ____________

(*b*) ____________

(*c*) ____________

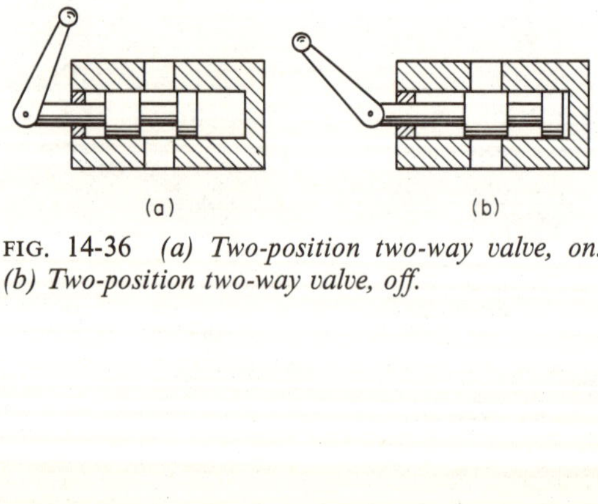


FIG. 14-36 *(a) Two-position two-way valve, on. (b) Two-position two-way valve, off.*

 

## 9 • TWO-POSITION THREE-WAY VALVE

PURPOSE: To show the operation of a two-position three-way valve.

COMPONENTS: Two-position three-way valve (Fig. 14-37). (This valve is installed on all training units manufactured after July, 1965.)

PROCEDURE

**1.** Connect pressure supply 1 to the inlet port on the valve.

**2.** Connect one control port on the valve to the metered return line and the second control port to the auxiliary return line.

**3.** Set the panel flow control to pressure supply 1 at "full flow."

**4.** Set pump flow control to maximum flow.

**5.** Start the unit. Observe the direction of flow through the valve at each of its two positions.

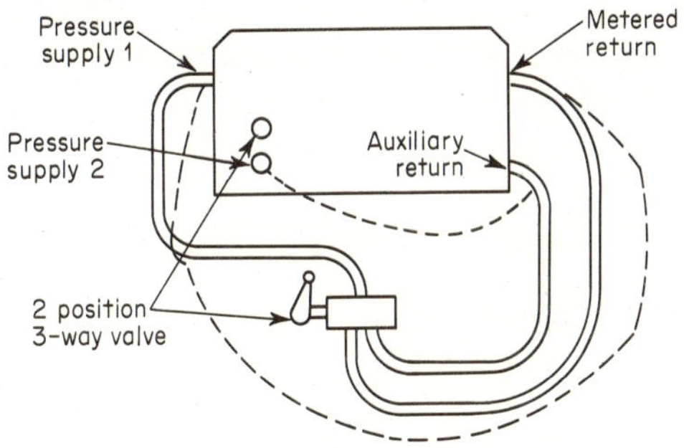


FIG. 14-37 *Circuit.*

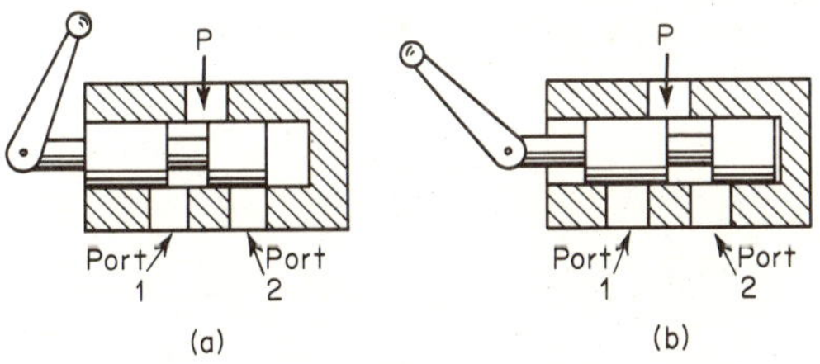


FIG. 14-38 *Three-position two-way valve.*

PROBLEMS

**6.** Draw flow lines on Fig. 14-38 for each of the two positions of the valve.

**7.** Draw symbols for a two-position three-way valve:

(*a*) Manually operated.

(*b*) Manually operated spring return; normally open.

(*c*) Manually operated spring return; normally closed.

**8.** Could this valve operate two hydraulic motors alternately? ______

**9.** Could this valve operate a hydraulic motor in two directions? ______

**10.** Would this valve operate two hydraulic single acting cylinders (*a*) alternately? ______ (*b*) simultaneously? ______

**11.** Could this cylinder be used to operate one hydraulic cylinder in two directions? (double acting) ______

PURPOSE: To show the operation of a two-position four-way valve.

COMPONENTS: Two-position four-way valve. Low-pressure return line tee fitting.

PROCEDURE

**1.** Set panel flow control to deliver 3 gpm to pressure supply (Fig. 14-39*a*).

**2.** Connect pressure supply 1 to the inlet port, and connect the tank, or outlet port, to the metered return line.

**3.** Connect port 1 and port 2 to the auxiliary return line using a tee.

**4.** Operate the unit and note the flow on the metered return line. Is there flow in position 1? ______ In position 2? ______

**5.** Draw flow arrows on Fig. 14-39*a*.

**6.** Change the tank or outlet port to the auxiliary return and port 1 to the metered return line. Operate and note the flow (Fig. 14-39*b*). (*a*) Is there a flow in position 1? ______ (*b*) In position 2? ______

**7.** Draw flow arrows on Fig. 14-39*b*.

**8.** Change port 1 from the metered return line to the auxiliary return line and connect 2 to the metered return line. (*a*) Is there any flow in position 1? ______ (*b*) In position 2? ______

**9.** In one position no flow will be recorded at port 1. To determine if this port will permit back flow from a source to the tank, connect this line to pressure supply 2 and connect the return or tank port to the metered return line. Port 2 will then be connected to the auxiliary return line. Actuate pressure supply 2 for this exercise. Is there a flow to metered return line? ______

**10.** Reverse the position of the spool and reverse port 1 and port 2. (Port 2 will be at pressure supply 2, and port 1 is connected to the auxiliary return line.) Actuate pressure supply 2. Is there a flow to metered return line? ______

CONCLUSION: A two-position two-way valve will direct the flow to either of the two ports depending on its position. The return flow from the second port will be directed back to the tank.

*Example:* In a double-acting hydraulic-cylinder operation the valve would direct flow to either of the two ports depending on its position. The return flow from the other end of the cylinder would be returned to the tank.

PROBLEMS

**11.** Indicate the flow lines on Fig. 14-40 for the two-position four-way valve.

**12.** If valve is directing flow to port 1, is it necessary to have return flow through the second port? ______

**13.** Draw the symbol for the valve used.

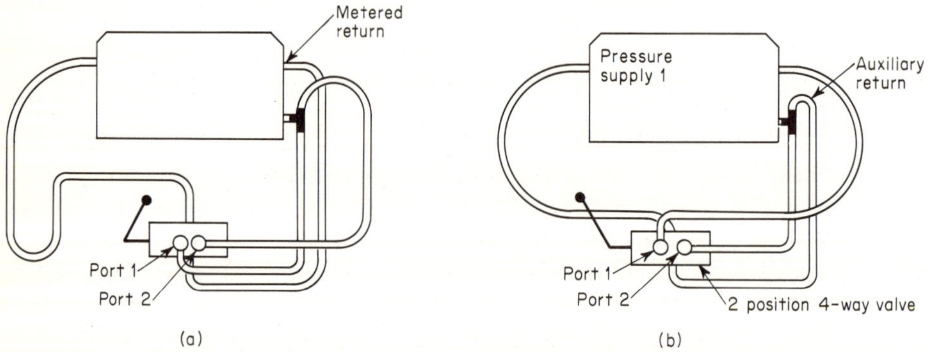


FIG. 14-39 *(a) Actuated position 1. (b) Actuated position 2.*

 

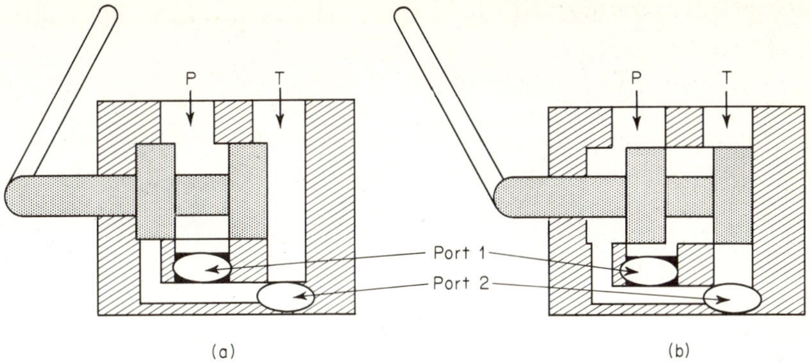


FIG. 14-40 *Two-position four-way valve.*

## 11 • THREE-POSITION FOUR-WAY VALVE

PURPOSE: To show the operation of a three-position four-way valve.

COMPONENTS: Three-position four-way valve. Low-pressure tee.

PROCEDURE

**1.** Connect pressure supply 1 to the inlet port of the valve and outlet port to metered return line (Fig. 14-41*a*).

**2.** Set the panel flow control to provide total flow to pressure supply 1.

**3.** Connect ports 1 and 2 to the auxiliary return line, and fill out Table 14-1, using conditions as shown in the table. Refer to Fig. 14-41. For the last two lines, set panel flow control to direct 4 gpm to pressure supply 1, and permit the balance of flow to pressure supply 2. Actuate pressure supply 2 by using two-position three-way valve on panel. (Circuit not shown in Fig. 14-41.)

**4.** From the information obtained, draw the

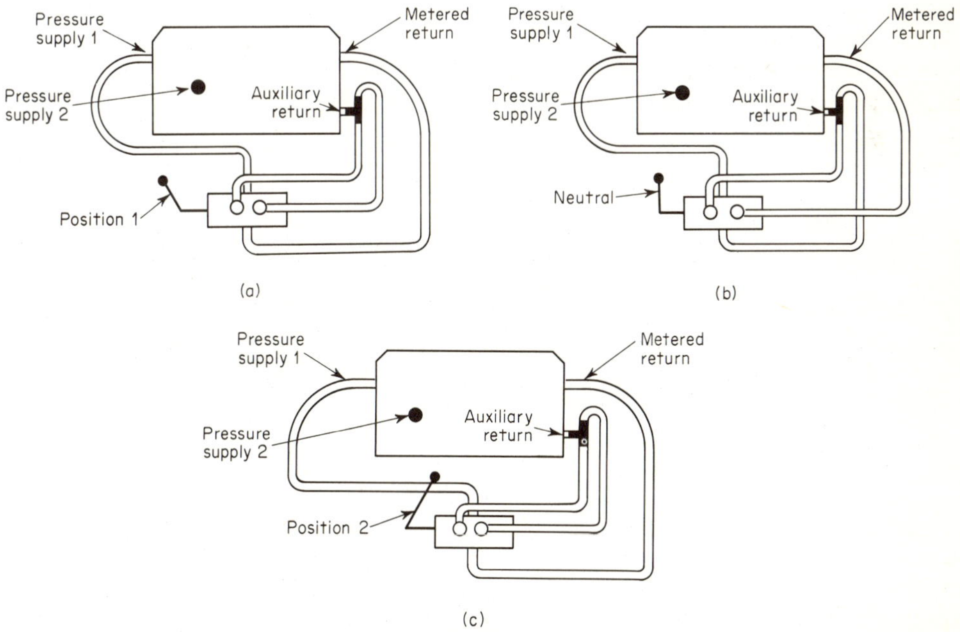


FIG. 14-41 *Demonstration circuit.*

flow lines for all ports for each of the three spool positions given in Fig. 14-42.

From the first nine lines in Table 14-1 some flow facts have been determined.

**5.** Is it an open or closed center valve?

______________________________

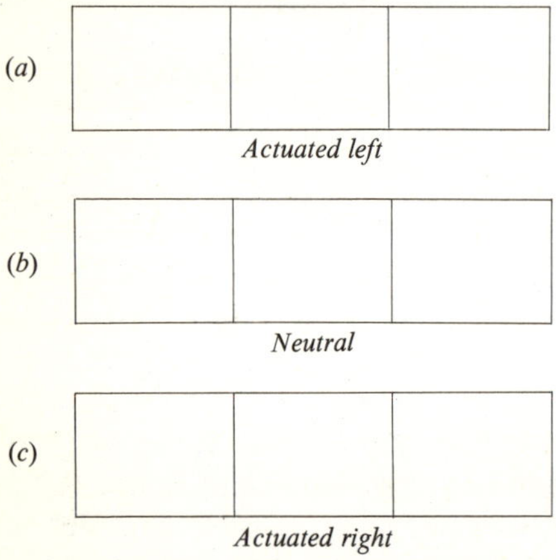


FIG. 14-42 *Valve-analysis results.*

**6.** In what position will it direct flow to port 1?

______________________________

**7.** In what position will it direct flow to port 2?

______________________________

**8.** Is there any flow from the outlet or tank port while flow is directed to port 1 or 2?

______________________________

To fully classify a four-way valve, additional information is needed. This is obtained from the last two lines of Table 14-1.

**9.** (*a*) In the neutral position, what is the condition of ports 1 and 2?

______________________________

______________________________

______________________________

______________________________

TABLE 14-1 THREE-POSITION FOUR-WAY VALVE ANALYSIS

| *Fig. 14-41* | SPOOL POSITION AND PORT CONNECTIONS | | | | | FLOW, GPM | | |
|---|---|---|---|---|---|---|---|---|
| | *Spool position* | *Pressure supply 1* | *Tank or outlet* | *Port 1* | *Port 2* | *Outlet* | *Port 1* | *Port 2* |
| (*a*) | Neutral | Inlet | Metered return | Auxiliary return | Auxiliary return | | | |
| (*a*) | 1 | " | " | " | " | | | |
| (*a*) | 2 | " | " | " | " | | | |
| (*b*) | Neutral | " | Auxiliary return | Metered return | " | | | |
| (*b*) | 1 | " | " | " | " | | | |
| (*b*) | 2 | " | " | " | " | | | |
| (*c*) | Neutral | " | " | Auxiliary return | Metered return | | | |
| (*c*) | 1 | " | " | " | " | | | |
| (*c*) | 2 | " | " | " | " | | | |
| Not shown | Neutral | " | " | Pressure 2 | " | | | |
| Not shown | Neutral | " | " | Metered return | Pressure 2 | | | |

 

(*b*) Will flow entering port 1 exhaust out of port 2? ______

(*c*) Will the flow enter port 1 or 2 and flow out the tank port? ______

(*d*) While the valve directs flow to port 1, is flow permitted to enter port 2 and flow out the return or tank port? ______

By completing Table 14-1 these features can be determined. Use 4 gpm on the panel flow control, with the balance to the tank, except when this excess flow is needed (as in the last two lines on Table 14-1). When pressure supply 2 is needed, use the two-position three-way valve to provide pressure flow to the ports on the valve to determine the characteristics of the valve.

CONCLUSION: A wide variety of spool configurations are available; the type used depends on the purpose the valve is to serve. There are various methods of operating the spool: (1) manually; (2) by solenoid; (3) by pilot; and (4) by solenoid-controlled pilot. Each is a method of moving the spool to one of its three positions. Springs may be installed to return the spool to its center position.

PROBLEMS

**10.** What was the type of spool installed in this valve—open or closed?

______

**11.** In the neutral position, are both port 1 and 2 blocked off? ______

**12.** How was the spool actuated?

______

**13.** If the features of the valve were not known, would the exercise determine the features of the valve? ______

**14.** What is the symbol of the valve used?

**15.** Can this valve be used to operate a bidirectional hydraulic motor? ______

**16.** Is this a good valve to use in a double-acting cylinder moving a load horizontally? ______

**17.** How could this valve be changed (without changing spools) into a two-position three-way valve?

______

______

______

______

**18.** Draw flow arrows for circuit on Fig. 14-41.

**19.** What are the four methods of operating a control valve?

______

______

______

______

## 12 • PRESSURE-REDUCING VALVE

PURPOSE: To show the operation of a pressure-reducing valve.

COMPONENTS: Two needle valves, a pressure-reducing valve, a pressure gauge, and a tee.

PROCEDURE

**1.** Connect pressure supply 1 to the metered line with a tee connection and needle valve 1. (Install needle valve after the tee as shown in Fig. 14-43.)

**2.** Connect a secondary line to the tee and

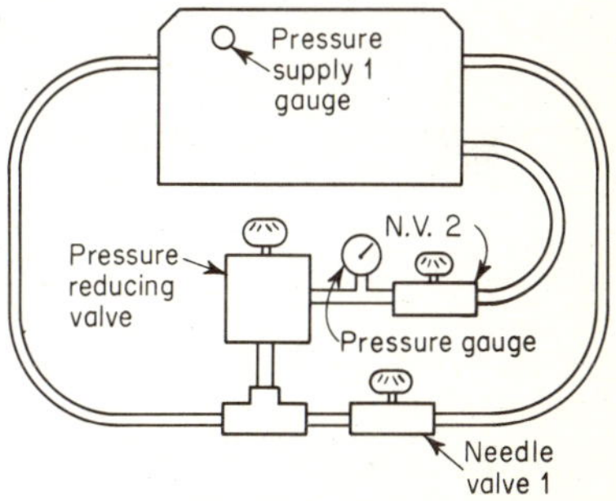


FIG. 14-43 *Circuit.*

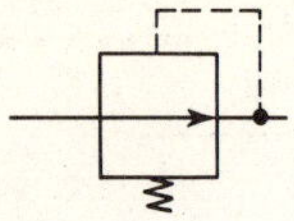

FIG. 14-44 *Symbol for pressure-reducing valve.*

inlet port on the pressure-reducing valve. From the outlet port of the pressure-reducing valve, connect a line to the auxiliary return with needle valve 2 installed. Connect a pressure gauge between the reducing valve and the needle valve.

**3.** Set the pressure-reducing valve at approximately 500 psi. (The maximum pressure in the secondary lines.)

**4.** Increase the pressure by the use of the needle valves until 400 psi is recorded on both the primary and the secondary system with a flow of approximately 4 gpm in the primary line and the balance through the secondary line.

**5.** Increase the pressure in the primary and secondary lines to the values listed on Table 14-2. Complete values that are not shown on the table.

RESULTS: The pressure-reducing valve determined the maximum pressure available in the secondary circuit. The flow took the path of least resistance at lower pressures. Pressures in the primary line above the setting of the valve had no effect on the pressure in the secondary line. It stayed at the maximum set on the valve, but at pressures below the valve setting, it fluctuated with the pressures of the primary line.

CONCLUSION: This valve provides a means of determining the maximum pressure in a secondary line, regardless of pressure variation in the primary line (provided the pressure in the primary line operates at pressures above the maximum pressure in the secondary line).

TABLE 14-2 PRESSURE-REDUCING-VALVE DATA

| *Primary-line press., psi* | *Secondary-line press., psi* | *Primary-line flow, gpm* | *Secondary-line flow, gpm* |
|---|---|---|---|
| (*a*) 300 | 100 | 4+ | 2− |
| (*b*) 300 | 300 | | |
| (*c*) 500 | 500 | | |
| (*d*) 600 | | 4 | 2 |
| (*e*) 700 | | 2 | 4 |
| (*f*) 800 | | | 2 |

PROBLEMS

**6.** At what maximum pressure was the pressure-reducing valve set? ______

**7.** At pressures below the valve setting, what determines the flow in the primary and the secondary lines?

______

**8.** At flows above the valve settings what determines the volume of flow to each line?

______

**9.** What resulted when the needle valve in the secondary line was closed and the primary line needle valve opened?

______

______

**10.** Is this pressure-reducing valve a flow control valve? ______

**11.** What is the symbol for a pressure-reducing valve?

**12.** Complete the circuit, using symbols:

 

NAME .......................................... DATE .............. CLASS ...........

## 13 • SEQUENCE VALVE

PURPOSE: To show the operation of a sequence valve.

COMPONENTS: Two needle valves, a sequence valve, and a pressure gauge.

PROCEDURE

**1.** Connect pressure supply 1 to the inlet port of the sequence valve.

**2.** Connect the primary supply from the sequence valve to the metered return line and a needle valve downstream from the sequence valve.

**3.** Connect secondary line from the sequence valve to the auxiliary return line with a needle valve and pressure gauge downstream as shown on Fig. 14-45.

**4.** Set the sequence valve to approximately 500 psi and operate the circuit under the various conditions listed on Table 14-3. Complete the values on this table.

**5.** If the sequence valve is pilot operated, connect a pilot line from the pressure line (pressure supply 1) to the pilot port of the valve.

**6.** If the sequence valve is externally drained, connect a low-pressure drain line from the drain port to any convenient return line or drain port on the panel.

To determine the flow rate of each circuit, either subtract the metered flow from the 5 gpm set on the panel flow control or alternate the return lines and verify the flow by the use of the flowmeter.

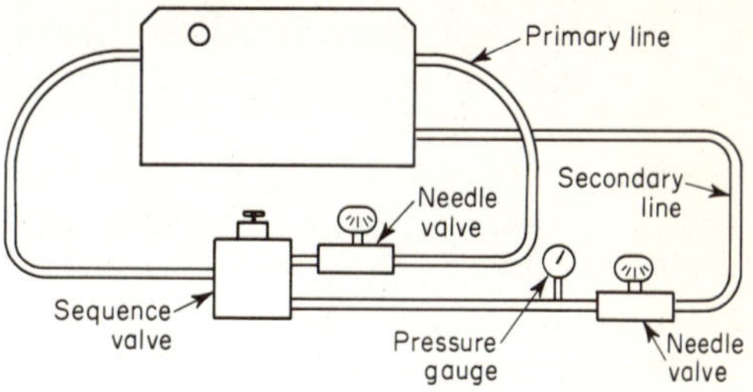


FIG. 14-45 *Circuit.*

RESULTS: Flow will be directed to the primary circuit until the pressure reaches the value set on the sequence valve. At this time the flow will be directed to the secondary circuit. When the pressure drops below the valve setting, all the flow will again be directed to the primary circuit.

CONCLUSION: This valve permits operation of two circuits in sequence (or consecutively) when the first operates at less pressure than is set on the valve.

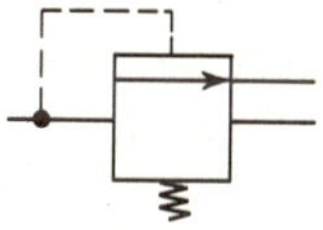

FIG. 14-46 *Symbol for sequence valve.*

TABLE 14-3 SEQUENCE-VALVE DEMONSTRATION

| *Condition* | *Pressure supply 1* | *Primary circuit* | *Secondary circuit* | *Primary pressure* | *Primary flow* | *Secondary pressure* | *Secondary flow* |
|---|---|---|---|---|---|---|---|
| 1 | Inlet | Metered | Auxiliary | 200 | 5* | 0 | 0 |
| 2 | " | " | " | 400 | | | |
| 3 | " | " | " | 500 | | | |
| 4 | " | | " | 600 | | | |
| 5 | | Auxiliary | Metered | † | | 500 | |
| 6 | " | " | " | † | | 600 | |
| 7 | " | " | " | † | | 700 | |
| 8 | " | " | " | 600 | | | |
| 9 | " | " | " | 500 | | | |
| 10 | " | " | | 400 | | | |
| 11 | " | " | " | 200 | | | |

* Use panel flow control.

† As required to obtain secondary line pressures.

PROBLEMS

**7.** At what pressure in the primary circuit was the secondary circuit energized? ______

**8.** What was the total initial flow through the primary circuit at 200 psi pressure? ______

**9.** (*a*) As the pressure was increased in the primary circuit (at pressures less than the valve setting), what happened to the flow in the primary circuit?

______

______

(*b*) What happened to the flow in the secondary circuit under the above conditions?

______

______

**10.** At pressures greater than the valve setting, what happened to the flow in the primary circuit as the pressure was increased in this circuit?

______

**11.** Does this valve determine direction of flow below its setting? ______

**12.** Does this valve determine direction of flow above its setting? ______

**13.** Is this a good valve for two hydraulic cylinders where they have to operate in sequence? ______

**14.** What is the symbol for a sequence valve?

**15.** Start with the quick disconnects, as shown on Fig. 14-47. Complete the circuit as shown on Fig. 14-46, using symbols.

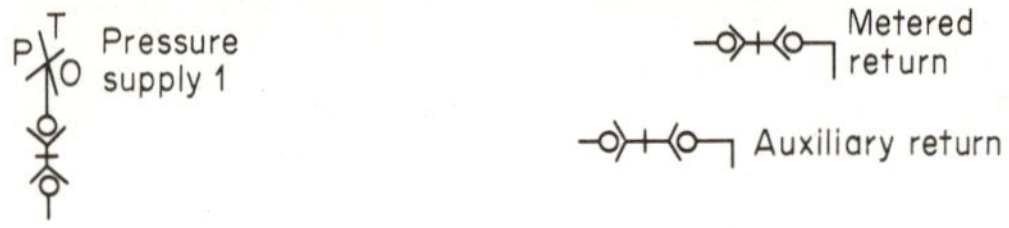


FIG. 14-47 *Partial circuit.*

## 14 • UNLOADING VALVE

PURPOSE: To show the operation of an unloading valve.

COMPONENTS: Needle valve, unloading valve, two pressure gauges, check valve, and three tees.

PROCEDURE

**1.** Set the panel flow control with approximately 50 percent of the total flow to pressure supply 1 and 50 percent to pressure supply 2.

**2.** Using a tee at the metered return port connect pressure supply 1 to one side of this tee with a needle valve between the tee and the metered return port.

**3.** Connect pressure supply 2 to the inlet port on the unloading valve.

**4.** Connect the outlet port from the valve to the other side of the tee on the metered return line with a check valve and pressure gauge installed as shown on Fig. 14-48.

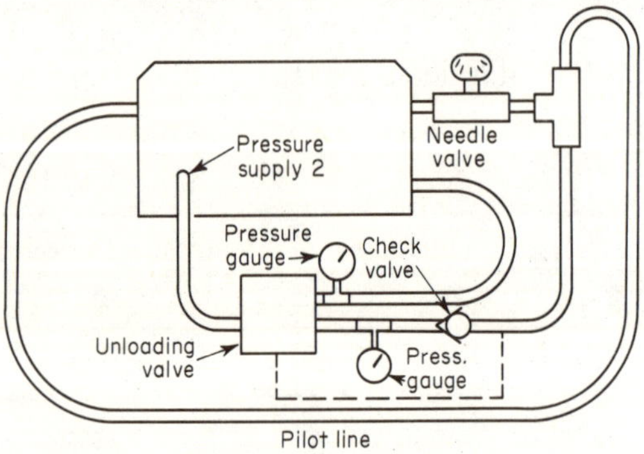


FIG. 14-48 *Circuit.*

Note that the needle valve will govern the pressure from the valve and from pressure supply 1.

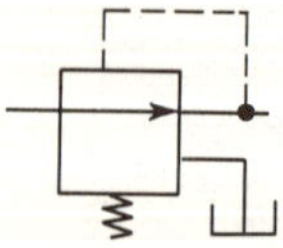

FIG. 14-49 *Symbol for unloading valve.*

 

**5.** Connect the discharge port on the valve to the auxiliary return line, and install the pressure gauge.

**6.** With the unloading valve set at approximately 500 psi and the panel flow control set as indicated in procedure 1, start the unit and record on Table 14-4 the results of the flow with pressure indicated on the table. Flow values will vary slightly depending on the type of pump used and its individual characteristics.

RESULTS: Total flow provided by the pump from the two sources on the training unit will be delivered to the actuator (in this case, the metered return line). Part of it, supply 1, is delivered direct; the balance, supply 2, is delivered through the unloading valve. As the pressure increases, the valve will divert one source, supply 2, directly to the tank, thus reducing the flow to the metered return.

CONCLUSION: This valve will permit a large volume of flow to be delivered until a pressure, set on the unloading valve, is reached. At this point it will divert one source of the total flow back to the tank with a minimum of pressure drop. This results in a fast movement of the actuator until the valve operates at its set pressure, and at that point a slow movement is attained. Consequently, if used in a high-low-volume pump (double pump), the high-volume pump would be unloaded. The total horsepower is now available to drive the low-volume portion of the pump to the actuator at high pressures. It now delivers a high pressure at slow speeds to the actuator.

PROBLEMS

**7.** At 200 psi what was the total gpm delivered to the metered return line? _______

**8.** At 200 psi what gpm in supply 2 was delivered to the metered return line through the valve? _______

**9.** At this same pressure on supply 1 what was the gpm delivered directly to the metered return line? _______

**10.** At what pressure did the valve unload (valve setting)? _______

**11.** What was the gpm at the metered return line after the valve unloaded? _______

**12.** How many gpm did the valve unload?

_______

**13.** How did the increased pressure affect the flow (*a*) to the metered return line (above valve setting)?

_______________________________________

_______________________________________

(*b*) to the auxiliary return line?

_______________________________________

_______________________________________

**14.** With a pressure of 400 psi in pressure supply line 1 what is the pressure in line 2 (*a*) upstream of the valve? _______ psi; (*b*) downstream of the valve? _______ psi.

TABLE 14-4 UNLOADING-VALVE ANALYSIS

| *Condition* | *Pressure supply 1, psi* | *Pressure supply 2, psi* | *Pressure at auxiliary return, psi* | *Flow at metered return, gpm* | *Flow at auxiliary return, gpm* |
|---|---|---|---|---|---|
| 1 | 200 | 200 | | Maximum | 0 |
| 2 | 400 | | | | |
| 3 | 500 | | | | |
| 4 | 600 | | | | |
| 5 | 700 | | | | |
| 6 | 800 | | | | |
| 7 | 700 | | | | |
| 8 | 600 | | | | |
| 9 | 500 | | | | |
| 10 | 400 | | | | |
| 11 | 200 | | | | |

**15.** With a pressure of 800 psi in pressure supply 1 and the valve unloading, what is the pressure in the secondary line (*a*) upstream of the valve? ______ psi; (*b*) downstream? ______ psi.

**16.** Using the fluid hp supplied formula, (pressure × gpm)/1714, give (*a*) the hp supplied at the needle valve at 400 psi ______ and (*b*) that supplied at 800 psi (after unloading) ______

**17.** If a valve is set to unload 50 percent of the flow, would the hp delivered be nearly one-half just before and just after unloading (assume a small pressure change)? ______

**18.** If a valve is set to unload 50 percent of flow at 500 psi, would fluid hp delivered be the same at 450 psi and 900 psi? ______ Why?

______________________________

______________________________

______________________________

______________________________

**19.** (*a*) Observing the wattmeter just before and just after unloading, there is no change in the hp that the electric motor supplies. Why does this hp not drop when the valve unloads?

______________________________

______________________________

______________________________

______________________________

______________________________

(*b*) Would it drop if the pump were a dual pump and one flow from it were unloaded? ______ Why?

______________________________

______________________________

______________________________

______________________________

______________________________

## 15 • LOAD-LOCK VALVE

PURPOSE: To show the operation of load-lock valve (counterbalance valve).

COMPONENTS: Counterbalance valve, or load-lock valve, needle valve, and pressure gauge.

PROCEDURE

**1.** Set the maximum pressure (relief valve) on the panel to by-pass flow at 100 psi.

**2.** Connect pressure supply 1 to the pressure port of the counterbalance valve. Set this valve at approximately 500 psi (solid line in Fig. 14-50).

**3.** Connect the outlet port to the metered return line (solid line in Fig. 14-50) with needle valve and pressure gauge installed.

**4.** Open needle valve.

**5.** Start the unit and record the flow and the return line pressure on Table 14-5.

**6.** Close the maximum-pressure valve (relief) until the supply pressure equals the values on Table 14-5 and fill in the rest of the data. Use needle valve to obtain pressures when pressure supply is connected to discharge port (free reverse flow inlet).

**7.** Upon reaching valve setting, note that

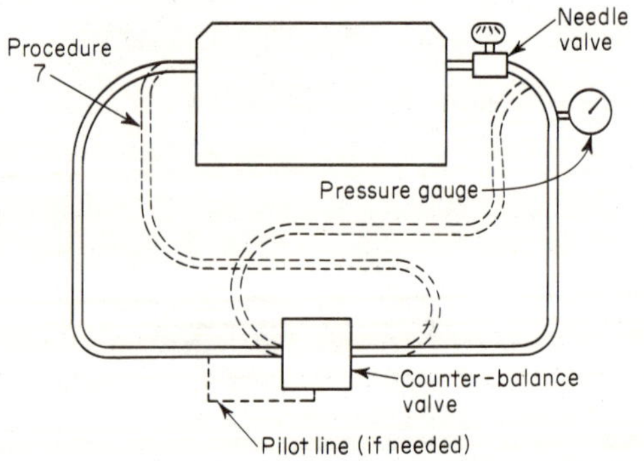


FIG. 14-50 *Circuit.*

 

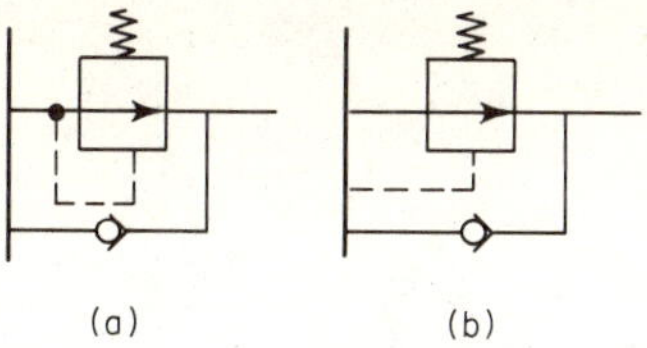

FIG. 14-51 *Symbol for load-lock valve (counterbalance).*

by closing the relief valve the pressure will not rise. Pressure is now governed by counterbalance valve setting. To obtain value of 800 psi as required on Table 14-5 use the counterbalance valve adjustment.

**8.** If counterbalance valve is remotely operated, install pilot line as shown on Fig. 14-50 (dotted line) and provide drain line if valve is externally drained.

**9.** Reverse the connections at the valve by connecting the pressure supply to the discharge port and the pressure port to the metered return line (dotted line in Fig. 14-50).

**10.** Open the needle valve.

**11.** Start the unit and by using the needle valve, increase the pressures as listed on Table 14-5, and record the balance of the data.

TABLE 14-5 COUNTERBALANCE VALVE ANALYSIS

| *Location at press. supply* 1 | *Supply press.* | *Flow, gpm* | *Return-line press.* |
|---|---|---|---|
| (*a*) Pressure port | 200* | | |
| (*b*) Pressure port | 600* | | |
| (*c*) Pressure port | 800* | | |
| (*d*) Discharge port | 200† | | |
| (*e*) Discharge port | 400† | | |
| (*f*) Discharge port | 600† | | |
| (*g*) Discharge port | 800† | | |

* Pressures obtained by relief valve.
† Pressures obtained by needle valve.

RESULTS: This valve requires 500 psi before it will permit flow in one direction. In the other direction it permits free flow.

CONCLUSION: The counterbalance valve will hold a load in one direction if it does not exceed the valve setting. Additional pressure is needed from the system to actuate the valve in the same direction. This results in controlled load lowering in a circuit where the hydraulic cylinder supports a load. If back pressure were not available to hold the load, it would cavitate the cylinder and rapidly drop. With this valve the load is lowered by the pump pressure. The load speed can be controlled with a variable-volume pump or a flow control valve. Note that this is an excellent safety feature if installed on the cylinder.

1. If the pressure line were to break, the pressure, of course, would drop and lock the cylinder in its position.
2. If the return line were to break, the load would still be lowered as required, and the oil would simply spill on the ground.

PROBLEMS

**12.** What pressure is required to permit flow in a free flow direction with the needle valve open and the oil supply entering the discharge port? ______

**13.** What is the minimum pressure required to allow flow in the controlled direction (oil flowing into the pressure port)?

______

**14.** At what pressure would the valve be set to afford load locking if the load caused a back pressure of 800 psi? ______

**15.** What maximum load would the counterbalance valve hold in a locked position? (Assume it is used with a 4-in.-dia. cylinder which has a 2-in.-dia. piston under the two conditions shown and assume that the load tends to retract the cylinder.) ______

**16.** What is the maximum load that the valve will hold if cylinder is reversed? ______

**17.** Explain why this valve is needed for the hoist cylinder in a front-end-loader application.

*Note:* This feature is often constructed into a four-way valve. This would eliminate the necessity for a separate load-locking valve.

## 16 • PRESSURE-COMPENSATED FLOW CONTROL VALVE

PURPOSE: To show operation of a pressure-compensated flow control valve (panel flow control).

COMPONENTS: Two needle valves, pressure gauge, pressure compensated flow, and control valve.

PROCEDURE

**1.** Connect pressure supply 1 to metered return line with a needle valve installed.

**2.** Connect pressure supply 2 to auxiliary return line with a needle valve and a pressure gauge installed.

**3.** By the use of the needle valves and flow control lever, set conditions as indicated on Table 14-6 and complete remainder of data.

RESULTS: This valve will deliver to either pressure supply a set amount of the total flow or the entire flow, depending on valve setting. The pressure of either supply source does not appreciably affect the flow at the other source. Note that it is a priority flow divider, and almost all the pump loss, as pressure is increased, is reflected at pressure supply 2. Thus, in this divider the priority port is 1. This can be further demonstrated by reducing the flow by the use of the pump flow control at condition 13 in the table.

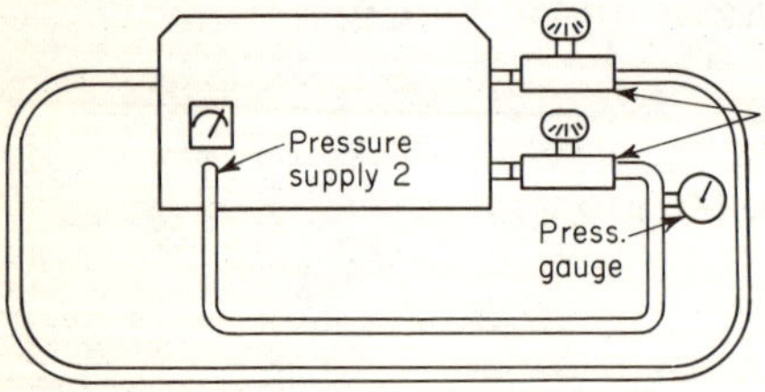


FIG. 14-52 *Circuit.*

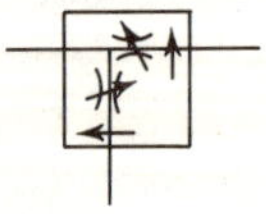

FIG. 14-53 *Symbol for pressure-compensated flow control valve.*

TABLE 14-6 PRESSURE-COMPENSATED FLOW CONTROL VALVE

| Condition no. | SUPPLY 1 | | SUPPLY 2 | |
|---|---|---|---|---|
| | *Pressure, psi* | *Flow, gpm* | *Pressure, psi* | *Flow, gpm* |
| 1 | 0 | Max. 6+ | 0 | 0 |
| 2 | 200 | Max. | | |
| 3 | 400 | Max. | | |
| 4 | 600 | Max. | | |
| 5 | 800 | Max. | | |
| 6 | 0 | 4 | 200 | 2 |
| 7 | 200 | | | |
| 8 | 400 | | | |
| 9 | 600 | | | |
| 10 | 800 | | | |
| 11 | 0 | 3 | 300 | 3 |
| 12 | 200 | | | |
| 13 | 400 | | | |
| 14 | 600 | | | |
| 15 | 800 | | | |

 

Note that under this condition flow is reduced at supply 2, whereas 1 remains the same.

PROBLEMS

**4.** Is this a good valve for flow control to pressure supply 1? _______

**5.** Can this valve be used for flow control to pressure supply 2? _______

**6.** In the full flow to pressure supply 1 what was the flow from pressure supply 2 at 0 psi?

_______

**7.** What happened to flow in 2 as pressure at supply 1 was increased?

_______________________________________

**8.** (*a*) The pump will deliver less flow at higher pressures. Which pressure supply (1 or 2) will reflect this decrease in flow?

_______

(*b*) One is then the controlled flow supply and the other the excess flow supply. Which is controlled? _______

**9.** Assume that flow is set to deliver 50 percent to each supply, both needle valves are set at 500 psi, and the pump flow control is used to reduce flow from maximum to 5 gpm, then to 4 gpm, 3 gpm, 2 gpm, and 0 gpm.

(*a*) What happens to the flow at 1?

_______________________________________

(*b*) At 2?

_______________________________________

(*c*) Which is the first to lose flow? _______

(*d*) How does this prove that 1 is the controlled flow pressure supply?

_______________________________________

_______________________________________

_______________________________________

_______________________________________

_______________________________________

_______________________________________

## 17 • PILOT-OPERATED RELIEF VALVE

PURPOSE: To show operation of pilot-operated relief valve.

COMPONENTS: Needle valve and pilot-operated relief valve.

PROCEDURE

**1.** Connect pressure supply 1 to the metered return line either with a tee connection or directly through the valve if it is an in-line type of valve.

**2.** Install a needle valve downstream from the relief valve in this line.

**3.** Set panel flow control to full flow at pressure supply 1.

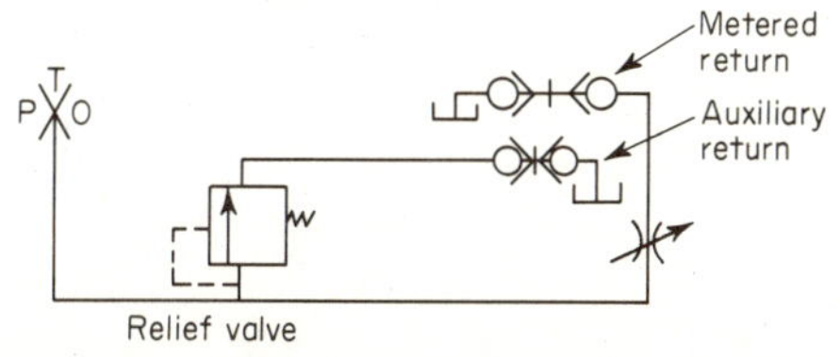


FIG. 14-54 *Circuit.*

**4.** Set panel relief valve to maximum bypass pressure.

**5.** Open needle valve, start unit, and slowly close the needle valve to determine the answers to the following questions. *Use procedures similar to demonstration 6 on panel relief valve or maximum pressure valve.*

PROBLEMS

**6.** What was the cracking pressure for the relief valve? _______

**7.** What was the full-flow pressure? _______

**8.** What was the reseat pressure for this valve? _______

**9.** What was the difference between cracking and reseat pressures? _______ psi.

**10.** Plot the results using the dotted line (cracking, full-flow, and reseat pressures) on Fig. 14-29 accompanying the discussion of simple relief valves.

**11.** Which valve had the flattest line between full-flow and cracking pressure? _______

**12.** What are the advantages of a pilot-operated relief valve?

**13.** Why are they most often used in a system?

**14.** Which valve, simple or pilot operated, will be most likely to cause malfunction if contaminated oil is used?

 

# CIRCUITS

# 15

## 1 ❯ HYDRAULIC CYLINDER WITH TWO-POSITION FOUR-WAY VALVE

PURPOSE: To operate a hydraulic cylinder with a two-position four-way valve.

COMPONENTS: Two needle valves, a hydraulic cylinder, and a two-position four-way valve. (A three-position four-way valve may be used if a two-position valve is not available. In this case, ignore the neutral position in the three-position valve.)

PROCEDURE

**1.** Connect pressure supply 1 to the inlet port of the valve and the outlet port to the metered return.

**2.** Connect the flow control ports to the blank end and rod ends of the hydraulic cylinder.

**3.** Mount the cylinder in a hydraulic cylinder testing device. If the device is not available, install a needle valve at each port on the hydraulic cylinder. These needle valves can be used to provide pressure in the system simulating pressures which would result from an actual cylinder load.

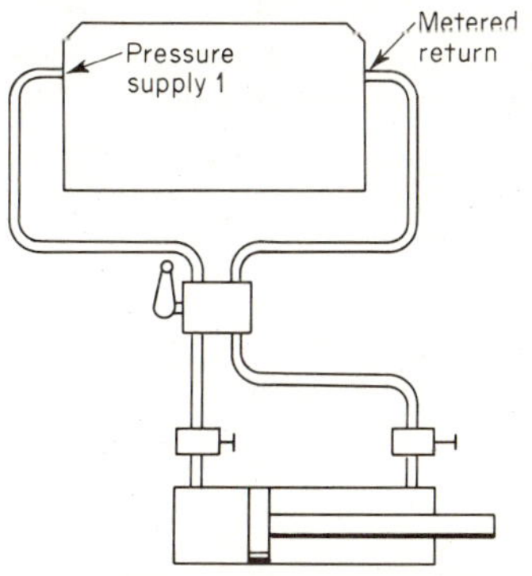


FIG. 15-1 *Schematic circuit.*

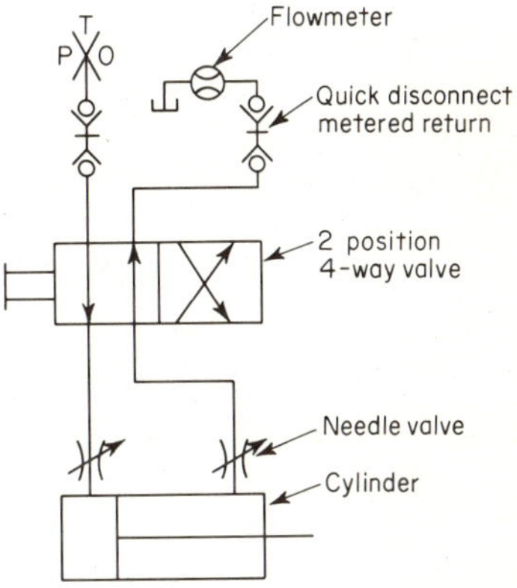


FIG. 15-2 *Circuit symbols.*

**4.** Use the panel flow control to provide flow to the circuit. Use the needle valve to induce pressure as required, and complete the values in Table 15-1.

**5.** Fill in the following valve data (for two-position four-way valve):

(*a*) How is it operated (manual, solenoid, or pilot operated)? ______

(*b*) Spring loaded (yes or no)? ______

(*c*) Manufacturer and model ______

______

(*d*) Intake port size ______

TABLE 15-1 CYLINDER PERFORMANCE DATA*

| Valve position | Supply pressure, psi | Effective area, sq in. | Force, lb | Rod travel speed, ft/min | Ft-lb/min | Delivered hp | Flow, gpm |
|---|---|---|---|---|---|---|---|
| (*a*) Extend 1 | 200 | | | | | | 5 |
| (*b*) Retract 2 | 200 | | | | | | 5 |
| (*c*) Extend 1 | 400 | | | | | | 5 |
| (*d*) Retract 2 | 400 | | | | | | 5 |
| (*e*) Extend 1 | 600 | | | | | | 5 |
| (*f*) Retract 2 | 600 | | | | | | 5 |
| (*g*) Extend 1 | 800 | | | | | | 5 |
| (*h*) Retract 2 | 800 | | | | | | 5 |

* Answers may be calculated or obtained from tables in Sec. 8.

(*e*) Tank port size ________

(*f*) Control port sizes ________

**6.** Fill in the following cylinder data (answers depend on cylinder used):

(*a*) Diameter ________ in.

(*b*) Stroke ________ in.

(*c*) Rod diameter ________ in.

(*d*) Manufacturer ________________

(*e*) Type of connection at blank end (welded, threaded, etc.) ________________

(*f*) Type of connection at rod end ________

(*g*) What is the effective pressure area on the piston when the rod is extending? ________

(*h*) When the rod is retracting? ________

**7.** Complete values not shown on Table 15-1.

*Example:*

Pressure/sq in. × area = force

Work = force × distance

If pressure in a cylinder is 200 psi and acts against an area of 15 sq in., the force is 200 × 15, or 3000 lb. The work done in a minute is the force × distance that the rod traveled in 1 min. Thus, if the rod traveled 10 ft/min with a force of 3000 lb (force × 10 ft), then the work done in 1 min = 30,000 ft-lb.

It is not possible to time the cylinder's travel speed for a full minute, but the problem can be solved by using proportion. If rod travel equals 2½ ft in 15 sec:

$$2\tfrac{1}{2}\text{ ft} : 15\text{ sec} = X\text{ ft} : 60\text{ sec}$$

$$15X = 150$$

$$X = 150/15 = 10\text{ ft/min}$$

This could be worked using inches of travel per minute and converting to feet per minute by dividing by 12.

By definition, one horsepower is the work done when 33,000 lb is lifted 1 ft in 1 min. Therefore, to find the hp delivered by the cylinder, we simply use the following proportion:

$$1\text{ hp} = 33{,}000\text{ ft-lb/min}$$

How many hp = 30,000 ft-lb/min?

$$X\text{ hp} = 30{,}000$$

$$X = \frac{30{,}000}{1\text{ hp}} \quad \text{or} \quad \frac{30{,}000}{33{,}000}$$

or $\quad 0.91$ hp

RESULTS: This circuit will deliver linear force at the work end of a circuit. The amount of force depends on cylinder area and pressure. The speed of the rod varies with the rate of flow. (Note that pressure does not affect the rate of rod travel.)

PROBLEMS

**8.** How was the four-way valve actuated?

________________________________

**9.** With this valve, can one stop a cylinder in any mid-position? ________

**10.** What happens to the circuit pressure when the cylinder bottoms at either end of its stroke?

________________________________

________________________________

 

**11.** What force would be delivered as the rod extends in a 12-in.-dia. cylinder with 6000 psi pressure while extending? ______

**12.** What force would be delivered in a 12-in.-dia. cylinder with a 4-in.-dia. rod and a pressure of 6000 psi pressure while retracting? ______

**13.** What is the travel speed of the rod, while extending, at 5 gpm? ______ ft/min

**14.** What is the travel speed of the rod, while retracting, at 5 gpm? ______ ft/min

**15.** At what speed would the load move up at 4 gpm if (*a*) the cylinder were operating vertically, with rod extending up? ______; (*b*) the cylinder were mounted vertically, with rod extending down? ______; (*c*) the cylinder were mounted and operated horizontally? ______

**16.** Why is a load-lock valve or load-lock feature in the control valve required where a heavy load is moved vertically intermittently?

______

______

______

______

______

**17.** What is cylinder cavitation?

______

______

______

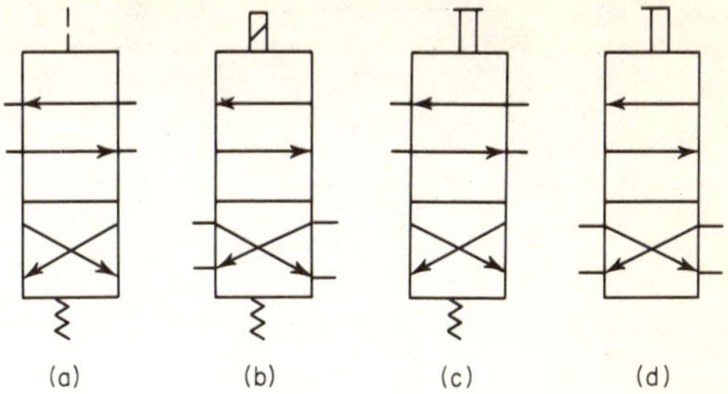


FIG. 15-3 *Two-position four-way valve symbols.*

**18.** Could cavitation occur in a cylinder mounted horizontally if a heavy load were moved rapidly and then suddenly stopped? ______

**19.** To solve the above problem, what device or valve is used?

______

______

**20.** Describe the two-position four-way valves shown in Fig. 15-3.

| *Valve* | (*a*) | (*b*) | (*c*) | (*d*) |
|---|---|---|---|---|
| How operated? | | | | |
| Spring loaded? | | | | |
| Position actuated or normal? | | | | |

## 2 • HYDRAULIC CYLINDER WITH THREE-POSITION FOUR-WAY VALVE

PURPOSE: To show the operation, application, and function of a hydraulic cylinder controlled by a three-position four-way valve.

COMPONENTS: Three-position four-way valve, double-acting hydraulic cylinder, and two needle valves.

PROCEDURE

**1.** Connect pressure supply 1 to the inlet port on the four-way valve and the outlet to the metered return line.

**2.** Connect the control ports on the four-way valve to opposite ends of the hydraulic cylinder.

**3.** Install the cylinder in a testing device, if available, or install a needle valve at each cylinder port. Use the needle valves to induce load on the system as required in Table 15-2.

**4.** Set pump flow control "full flow," and govern flow by use of the panel flow control.

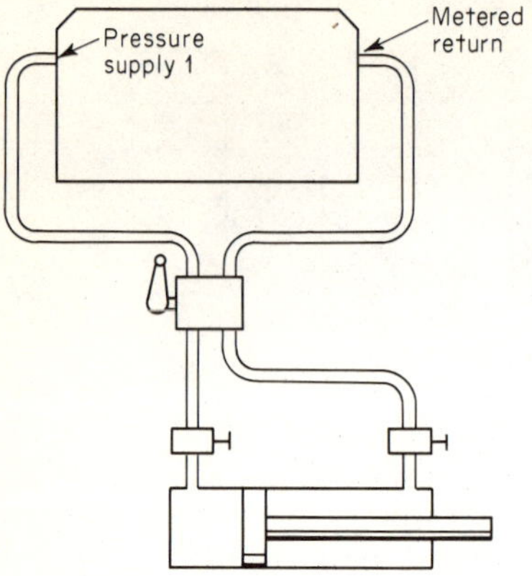


FIG. 15-4 *Schematic circuit.*

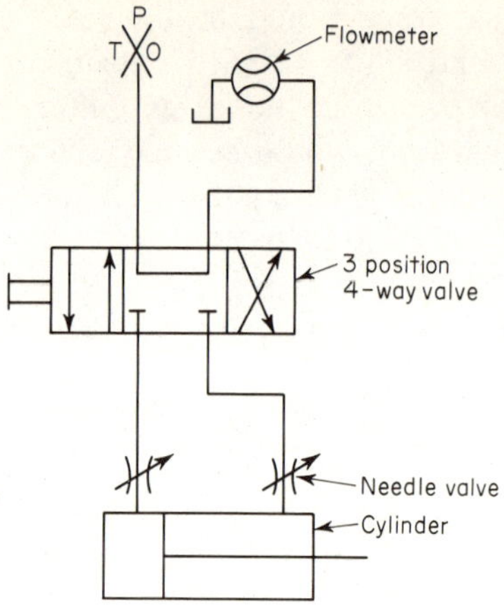


FIG. 15-5 *Symbol circuit.*

**5.** Fill in the following cylinder data:

(*a*) Cylinder diameter ________ in.

(*b*) Rod diameter ________ in.

(*c*) Stroke ________ in.

(*d*) Manufacturer ________________

(*e*) Pressure rating ________

(*f*) Port thread type ________

(*g*) Port size ________

**6.** Complete the values in Table 15-2.

**7.** Fill in the following information concerning a four-way valve.

(*a*) Is it possible to move the rod when the valve is in the neutral position? ________

(*b*) How is the valve actuated (manually, solenoid, or pilot operated)?

________________________

(*c*) Intake port size and thread type

________________________

(*d*) Tank port size and thread type

________________________

(*e*) Control port size and thread type

________________________

(*f*) Does the valve have a load-lock or a counterbalance feature? ________________

(*g*) Manufacturer ________________

(*h*) Model ________________

(*i*) Maximum pressure rating ________

PROBLEMS

**8.** Figure 15-6 shows the symbols for the center position of a three-position four-way valve. Describe the flow in each of the conditions.

(*a*) ________________________

________________________

TABLE 15-2 HYDRAULIC CYLINDER ANALYSIS

| *Control valve position* | *Supply pressure, psi* | *Rod movement (in, none, out)* | *Rod speed, ft/min* | *Flow, gpm* | *Rod force (retracting), lb* | *Rod force (extending), lb* | *Delivered hp* |
|---|---|---|---|---|---|---|---|
| (*a*) Center | | | | | | | |
| (*b*) No. 1 | 100 | | | 5 | | | |
| (*c*) No. 2 | 100 | | | 5 | | | |
| (*d*) Center | | | | | | | |
| (*e*) No. 1 | 400 | | | 4 | | | |
| (*f*) No. 2 | 400 | | | 4 | | | |
| (*g*) Center | | | | | | | |
| (*h*) No. 1 | 600 | | | 3 | | | |
| (*i*) No. 2 | 600 | | | 3 | | | |

 

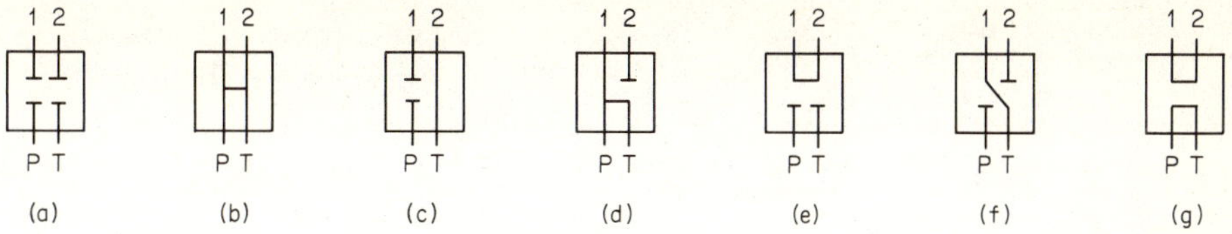


FIG. 15-6 *Center position symbols for three-position four-way valve.*

(*b*) ______________________________________

(*c*) ______________________________________

(*d*) ______________________________________

(*e*) ______________________________________

(*f*) ______________________________________

(*g*) ______________________________________

(*h*) ______________________________________

**9.** Figure 15-7 shows four methods of operating a four-way valve. Describe the method by which each is actuated.

(*a*) ______________________________________

(*b*) ______________________________________

(*c*) ______________________________________

(*d*) ______________________________________

**10.** Draw the symbol for the four-way valve used in this circuit.

**11.** What is the inside diameter of the cylinder? _______

**12.** What is the area of the cylinder or piston on the blank side? _______

**13.** What is the diameter of the rod? _______

**14.** What is the area of the rod? _______

**15.** What is the difference between the cylinder area and the rod area? _______

**16.** What is the effective area for pressure while the rod is extending? _______

**17.** What is the effective area for pressure while the rod is retracting? _______

**18.** What was the speed of the rod while extending at 5 gpm? _______ ft/min.

**19.** What was the speed of the rod while retracting at 5 gpm? _______ ft/min.

**20.** What force could the rod deliver while extending at 100 psi? _______

**21.** What force could the rod deliver while retracting at 100 psi? _______

**22.** What force could this rod deliver while extending at 6000 psi? _______

**23.** What force could this rod deliver while retracting at 6000 psi? _______

**24.** What would the speed of rod travel be if it were retracting and 50 gpm were delivered? _______

**25.** What would the speed of rod travel be if it were extending and 50 gpm were delivered? _______

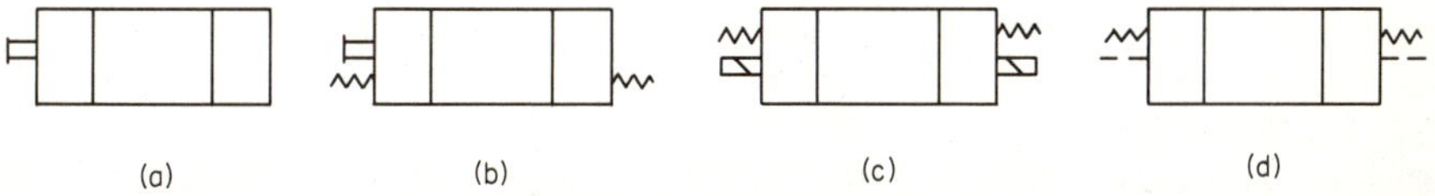


FIG. 15-7 *Methods of valve actuation.*

TABLE 15-3 CIRCUIT RESULTS

| Pressure, psi | Cylinder area, sq in. | Force, lb | Travel speed, ft/min | Work, ft-lb/min | hp |
|---|---|---|---|---|---|
| EXTENDING | | | | | |
| 100 | | | | | |
| 400 | | | | | |
| 600 | | | | | |
| RETRACTING | | | | | |
| 100 | | | | | |
| 400 | | | | | |
| 600 | | | | | |

**26.** Using the following formula fill out Table 15-3 (for a 4-in.-dia. cylinder with a 2-in. rod and a flow of 10 gpm).

$$\text{Area} = \frac{3.14 \times \text{diameter squared}}{4}$$

*Example:*

Rod travel speed in inches per minute

$$= \frac{231 \times \text{gpm}}{\text{area}}$$

Effective area rod end
= area blank end minus rod area

Force = pressure × effective piston area

Work = force × distance

Horsepower = work done in a given time

$$\text{Delivered horsepower} = \frac{\text{ft-lb/min}}{33{,}000}$$

If the pressure is 100 psi and the area is 10 sq in., then the force is 100 × 10, or 1000 lb.

Given a force of 1000 lb with a rod speed of 20 ft/min, the work done in a minute is 1000 × 20, or 20,000 ft-lb.

To find the hp delivered, we can use a simple proportion:

$$1 \text{ hp} = 33{,}000 \text{ ft-lb/min}$$
$$X \text{ hp} = 20{,}000$$
$$X = \frac{20{,}000}{1 \text{ hp}} \quad \text{or} \quad \frac{20{,}000}{33{,}000}$$

therefore

$$X = \frac{20}{33} \quad \text{or} \quad 0.6 \quad \text{or} \quad 2/3 \text{ hp approx.}$$

**27.** The displacement of a cylinder is the total volume of fluid that is required to move the rod its entire stroke. (The displacement would be calculated by multiplying the area by the stroke.) What is the extending displacement of a cylinder 2 in. × 24 in. (*a*) in cublic inches? ________; (*b*) in cubic feet? ________

**28.** What is the retracting displacement of the cylinder (*a*) in cubic inches? ________; (*b*) in cubic feet? ________

**29.** What percent greater force is the cylinder capable of exerting (at constant pressure) extending over retracting? ________

## 3 • HYDRAULIC MOTOR

PURPOSE: To show the operation of a hydraulic motor.

COMPONENTS: Training unit, hydraulic motor, Prony brake, and tachometer.

PROCEDURE

**1.** Connect pressure supply 1 to the pressure port on the motor and the outlet port to the metered return line.

**2.** Attach the Prony brake to the output shaft of the motor.

**3.** On the panel flow control set full flow to pressure supply 1.

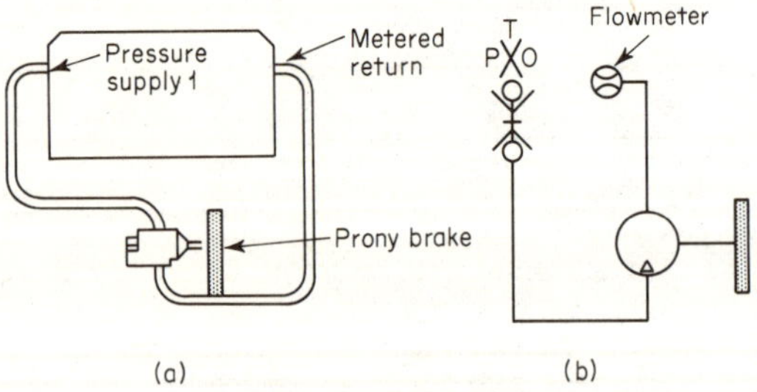


FIG. 15-8 *Motor circuit.*

**4.** Use the pump flow control to regulate flow to the circuit and the Prony brake to induce a load on the motor. Make adjustments to meet the conditions in Table 15-4.

**5.** Fill in the following motor data.

(*a*) Type ______

(*b*) Manufacturer ______

(*c*) Model ______

(*d*) Displacement ______

(*e*) Manufacturer's maximum gpm rating ______

(*f*) Manufacturer's maximum pressure rating ______

(*g*) Pressure port size ______

(*h*) Thread type ______

(*i*) Tank port size ______

(*j*) Thread type ______

(*k*) Birotational? ______

(*l*) Internal or external drain? ______

PROBLEMS

**6.** Calculate the hp for each pressure flow combination in Table 15-4. If a Prony brake is not available, obtain the torque data from the manufacturer's performance curves. The horsepower output formula is:

$$\frac{\text{torque ft-lb} \times \text{rpm}}{5252}$$

For example, if torque = 7 ft-lb and rpm = 1000, then

$$\frac{7 \times 1000}{5252} = \frac{7000}{5252} = 1.3 \text{ output hp}$$

**7.** If the pressure doubles, does the value of the torque double? ______

**8.** If the flow to a pump doubles, does the rpm double? ______

**9.** From the manufacturer's data, is this motor a bidirectional or unidirectional motor (bidirectional is a two-directional motor)? ______

**10.** From the data, plot the motor rpm and gpm data for 300 psi and 700 psi on the chart (Fig. 15-9). This will result in two lines (one for each pressure). Label each line, indicating the pressure for the line.

TABLE 15-4 MOTOR PERFORMANCE CHART*

| *Supply pressure, psi* | *Flow, gpm* | *Motor, rpm* | *Torque, in.-lb* | *Torque, ft-lb* | *Output, hp* |
|---|---|---|---|---|---|
| (*a*) 100 | 3 | | | | |
| (*b*) 300 | 3 | | | | |
| (*c*) 500 | 3 | | | | |
| (*d*) 700 | 3 | | | | |
| (*e*) 800 | 3 | | | | |
| (*f*) 100 | 4 | | | | |
| (*g*) 300 | 4 | | | | |
| (*h*) 500 | 4 | | | | |
| (*i*) 700 | 4 | | | | |
| (*j*) 800 | 4 | | | | |
| (*k*) 100 | 5 | | | | |
| (*l*) 300 | 5 | | | | |
| (*m*) 500 | 5 | | | | |
| (*n*) 700 | 5 | | | | |
| (*o*) 800 | 5 | | | | |
| (*p*) 100 | † | | | | |
| (*q*) 300 | † | | | | |
| (*r*) 500 | † | | | | |
| (*s*) 700 | † | | | | |
| (*t*) 800 | † | | | | |

* 1.07 cu in. displacement.
† Maximum flow.

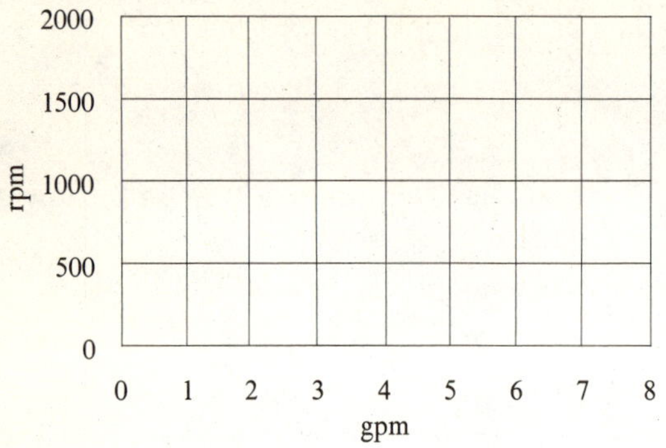


FIG. 15-9 *Demonstration graph.*

**11.** Fill in the missing data on Table 15-4.

**12.** From the data, plot the pressure and output horsepower for 3 gpm and 5 gpm (Fig. 15-10).

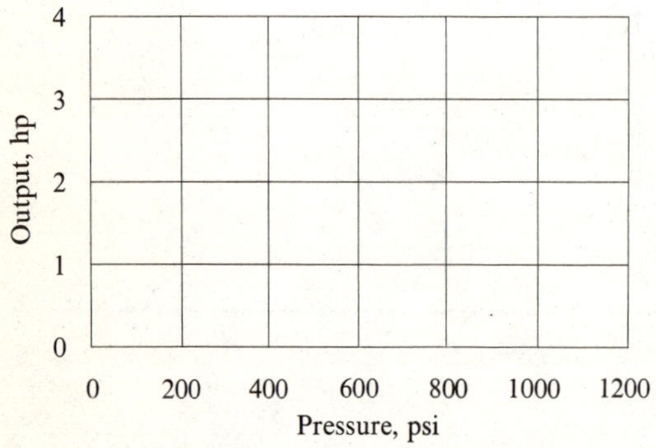


FIG. 15-10 *Demonstration graph.*

**13.** What displacement value do the test results indicate at 500 psi for the motor used? ________ cu in./rev

*Hint:* Use the following formula:

$$\frac{\text{gpm} \times 231}{\text{rpm}} = \text{displacement at a specific pressure}$$

**14.** On the graph (Fig. 15-11), draw the hp output line for the motor at 5 gpm and 0 to 800 psi, and label the line "hp output at 5 gpm." (This information can be found on Table 15-4.)

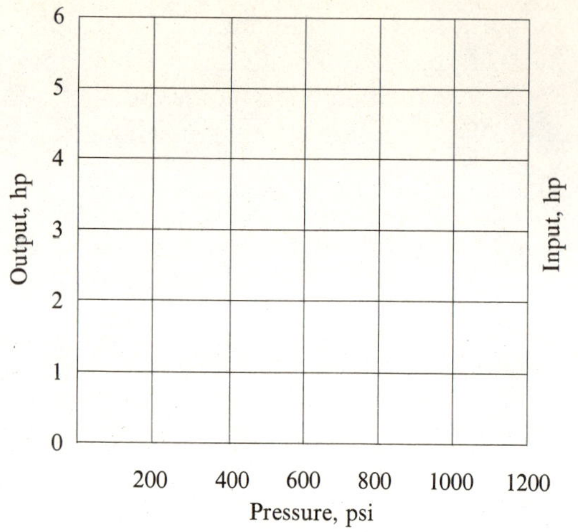


FIG. 15-11 *Demonstration graph.*

**15.** On Fig. 15-11 also draw the input hp, using the following formula:

$$\text{input hp} = \frac{\text{pressure} \times \text{gpm}}{1714}$$

*Note:* Pressure for this formula is the differential pressure between the input and output ports of the motor. This is the supply pressure minus the return-line pressure on the unit.

*Example:* Pressure = 800; gpm = 5

$$\frac{800 \times 5}{1714} = \frac{4000}{1714} = 2.33 \text{ input hp}$$

**16.** (*a*) Using the information in Fig. 15-11, give the mechanical efficiency (in percent) of the motor (*a*) at 300 psi ________

(*b*) At 800 psi ________

*Hint:* Use the following formula:

$$\frac{\text{Output hp}}{\text{Input hp}} = ? \times 100 = \text{percent efficiency}$$

**17.** How does the actual output hp compare with that claimed by the manufacturer?

________________________________________

 

## 4 • BIDIRECTIONAL OPERATION OF HYDRAULIC MOTOR

PURPOSE: To show bidirectional operation of a hydraulic motor.

COMPONENTS: Hydraulic motor, three-position four-way valve, Prony brake, and tachometer.

PROCEDURE

**1.** Connect pressure supply 1 to the inlet port of the four-way valve. Connect outlet, or tank port, to the metered return and the motor to the two flow control ports.

**2.** Mount the motor with the Prony brake as in the discussion of the hydraulic motor (shown on Figs. 15-12 and 15-13).

**3.** Set pump flow control to maximum.

**4.** Use panel flow control to control flow.

**5.** Operate the motor in two directions using a four-way valve, and record the pressure flow and torque data as required to complete Table 15-5.

PROBLEMS

**6.** Analyze the valve used and record its features.

(*a*) Two- or three-position? ______

(*b*) How operated?

______________________

(*c*) Type of center (if three-position)?

______________________

(*d*) Does it have the float feature (common flow from control ports to tank while valve is in neutral position)? ______

(*e*) Does it have the load check feature? ______

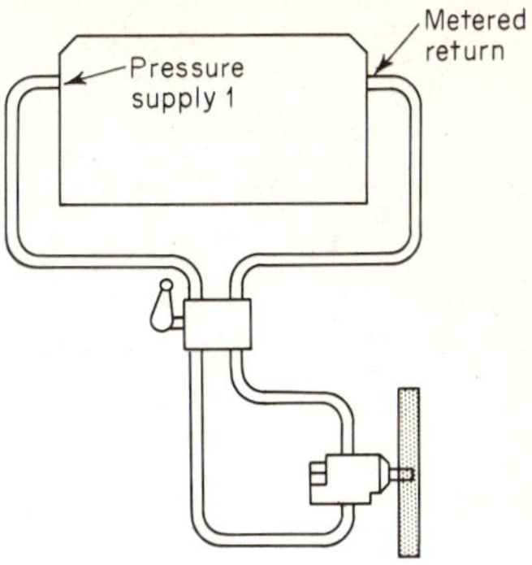


FIG. 15-12 *Motor circuit.*

(*f*) Is it a spring return type? ______

(*g*) Intake port size? ______ Thread type? ______

(*h*) Tank port size? ______ Thread type? ______

(*i*) Control port sizes? ______ Thread type? ______

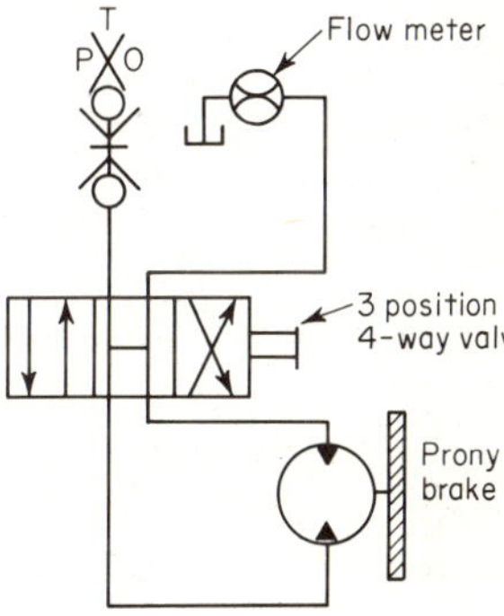


FIG. 15-13 *Motor circuit.*

TABLE 15-5 CIRCUIT RESULTS

| *Motor rotation** | *Supply pressure* | *gpm* | *rpm* | *Torque, ft-lb* | *Input, hp* | *Output, hp* |
|---|---|---|---|---|---|---|
| (*a*) cw | 300 | 4 | | | | |
| (*b*) ccw | 300 | 4 | | | | |
| (*c*) cw | 500 | 4 | | | | |
| (*d*) ccw | 500 | 4 | | | | |
| (*e*) cw | 700 | 4 | | | | |
| (*f*) ccw | 700 | 4 | | | | |
| (*g*) cw | 800 | 4 | | | | |
| (*h*) ccw | 800 | 4 | | | | |

* Rotation as you look at output shaft of motor; cw = clockwise, ccw = counterclockwise.

**7.** At 700 psi and 4 gpm:
(*a*) What was the hp while rotating cw? (input) ______ (output) ______
(*b*) What was the hp while rotating ccw? (input) ______ (output) ______

**8.** Does this motor deliver the same hp in either direction of rotation at the same pressure and flow volume? ______

**9.** What are the advantages of delivering hp or torque to a machine by using fluid power in preference to
(*a*) Electrical power?

______

______

______

______

(*b*) Mechanical power?

______

______

______

______

**10.** Complete the accompanying 4 gpm-rpm pressure graph (Fig. 15-14*a*) using data obtained in Table 15-5.

**11.** Complete the following 4 gpm-hp output pressure graph (Fig. 15-14*b*) using data obtained in Table 15-5.

**12.** What is motor slippage?

______

______

**13.** Where does it occur in a motor?

______

______

**14.** Does slippage increase with pressure? ______ Why?

______

______

**15.** How does slippage affect the rpm's delivered by the motor?

______

______

______

______

**16.** Using motor in circuit, what is displacement at 0 ft-lb torque delivery [formula: (gpm × 231)/rpm]?

______

**17.** What is displacement of motor at 500-psi inlet pressure? ______

**18.** What is torque at 500-psi pressure? ______ ft-lb.

**19.** What is displacement of motor at 800-psi inlet pressure? ______

**20.** What is torque at 800-psi pressure? ______ ft-lb.

**21.** What is volumetric efficiency at 4 gpm based on 0 ft-lb torque and torque delivered at 500-psi inlet pressure? ______

*Note:* Volumetric efficiency is the motor's actual rpm's delivered at a given pressure

(*a*) (*b*)

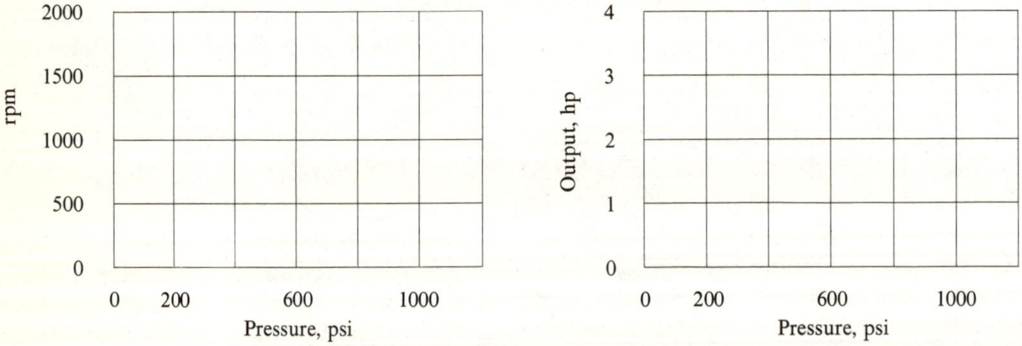


FIG. 15-14 *Demonstration graphs.*

divided by theoretical rpm's at the same flow rate at 0 pressure. For your purpose the theoretical rpm's are those received at 0 ft-lb torque. Therefore, volumetric efficiency for a motor at 4 gpm would be

$$\frac{\text{rpm at 500 psi at 4 gpm}}{\text{rpm at 0 psi at 4 gpm}} = \text{efficiency}$$

(Percent of efficiency equals efficiency multiplied by 100.)

**22.** What is the volumetric efficiency based on 4 gpm and 0 ft-lb torque compared with 4 gpm and the torque delivered at an 800-psi inlet pressure? ______

## 5 • PARALLEL CIRCUIT OF TWO HYDRAULIC CYLINDERS

PURPOSE: To show the parallel circuit of two hydraulic cylinders.

COMPONENTS: Four-way valve, two hydraulic cylinders, and two needle valves.

PROCEDURE

**1.** Connect pressure supply 1 to the four-way valve, and connect the outlet port to the metered return line.

**2.** With a tee at each control port on the valve, connect the blank end ports of the cylinders to one tee and the rod end ports to the other tee (Figs. 15-15 and 15-16).

**3.** Operate the cylinders through their strokes, imposing loads on the cylinders as shown in Table 15-7. Use needle valves to simulate loaded conditions if testing devices are not available.

**4.** To meet the requirements on Table 15-7 it will be necessary to use appropriate needle valves to induce loads as they are required.

TABLE 15-6 CYLINDER DATA

| | *1* | *2* |
|---|---|---|
| Cylinder diameter | | |
| Cylinder stroke | | |
| Rod diameter | | |
| Effective cylinder area (ext.) | | |
| Effective cylinder area (ret.) | | |

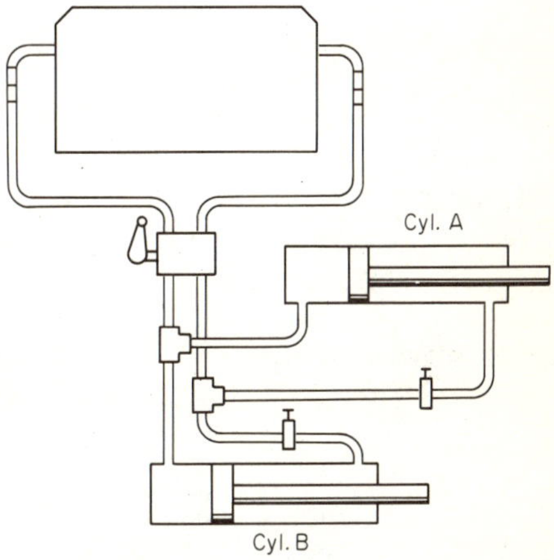


FIG. 15-15 *Two-cylinder circuit.*

PROBLEMS

**5.** Under no load which cylinder moved through its stroke first? ______

**6.** When cylinder *A* was loaded, why did cylinder *B* move first?

______

______

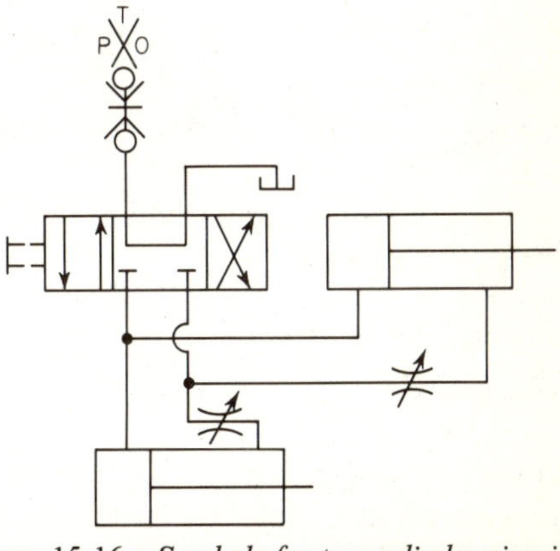


FIG. 15-16 *Symbols for two-cylinder circuit.*

TABLE 15-7 PARALLEL CYLINDER PERFORMANCE DATA*

| *Flow, gpm* | *Pressure supply,† psi* | *Load conditions* | | *Extending cylinder travel speed, ft/min* | | *Retracting cylinder travel speed, ft/min* | |
|---|---|---|---|---|---|---|---|
| | | *Cyl. A* | *Cyl. B* | *Cyl. A* | *Cyl. B* | *Cyl. A* | *Cyl. B* |
| 2 | | No load | No load | | | | |
| 5 | | No load | No load | | | | |
| 2 | | Loaded | No load | | | | |
| 5 | | Loaded | No load | | | | |
| 2 | | No load | Loaded | | | | |
| 5 | | No load | Loaded | | | | |
| 2 | 300 | Loaded | Loaded | | | | |
| 5 | 300 | Loaded | Loaded | | | | |

* Two 2-in.-dia. cylinders; 1⅛-in.-dia. rods.
† Pressures before bottoming.

**7.** At 500 psi what force was cylinder *A* capable of exerting while (*a*) extending? _______; (*b*) retracting? _______

**8.** What was force capacity of cylinder *B* at 500 psi (*a*) extending? _______; (*b*) retracting? _______

**9.** What was extending speed of travel for cylinder *A* at 5 gpm? _______

**10.** What was retracting speed of travel for cylinder *A* at 5 gpm? _______

**11.** What was travel speed of cylinder *B* at 5 gpm extending? _______

**12.** What was travel speed of cylinder *B* at 5 gpm retracting? _______

**13.** Would this be a good circuit if it were immaterial which cylinder completed its stroke first? _______ Why?

________________________________________

________________________________________

________________________________________

**14.** Would this be a good circuit if both cylinders were of the same diameter, had an equal load, and were expected to complete their cycles simultaneously? _______ Why?

________________________________________

________________________________________

________________________________________

________________________________________

________________________________________

## 6 • PARALLEL CIRCUIT OF TWO HYDRAULIC MOTORS

PURPOSE: To show the parallel circuit of two hydraulic motors.

COMPONENTS: Two hydraulic motors, a four-way valve, two Prony brakes, and two pressure gauges.

PROCEDURE

**1.** Connect pressure supply 1 to the inlet of the four-way valve and the outlet to the metered return line.

**2.** Using a tee at each control port on the four-way valve, connect the inlet ports of the motor to one tee and the return lines to the other tee.

**3.** Operate the hydraulic motors as indicated in Tables 15-8 and 15-9 and complete the information. Induce load by use of Prony brakes.

**4.** Fill in the following motor data.

 

NAME .............................................. DATE .............. CLASS ...........

| | Motor 1 | Motor 2 |
|---|---|---|
| (*a*) Manufacturer, type | | |
| (*b*) Model | | |
| (*c*) Displacement (cu in. at 500 psi) | | |
| (*d*) Torque (ft-lb at 500 psi) | | |
| (*e*) Rpm at 500 psi and 6 gpm | | |

PROBLEMS

**5.** Will two motors with the same displacement in a parallel circuit operate at equal rpm's if they are under no load condition? ______ Why?

______________________________

______________________________

______________________________

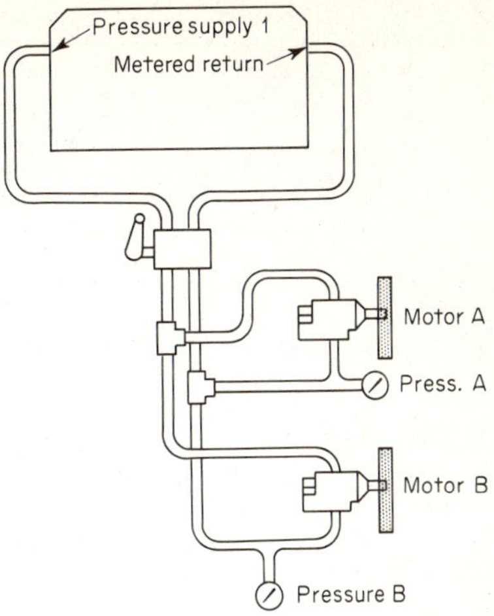


FIG. 15-17 *Two-motor parallel circuit.*

**6.** Will hydraulic fluid take the path of least resistance in this circuit? ______

TABLE 15-8 MOTOR A: PARALLEL MOTOR CIRCUIT PERFORMANCE*

| *Condition number* | *Pressure supply, psi* | *Total delivered gpm* | *"A" pressure, psi*† | *Input, hp* | *rpm* | *Torque, ft-lb* | *Output, hp* | *Motor A, gpm* |
|---|---|---|---|---|---|---|---|---|
| 1 | | 2 | | | | 0 | | |
| 2 | | 4 | | | | 0 | | |
| 3 | | 6 | | | | 0 | | |
| 4 | | 2 | 200 | | | | | |
| 5 | | 2 | 400 | | | | | |
| 6 | | 2 | 600 | | | | | |
| 7 | | 4 | 200 | | | | | |
| 8 | | 4 | 400 | | | | | |
| 9 | | 4 | 600 | | | | | |
| 10 | | 6 | 200 | | | | | |
| 11 | | 6 | 400 | | | | | |
| 12 | | 6 | 600 | | | | | |
| 13 | | | 600 | | | | | |
| 14 | | | 600 | | | | | |
| 15 | | | 600 | | | | | |

* 1.07 cu in. displacement motor.

† Connect circuits so that pressure gauges (A and B) are connected to the inlet side of the motors. (Pressure is the differential pressure—supply or motor inlet pressure minus return-line pressure.)

*Note:* Conditions 13, 14, and 15 will be impossible to obtain. Fluid in a system will take the path of least resistance. The motor requiring the least pressure will receive the fluid, and its rpm will increase proportionately. That is,

$$\text{Input hp} = \frac{\text{gpm} \times \text{psi}}{1714}$$

$$\text{Output hp} = \frac{\text{ft-lb (torque)} \times \text{rpm}}{5252}$$

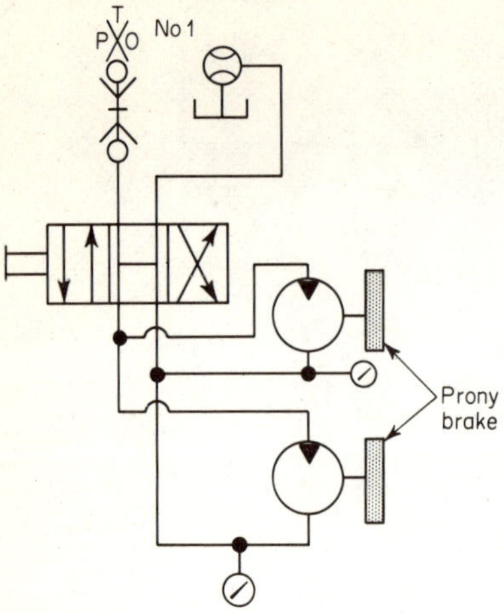


FIG. 15-18 *Two-motor parallel symbols circuit.*

**7.** What is the input hp under conditions 5 and 7 in (*a*) motor 1? ______; (*b*) motor 2? ______; (*c*) Are they the same? ______ Why?

______________________________

______________________________

______________________________

______________________________

______________________________

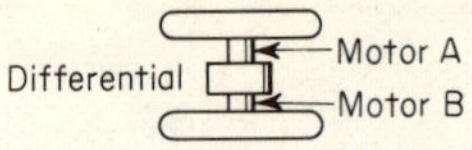


FIG. 15-19 *Two-motor connection.*

**8.** What is the total output hp by the two motors at (*a*) condition 6? ______; (*b*) condition 10? ______; (*c*) Are they the same for both motors? ______ Why?

______________________________

______________________________

______________________________

______________________________

**9.** Would Fig. 15-19 be a good circuit with (*a*) tractor drive wheels, dry conditions? ______; (*b*) tractor drive wheels, icy conditions? ______

**10.** Would a parallel circuit to the two wheels in Fig. 15-20 be a good, fair, or poor circuit? ______ Why?

______________________________

______________________________

______________________________

______________________________

TABLE 15-9 MOTOR B: PARALLEL MOTOR CIRCUIT PERFORMANCE*

| *Condition number* | *Pressure supply, psi* | *Delivered gpm* | *"B" pressure, psi* | *Input, hp* | *rpm* | *Torque, ft-lb* | *Output, hp* | *Motor B, gpm* |
|---|---|---|---|---|---|---|---|---|
| 1 | | 2 | | | | 0 | | |
| 2 | | 4 | | | | 0 | | |
| 3 | | 6 | | | | 0 | | |
| 4 | | 2 | 200 | | | | | |
| 5 | | 2 | 400 | | | | | |
| 6 | | 2 | 600 | | | | | |
| 7 | | 4 | 200 | | | | | |
| 8 | | 4 | 400 | | | | | |
| 9 | | 4 | 600 | | | | | |
| 10 | | 6 | 200 | | | | | |
| 11 | | 6 | 400 | | | | | |
| 12 | | 6 | 600 | | | | | |
| 13 | | | 550 | | | | | |
| 14 | | | 500 | | | | | |
| 15 | | | 300 | | | | | |

* 1.07 cu in. displacement motor.

 

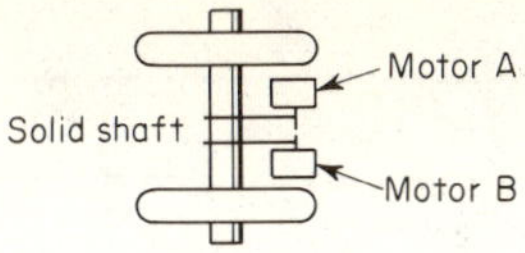


FIG. 15-20 *Two-motor connection.*

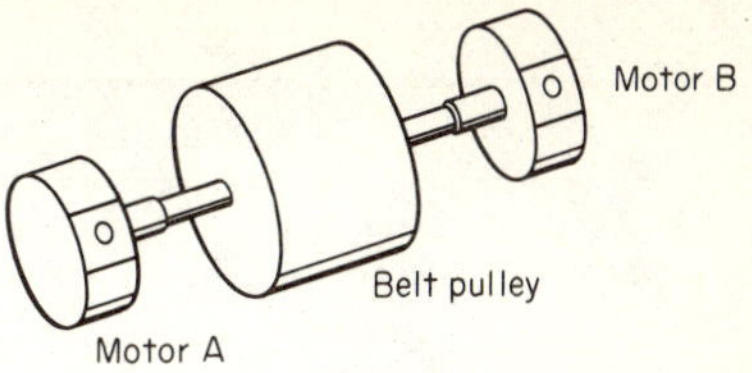


FIG. 15-21 *Two-motor to pulley connection.*

**11.** For a head pulley on a conveyor (Fig. 15-21) would a parallel circuit be good, fair, or poor? ____________ Why?

______________________________

______________________________

______________________________

______________________________

______________________________

**12.** What would happen to pulley speed if these motors, with the same gpm, were changed to a series connection?

______________________________

**13.** What would happen to the pressure required?

______________________________

______________________________

## 7 • TWO HYDRAULIC CYLINDERS IN SERIES

PURPOSE: To show the operation of two hydraulic cylinders connected in a series system.

COMPONENTS: Two hydraulic cylinders, a four-way valve, three needle valves, and four pressure gauges.

PROCEDURE (PART I)

**1.** Connect pressure supply to the inlet port of the four-way valve.

**2.** Connect the outlet of the four-way valve to the metered return line.

**3.** Connect one flow control port on the four-way valve to the port on the blank end of hydraulic cylinder 1.

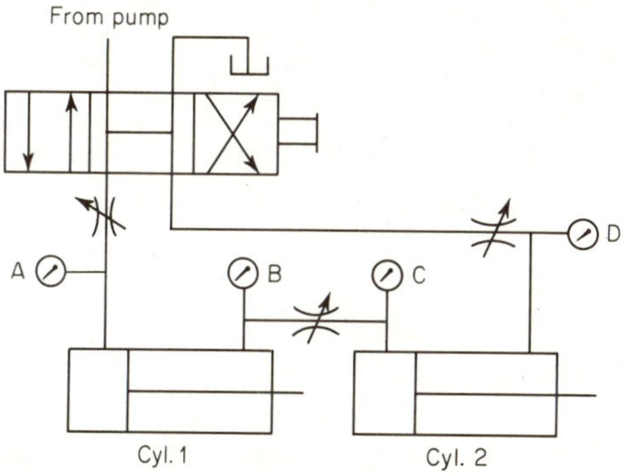


FIG. 15-22 *Symbols for two-cylinder series connection.*

**4.** Connect the rod end port of cylinder 1 to the blank end of cylinder 2.

**5.** Connect the rod end port of cylinder 2 to the second control port of the four-way valve.

**6.** Use needle valves to induce loads. Operate the circuit and complete missing data on Table 15-10. Before completing the table, determine for each cylinder the pressure required to move a 500- and 1100-lb load. Answer questions 7, 8, 9, and 10.

PROBLEMS (PART I)

**7.** (*a*) What is the effective area of cylinder 1 while extending? ________ sq in.
(*b*) Rod diameter? ________ in.
(*c*) Rod area? ________ sq in.
(*d*) Effective area retracting? ________

**8.** (*a*) What is effective area of cylinder 2 extending? ________ sq in.
(*b*) Rod diameter? ________ in.
(*c*) Rod area? ________ sq in.
(*d*) Effective area retracting? ________

**9.** (*a*) What pressure is required for cylinder 1 to move a load which requires a 500-lb force while extending? ________ psi.
(*b*) While retracting? ________ psi.
(*c*) 1000 lb extending? ________ psi.
(*d*) 1000 lb retracting? ________ psi.

TABLE 15-10*

| | CYLINDER 1 | | | | CYLINDER 2 | | | |
|---|---|---|---|---|---|---|---|---|
| *gpm* | *Pressure gauge A* | *Pressure gauge B* | *Load, lb* | *Direction of travel* | *Pressure gauge C* | *Pressure gauge D* | *Load, lb* | *Direction of travel* |
| 4 | | | 1000 | Extending | | | 0 | |
| 4 | | | 0 | | | | 1000 | Extending |
| 4 | | | 1000 | Retracting | | | 0 | |
| 4 | | | 0 | | | | 1000 | Retracting |
| 4 | | | 500 | Extending | | | 500 | |
| 4 | | | 500 | | | | 500 | Retracting |

* 2-in. cylinders with 1⅛-in.-dia. rods.

**10.** (*a*) What pressure is required for cylinder 2 to move a load which requires a 500-lb force while extending? _______ psi.

(*b*) While retracting? _______ psi.

(*c*) 1000 lb extending? _______ psi.

(*d*) 1000 lb retracting? _______ psi.

**11.** At 4 gpm what is the speed of travel, extending, for (*a*) cylinder 1? _______ ft/min; (*b*) cylinder 2? _______ ft/min. (*c*) Are they the same? _______ Why?

_______________________________________

_______________________________________

_______________________________________

**12.** Assume the following:

(*a*) Both cylinders identical

(*b*) 10 sq-in. area on piston (blank end)

(*c*) 5 sq-in. rod area

(*d*) 5 sq-in. effective piston area (rod end)

(*e*) Flow 500 cu in./min = 500/231 or 2.12 gpm

What would be the extending speed of cylinder 1? _______ in./min.

**13.** What would be the extending speed of cylinder 2? _______ in./min.

**14.** What would be the retracting speed of cylinder 1? _______ in./min.

**15.** What would be the retracting speed of cylinder 2? _______ in./min.

**16.** If cylinder 1 had a load of 1000 lb while extending (actual load, not pressure induced by a needle valve), and cylinder 2 had no load (Fig. 15-22), what would pressure be at: *A*? _______ psi; *B*? _______ psi; *C*? _______ psi; *D*? _______ psi.

**17.** If cylinder 2 had a load of 1000 lb while extending (cylinder 1 no load), what would pressure be at: *A*? _______ psi; *B*? _______ psi; *C*? _______ psi; *D*? _______ psi.

**18.** If cylinder 1 had a load of 1000 lb while retracting (cylinder 2 no load), what would pressure be at: *A*? _______ psi; *B*? _______ psi; *C*? _______ psi; *D*? _______ psi.

**19.** If cylinder 2 had a load of 1000 lb while retracting (cylinder 1 no load), what would pressure be at: *A*? _______ psi; *B*? _______ psi; *C*? _______ psi; *D*? _______ psi.

**20.** If both cylinders had a load of 500 lb while extending, what would the pressures be at: *A*? _______ psi; *B*? _______ psi; *C*? _______ psi; *D*? _______ psi.

**21.** If both cylinders had a load of 500 lb while retracting, what would the pressure be at: *A*? _______ psi; *B*? _______ psi; *C*? _______ psi; *D*? _______ psi.

**22.** While extending cylinder 1, would cylinder 2 extend fully? _______

**23.** While retracting, which cylinder would make (*a*) a full stroke? _______; (*b*) less than a full stroke? _______

CONCLUSION: While these cylinders are extending and cylinder 2 is loaded, this circuit is a simple form of pressure intensification. Note from your answers that cylinder 2 is moving a 1000-lb load with a pressure on the blank side of 100 psi. The pressure from the pump, or on the blank side of cylinder 1, is 50 psi, or one-half. While the pressure has doubled, note that the speed of travel on cylinder 2 is one-half that of 1.

 

Thus with a pump delivering pressure of 50 psi we are moving a load which otherwise would take 100 psi. Also from your answers, you can see that this is *not* a pressure intensifier while the cylinders are retracting (load on cylinder 1, no load on 2). Pressure required to retract 1 is 1000 lb divided by 5 sq in., or 200 psi (Gauges *B* and *C*). The pressure on Gauge *D* (incoming) is 400 psi. While in this direction, the incoming pressure is doubled. Note, however, that the speed of both cylinders is greater; cylinder 1 has four times the speed it had while extending, and cylinder 2 has twice the speed it had while extending. Notice the correlation of pressure and speed of travel in the two directions of travel.

PROCEDURE (PART II) (shown on Fig. 15-23)

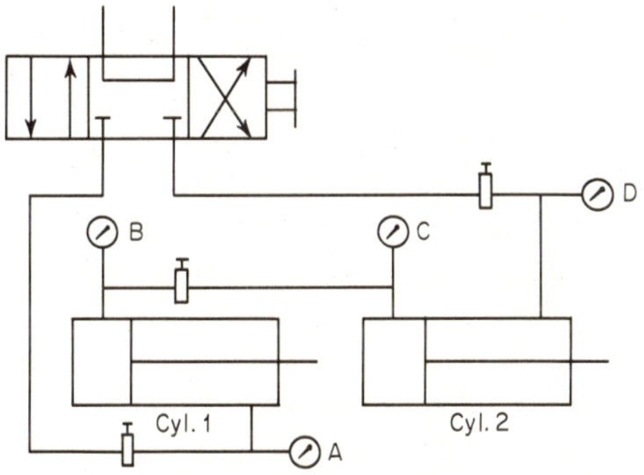


FIG. 15-23 *Symbols for two-cylinder series connection.*

**24.** Connect control port on four-way valve to rod end of cylinder 1 and its blank end to the blank end of cylinder 2.

**25.** Connect the rod end of cylinder 2 to the second control port on the four-way valve.

**26.** Operate system and complete missing information on Table 15-11.

PROBLEMS (PART II)

**27.** When cylinder 1 is extending, what is direction of travel of 2?

______________________________

**28.** When cylinder 2 is extending, what is direction of travel of 1?

______________________________

**29.** At 4 gpm what would be the extending speed of cylinder 1? ________ in./min.

**30.** At 4 gpm what would be the extending speed of cylinder 2? ________ in./min.

**31.** At 4 gpm what would be the retracting speed of cylinder 1? ________ in./min.

**32.** At 4 gpm what would be the retracting speed of cylinder 2? ________ in./min.

**33.** Would two cylinders connected in series remain synchronized under normal operating conditions? ________ Why?

______________________________

______________________________

______________________________

______________________________

______________________________

TABLE 15-11*

| | CYLINDER 1 | | | | CYLINDER 2 | | | |
|---|---|---|---|---|---|---|---|---|
| *gpm* | *Pressure gauge A* | *Pressure gauge B* | *Load, lb* | *Direction of travel* | *Pressure gauge C* | *Pressure gauge D* | *Load, lb* | *Direction of travel* |
| 4 | | | 1000 | Extending | | | 0 | |
| 4 | | | 0 | | | | 1000 | Extending |
| 4 | | | 1000 | Retracting | | | 0 | |
| 4 | | | 0 | | | | 1000 | Retracting |
| 4 | | | 500 | Extending | | | 500 | |
| 4 | | | 500 | | | | 500 | Retracting |

* 2 in. x 24 in. hydraulic cylinder.

**34.** What valve could be installed to keep cylinders synchronized (in opposite directions) so they both would bottom simultaneously in the event they would become unsynchronized? Explain your answer.

**35.** Using the previous circuit, assume the following data. Both cylinders identical; 10 sq-in. area on piston (blank end); 5 sq-in. rod area; 5 sq-in. effective piston area (rod end); flow 500 cu in./min = $500/231$ or 2.12 gpm (1 gpm – 231 cu in.). When cylinder 1 is extending, what is the direction of travel of 2?

**36.** When cylinder 2 is extending, what is the direction of travel of 1?

**37.** What would be the extending speed of cylinder 1? ______ in./min.

**38.** What would be the extending speed of cylinder 2? ______ in./min.

**39.** What would be the retracting speed of cylinder 1? ______ in./min.

**40.** What would be the retracting speed of cylinder 2? ______ in./min.

## 8 • TWO HYDRAULIC MOTORS IN SERIES

PURPOSE: To show the operation of two hydraulic motors connected in series.

COMPONENTS: Two hydraulic motors, two Prony brakes, and a pressure gauge.

PROCEDURE

**1.** Connect pressure supply 1 to the inlet port of motor 1.

**2.** From the outlet port of motor 1 connect to the inlet port of motor 2 with a pressure gauge.

**3.** Connect outlet port of motor 2 to the metered return line.

**4.** Fill in specifications for motor data sheet.

| | *Motor 1* | *Motor 2* |
|---|---|---|
| (*a*) Manufacturer, type | | |
| (*b*) Model | | |
| (*c*) Displacement, cu in. at 500 psi | | |
| (*d*) Max. pressure (mfg. recommended) | | |
| (*e*) Max. rpm (mfg. recommended) | | |
| (*f*) Bi or unidirectional | | |
| (*g*) Series installation permitted? | | |

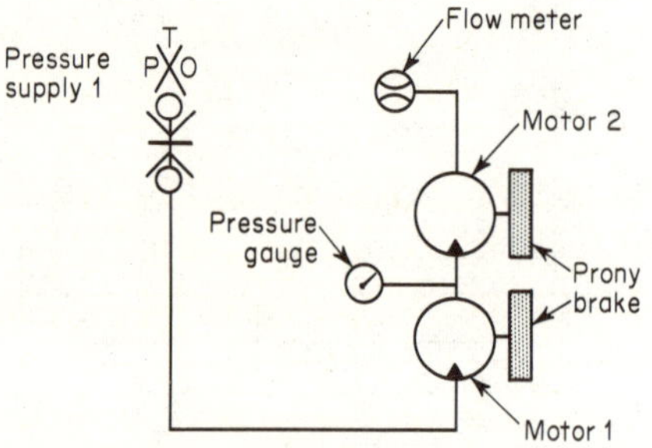


FIG. 15-24 *Two hydraulic motor series connections.*

**5.** Complete Table 15-12.

Note that only motor 1 should be loaded *unless* manufacturer's specifications state that motor 1 is capable of sustaining back pressures caused by loading motor 2. (Some motors are constructed by design with drains to facilitate back pressure which will be induced in motor 1 whenever 2 is loaded.)

PROBLEMS

**6.** Is the rotation of motor 2 dependent on the rotation of motor 1? ______

**7.** If the relief valve were set at 1000 psi and motor 1 required 1100 psi, would (*a*) motor 1 rotate? ______; (*b*) motor 2? ______

 

TABLE 15-12

| MOTOR 1 | | | | | MOTOR 2 | | | | |
|---|---|---|---|---|---|---|---|---|---|
| *Supply pressure, psi* | *gpm* | *Torque, ft-lb* | *rpm* | *Delivered hp* | *Supply pressure, psi* | *gpm* | *Torque, ft-lb* | *rpm* | *Delivered hp* |
| (*a*) 800 | 4 | | | | | 4 | 0 | | |
| (*b*) 400 | 4 | 0 | | | | 4 | | | |
| (*c*) 800 | 6 | | | | 800 | | 0 | | |
| (*d*) | 6 | 0 | | | 800 | 6 | | | |
| (*e*) 800 | 4 | | | | 400 | | | | |
| (*f*) 800 | 5 | | | | 400 | | | | |

**8.** If the relief valve were set at 1000 psi and motor 2 required 1100 psi to operate, would (*a*) motor 2 rotate? ______ ; (*b*) motor 1? ______

**9.** (*a*) Is the combined torque of the two motors essentially the same (at 800-psi supply pressure to 1 and 400 psi to 2) as is obtained from one motor operation at 800-psi supply pressure? ______

(*b*) Is there any difference? ______ Why?

______________________________________

______________________________________

______________________________________

**10.** Figure 15-25 is a similar circuit, without actual analysis on the power unit. Using the information in Table 15-13, complete its missing data.

*Note:* Assume that the relief valve is set at 1100 psi (cracking pressure). Disregard line losses, motor friction, and slippage (Table 15-13).

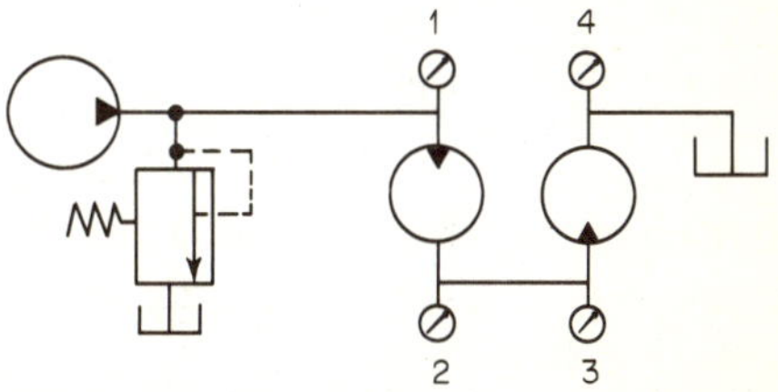


FIG. 15-25 *Two-motor series system.*

**11.** Two motors are connected in parallel, and each operates at 500 rpm with 25 gpm to each motor and a work load that requires 500 psi. Assume that these motors are changed to a series connection and are required to operate at 500 rpm and produce the same torque.

(*a*) How many gpm will be required? ______

(*b*) What pressure will be required at the inlet to motor 1? ______

(*c*) What pressure will be required at the inlet to motor 2? ______

TABLE 15-13

| MOTOR 1 | | | | | MOTOR 2 | | | | |
|---|---|---|---|---|---|---|---|---|---|
| *Gauge 1 pressure* | *Gauge 2 pressure* | *Motor gpm* | *rpm* | *Torque, ft-lb* | *Gauge 3 pressure* | *Gauge 4 pressure* | *Motor gpm* | *rpm* | *Torque, ft-lb* |
| 1000 | 500 | 4 | 500 | 100 | 500 | 0 | 4 | 500 | 100 |
| 1000 | | | | | 0 | | | | |
| 500 | 0 | 4 | | | | | | | |
| 750 | 750 | 2 | | | 750 | 0 | 2 | | |
| 500 | 250 | 2 | | | 250 | 0 | | | |

## 9 • DIFFERENTIAL CIRCUIT TO HYDRAULIC CYLINDER

PURPOSE: To show the differential circuit to a hydraulic cylinder.

COMPONENTS: Three-position four-way valve, hydraulic cylinder, needle valve, and pressure gauge.

PROCEDURE

**1.** Symbols to the left of pressure supply 1 are components of the power unit circuit. Label each with the appropriate letter.

- A Electric motor
- B Strainer
- C Pump
- D Relief valve
- E Tank

**2.** Connect pressure supply 1 to the inlet port on the four-way valve with a tee prior to the inlet port.

**3.** Connect the outlet to the metered return line.

**4.** To the tee connect a line to the rod end of the hydraulic cylinder.

**5.** Plug one control port on the four-way valve as shown on Fig. 15-26.

**6.** Connect the other control port of the valve to the blank end of the cylinder.

**7.** If a cylinder loading device is not available, install needle valve as shown in Fig. 15-26.

**8.** Determine the following cylinder data:

(*a*) Piston diameter ______

(*b*) Rod diameter ______

(*c*) Net area rod end ______

(*d*) Stroke ______ in.

**9.** Operate the cylinder in both directions and record data on Table 15-14.

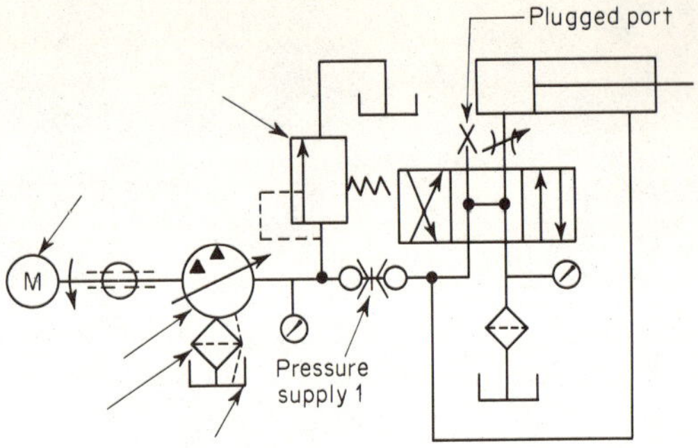


FIG. 15-26 *Differential circuit to hydraulic cylinder.*

RESULTS: The extending speed is increased by the additional fluid from the rod end. (The blank end received the pump flow plus the fluid from the rod end, since it had no other place to go.) While the rod is retracting, the pump flow is directed to the rod end. (The fluid from the blank end is returned to the tank.) While the rod is extending, the force it is capable of exerting is reduced because the pressure is equal on both sides. (The maximum force it can exert is the pressure times the differential area between the blank end and the rod end at the cylinder.) This will increase the speed of travel but reduce the maximum force that the cylinder is capable of exerting.

PROBLEMS

**10.** What is the area of the blank end of the cylinder used? ______ sq in.

**11.** What is the area of the rod of the cylinder used? ______ sq in.

**12.** What is the net effective area in the rod end? ______ sq in.

TABLE 15-14 PERFORMANCE RESULTS

| *gpm* | *Extending travel speed, ft/min* | *Extending pressure, psi* | *Extending force, lb* | *Retracting travel speed, ft/min* | *Retracting pressure, psi* | *Retracting force, lb* |
|---|---|---|---|---|---|---|
| 2 | | 500 | | | 500 | |
| 2 | | 1000 | | | 1000 | |
| 4 | | 500 | | | 500 | |
| 4 | | 1000 | | | 1000 | |

 

**13.** What is the differential area between blank and rod end? _______ sq in.

**14.** At 500-psi pressure what force is the cylinder capable of exerting while retracting? _______

**15.** At 500 psi what force is the cylinder capable of exerting while extending? _______

**16.** What was extending speed of travel at 4 gpm? _______ ft/min.

**17.** What was the retracting speed of travel at 4 gpm? _______ ft/min.

**18.** (*a*) If this cylinder had 4 gpm delivered to its blank end without additional fluid from the rod end, what would be its speed of travel? _______ (*b*) What would be its speed if 2 gpm were delivered? _______

**19.** What percent increase in extending speed was experienced as a result of this differential circuit? _______

**20.** Identify the components or explain the meaning of the symbols used in Fig. 15-27.

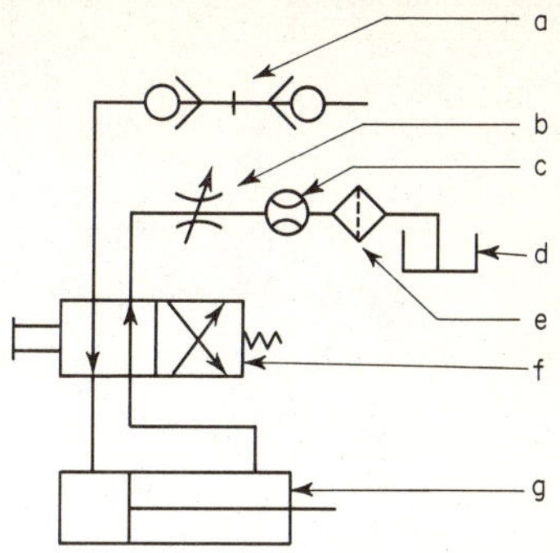


FIG. 15-27 *Cylinder, four-way valve circuit.*

(*a*) ______________________

(*b*) ______________________

(*c*) ______________________

(*d*) ______________________

(*e*) ______________________

(*f*) ______________________

(*g*) ______________________

## 10 • PRESSURE-COMPENSATED FLOW CONTROL

PURPOSE: To show the speed control of a hydraulic cylinder using a pressure compensated flow control.

COMPONENTS: Three-position four-way valve, hydraulic cylinder, and the pressure-compensated flow control on the basic unit.

PROCEDURE

1. On Fig. 15-28 label:

   A Pressure supply 1 quick disconnect
   B Pressure supply gauge
   C Three-position four-way valve
   D Hydraulic cylinder
   E Metered return quick disconnect
   F Flowmeter
   G Filter
   H Tank

2. Connect pressure supply 1 to the inlet of the four-way valve and the outlet of the valve to the metered return line.

3. Connect the control ports on the four-way valve to the rod and blank end of the hydraulic cylinder.

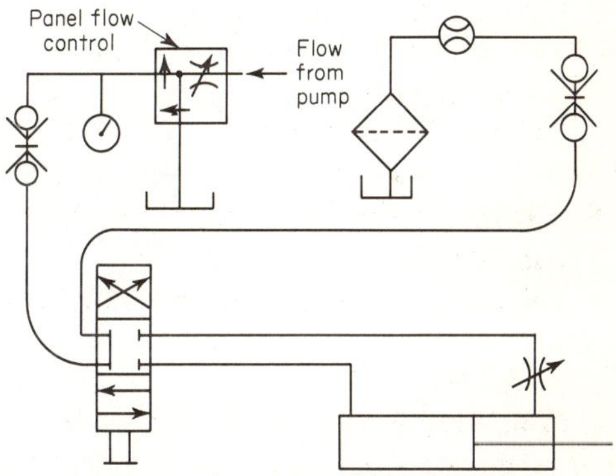


FIG. 15-28 *Cylinder and flow control circuit.*

4. Operate the cylinder at the values shown on Table 15-15 and load the cylinder if possible. If cylinder loading device is not available, induce the load by using the needle valve as shown in Fig. 15-28.

5. Set pump flow control to maximum gpm; control rate of flow by using panel flow control.

TABLE 15-15

| gpm | *Extending pressure, psi* | *Extending speed, ft/min* | *Force, lb* | *Retracting pressure, psi* | *Retracting speed, ft/min* | *Force, lb* |
|---|---|---|---|---|---|---|
| 2 | 400 | | | 400 | | |
| 2 | 600 | | | 600 | | |
| 4 | 400 | | | 400 | | |
| 4 | 600 | | | 600 | | |
| 5 | 400 | | | 400 | | |
| 5 | 600 | | | 600 | | |

**6.** Fill in the following cylinder data:
(*a*) Blank end diameter ________ in.
(*b*) Blank end area ________ sq in.
(*c*) Rod diameter ________ in.
(*d*) Rod area ________ sq in.
(*e*) Net area rod end ________ sq in.
(*f*) Stroke ________ in.

**7.** Fill in the performance data on Table 15-15. Control the gpm with the panel flow control. Control the pressure by loading the cylinder or by use of a needle valve.

PROBLEMS

**8.** When 2 gpm are used in the circuit, what gpm is the pump delivering to the panel flow control valve? ________

**9.** At what pressure is the total flow delivered to the panel flow control? ________

**10.** What happens to the energy in the balance of flow which is delivered to the flow control valve?

________________________

**11.** (*a*) Do changes in pressure required at the cylinder affect the flow rate? ________
(*b*) If so, what is percent of change at 4 gpm and 200 and 800 psi? ________

**12.** Does the pump deliver the same amount of energy to the flow control valve regardless of the gpm's delivered to the cylinder (assuming the pressure to be constant)? ________

## 11 • PANEL PRESSURE-COMPENSATED FLOW CONTROL VALVE

PURPOSE: To show speed control of the hydraulic motor using the panel pressure-compensated flow control valve.

COMPONENTS: Hydraulic motor and Prony brake.

PROCEDURE

**1.** Connect pressure supply 1 to the pressure port of the hydraulic motor. Connect the pump outlet to the metered return line.

**2.** Attach the Prony brake to the motor.

**3.** Determine the following motor specifications:
(*a*) Manufacturer and type ________
(*b*) Model ________
(*c*) Displacement 500 psi ________
(*d*) Maximum pressure ________
(*e*) Maximum rpm ________
maximum gpm ________
(*f*) Unidirectional or bidirectional ________

**4.** Operate the circuit as shown in the tables, and record the results on Table 15-16.

PROBLEMS

**5.** What was the rpm at 2 gpm and 300 psi? ________

**6.** What was the rpm at 2 gpm and 600 psi? ________

**7.** What was the rpm at 4 gpm and 600 psi? ________

**8.** What was the torque at 2 gpm and 300 psi? ________

**9.** What was the torque at 4 gpm and 300 psi? ________

**10.** What was the torque at 6 gpm and 300 psi? ________

**11.** If the torque stayed the same and the rpm doubled, would the output hp double? ________

 

TABLE 15-16

| *gpm* | *rpm* | *Supply pressure, psi* | *Torque, in.-lb* | *Torque, ft-lb* | *Output, hp* | *Input, hp* | *Return-line pressure* |
|---|---|---|---|---|---|---|---|
| (*a*) 2 | | 300 | | | | | |
| (*b*) 4 | | 300 | | | | | |
| (*c*) 6 | | 300 | | | | | |
| (*d*) 2 | | 600 | | | | | |
| (*e*) 4 | | 600 | | | | | |
| (*f*) 6 | | 600 | | | | | |

Output hp formula: $\frac{\text{Torque (ft-lb)} \times \text{rpm}}{5252}$

Input hp formula: $\frac{\text{Pressure (psi)*} \times \text{gpm}}{1714}$

* Differential pressure across motor.

**12.** If the rpm stayed the same and torque doubled, would the output hp double? ________

**13.** In a hydraulic motor the rpm is related theoretically directly to gpm. Why does the rpm not exactly follow the gpm in actual tests? ________________________________

________________________________

________________________________

**14.** Does the supply pressure determine the torque in a hydraulic motor? ________

**15.** How does the supply pressure affect rpm delivered by the motor (assuming constant flow rate)? ________________________________

________________________________

________________________________

**16.** Are kilowatts used proportionate to flow when pressure remains constant? ________ Why? ________________________________

________________________________

________________________________

**17.** Since there is no change in the electric motor power output as the flow to supply 1 is changed, what happens to the energy in the fluid that is returned to the tank?

________________________________

________________________________

________________________________

________________________________

**18.** Vary the flow to the motor with the pump flow control while panel flow control is set to deliver 100 percent to pressure supply 1, and note the kilowatts required while constant pressure is maintained. Does it fluctuate directly with flow? ________ Why?

________________________________

________________________________

________________________________

________________________________

________________________________

PURPOSE: To show input and output horsepower of an electric motor.

COMPONENT: Needle valve.

PROCEDURE

**1.** Connect pressure supply 1 to the needle valve and then to the metered return line as shown on Fig. 15-29.

**2.** Set the panel flow control to maximum (100 percent) to pressure supply 1.

**3.** Fully open the needle valve.

**4.** Set rate of flow to 5 gpm using the pump flow control.

**5.** Set the relief valve "full in" (maximum pressure).

**6.** Start the unit and record on Table 15-17 the kilowatts required for 5 gpm as the pressure is increased by the use of the needle valve. Use pump flow control to maintain 5 gpm.

**7.** Repeat the exercise using 4 gpm; again use pump flow control to maintain a constant flow. Increase pressure by using needle valve, and record values on Table 15-17.

**8.** (*a*) From the information recorded on Table 15-17, plot the kilowatts and pressures on the graph (Fig. 15-30). Label the 4- and 5-gpm lines.

(*b*) Plot on the graph (Fig. 15-31) the input and output hp for 5 gpm. Use a dotted line for input hp and label each line.

*Formula:*

$$\text{Input hp} = \frac{\text{watts}}{746} \qquad 1 \text{ kilowatt} = 1000 \text{ watts}$$

$$\text{Output hp} = \frac{\text{watts}}{746} \times \text{motor efficiency}$$

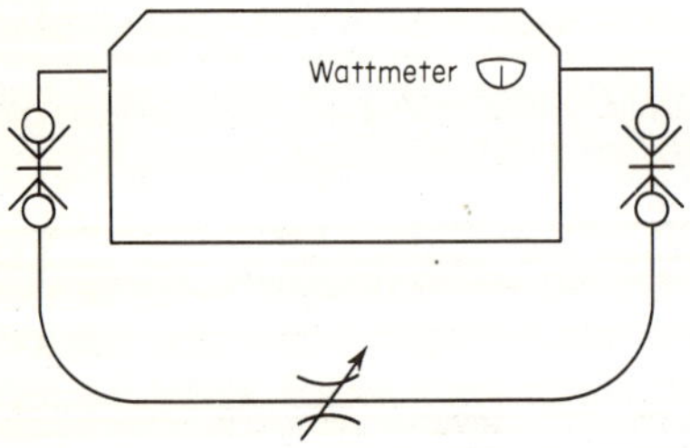


FIG. 15-29 *Circuit.*

TABLE 15-17 KILOWATTS

| FLOW | PRESSURE, PSI | | | | | | | |
|---|---|---|---|---|---|---|---|---|
| | *0* | *200* | *300* | *400* | *500* | *700* | *800* | *900* |
| 5 gpm | | | | | | | | |
| 4 gpm | | | | | | | | |

Two methods can be used to determine hp output of this electric motor: (1) Prony brake; (2) input hp multiplied by the efficiency of the motor. For the purpose of this circuit analysis, determine the output hp by using an efficiency of 80 percent.

Table 15-18 shows the approximate efficiencies for electric motors as given by the manufacturer. The power unit utilizes a 5-hp 220-volt "induction start and run" electric motor.

PROBLEMS

**9.** How many watts does the electric motor use to pump 5 gpm with the needle valve full open? ______

**10.** What is the input hp under the conditions in Prob. 9? ______

**11.** Correcting for electric motor efficiency, what is the output hp delivered by the electric motor at zero pressure?

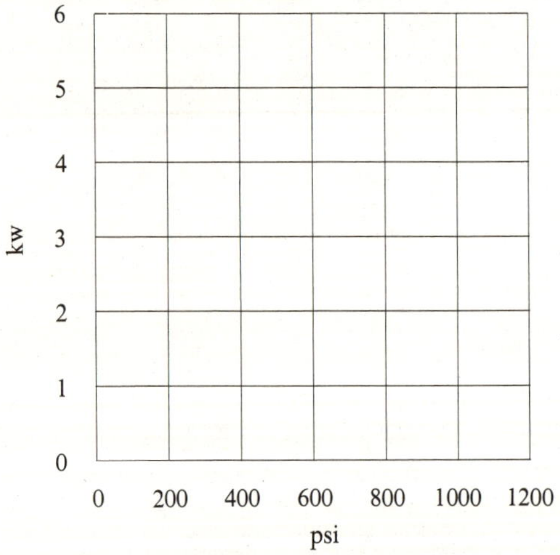


FIG. 15-30 *Demonstration graph.*

 

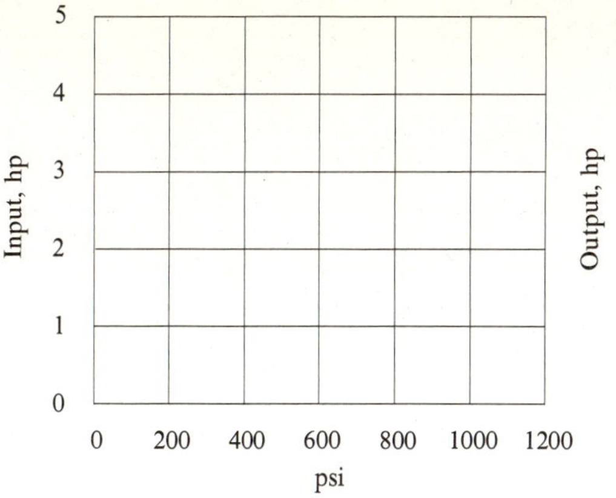


FIG. 15-31 *Demonstration graph.*

**12.** What is the electric motor input hp at 800 psi and 5 gpm? ______

**13.** What is the electric motor output hp at 800 psi and 5 gpm? ______

**14.** Would the electric motor output hp be the same as the input hp to the hydraulic pump that it drives? ______

**15.** Continue to increase the pressure, and note that the flow drops and is reduced to zero as the pump compensates for this increased pressure. Also, the watts, or power required, will be reduced. Since horsepower is the measure of work done in a specific time, it follows that as the flow is decreased, less work is done and, consequently, less horsepower is required from the electric motor even though a higher pressure is maintained. How many more watts are required to maintain maximum pressure at no-flow condition as compared with a full-flow condition at the minimum pressure (needle valve full open)?

(*a*) Maximum pressure—no flow:
______ watts; ______ input hp

(*b*) Minimum pressure—full flow:
______ watts; ______ input hp

**16.** Plot the rpm of the electric motor against the output hp on the graph (Fig. 15-32).

**17.** Figure 15-33 is the complete symbol drawing of the power unit as used in this circuit. Match the letters with the corresponding components. (See next page.)

TABLE 15-18

| MOTOR TYPE AND SIZE | PERCENT EFFICIENT | | TORQUE, FULL LOAD, FT-LB |
|---|---|---|---|
| | *½ load* | *Full load* | |
| 5 hp 3 p 220 v: run* start | 80 | 85 | 15 |
| 3 hp 3 p 220 v: run start | 79 | 83 | 9 |
| 5 hp 3 p 110 v: run start | 78 | 81 | 15 |
| 3 hp 3 p 110 v: run start | 75 | 79 | 9 |
| 1 hp 3 p 220 v: run start | 73 | 78 | 3 |
| ¼ hp 110 v: capac. | 50 | 58 | ¾ |

* Induction run and induction start.

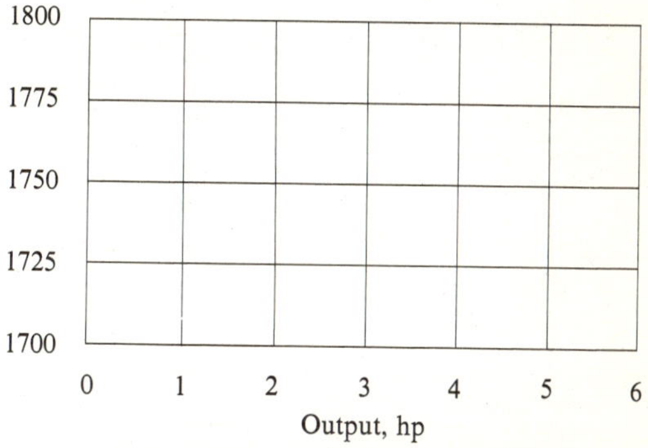


FIG. 15-32 *Demonstration graph.*

A Two-position three-way valve
B Quick disconnect
C Tank
D Flowmeter
E Strainer
F Filter
G Flow control
H Pump
I Motor (electric)
J Relief valve
K Supply pressure gauge
L Return-line pressure gauge
M Needle valve
N Supply pressure
O Return-line pressure gauge

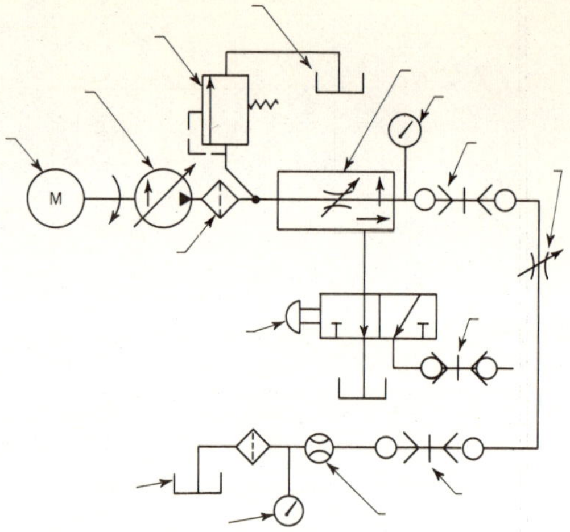


FIG. 15-33 *Power unit circuit.*

## 13 • INPUT AND OUTPUT OF HYDRAULIC PUMP

PURPOSE: To show input and output horsepower of a hydraulic pump.

COMPONENTS: Pressure-compensated vane pump on the power unit, and a needle valve.

PROCEDURE

**1.** Connect pressure supply 1 to the metered return line with the needle valve installed as shown on Fig. 15-34.

**2.** Set the pump flow control at 5 gpm, determine pump data, and record pump performance on Table 15-19.

| | *Vane* | *Piston* |
|---|---|---|
| (*a*) Manufacturer | | |
| (*b*) Model | | |
| (*c*) Maximum displacement at 500 psi | | |
| (*d*) Maximum gpm | | |

TABLE 15-19 PUMP PERFORMANCE

| *Press., psi* | *gpm* | *Watts* | *Input, hp* | *Output, hp* | *Mechanical eff, %* |
|---|---|---|---|---|---|
| 200 | 5 | | | | |
| 400 | 5 | | | | |
| 600 | 5 | | | | |
| 800 | 5 | | | | |

$$\text{Mechanical efficiency} = \frac{\text{pump output hp}}{\text{pump input hp}} \text{ or } \frac{\text{fluid hp delivered}}{\text{electric output hp}}$$

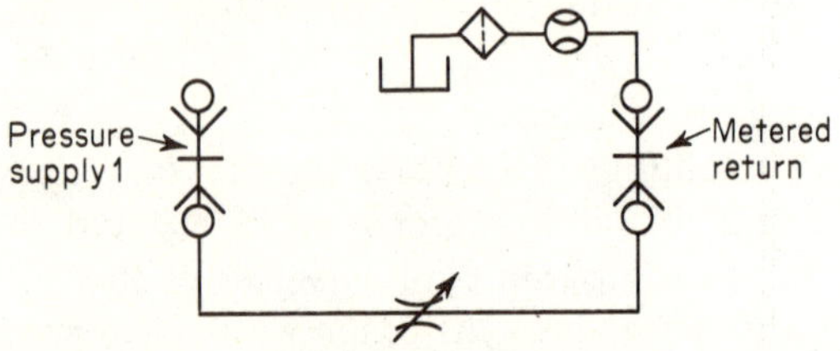


FIG. 15-34 *Circuit.*

It is known from the previous exercise that the electric motor output hp is the input hp to the hydraulic pump. Also, the hydraulic pump output hp is derived from this formula:

$$\frac{\text{gpm} \times \text{pressure (psi)}}{1714} = \text{output (hp)}$$

**3.** On the graph (Fig. 15-35) plot the input hp* and the output hp of the pump installed on the unit. Increase the pressure as required in Table 15-19. Use a dotted line for input and a solid line for output.

Hydraulic pump input hp

$$= \frac{\text{watts}}{746} \times \text{efficiency of electric motor}$$

Hydraulic pump output hp

$$= \frac{\text{gpm} \times \text{pressure (psi)}}{1714}$$

* See Table 15-18.

 

NAME .......................................... DATE .............. CLASS ...........

**Problems**

**4.** How many hp did the pump deliver at 5 gpm and 400 psi? ________

**5.** How many hp did the electric motor deliver at 5 gpm and 400 psi? ________

**6.** What is the mechanical efficiency of this pump at 5 gpm and 400 psi? ________

**7.** If the gpm delivered remains the same and the pressure is doubled, will the hp delivered double? ________

**8.** If the pressure remains the same and the gpm output is doubled, will the delivered hp be doubled? ________

**9.** What is the displacement of this pump in cubic inches per revolution and the maximum flow at 400 psi? [*formula:* displacement = (gpm × 231)/rpm] ________

**10.** Is the pump output hp that which will be the circuit input hp (with the panel flow control set at 100 percent to the circuit)? ________

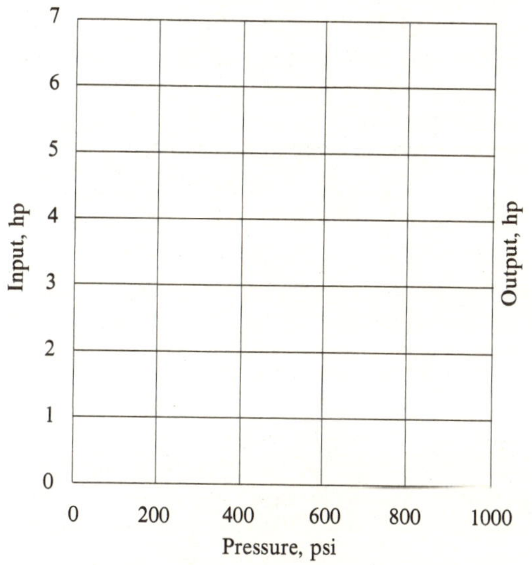


FIG. 15-35 *Demonstration graph.*

*Note:* When the panel flow control directs total flow to pressure supply 1, the entire electric output hp is used to deliver flow to supply 1. If the control is set to deliver 50 percent to pressure supply 1, the other 50 percent of the pump output hp will be wasted and returned to the tank. By reducing the flow to pressure supply 1, the watts, or power, will not change unless 100 percent of the flow is returned to the tank. (This can be demonstrated by using a needle valve to induce 800 psi and setting total flow to pressure supply 1.) Maintain 800 psi and reduce flow to 1 by using panel flow control and record the watts required.

**11.** As an illustration, complete the graph (Fig. 15-36).

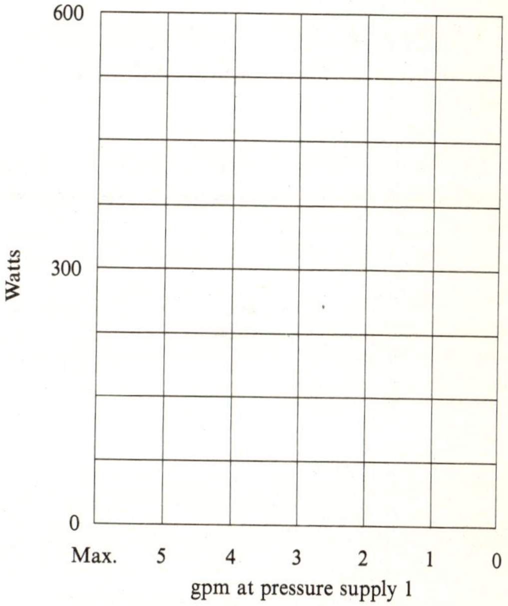


FIG. 15-36 *Demonstration graph.*

PURPOSE: To show the input and output horsepower of a hydraulic motor.

COMPONENTS: Hydraulic motor, Prony brake, and tachometer.

PROCEDURE

**1.** Connect the pressure port of the hydraulic motor to pressure supply 1 with the outlet connected to the metered return line.

**2.** Connect the Prony brake to the hydraulic motor.

**3.** Set the panel flow control to maximum flow position.

**4.** Increase the torque by using the Prony brake in such increments as to cause the pressures indicated in Table 15-20. Use the pump flow control to regulate the flow. Establish the conditions shown in Table 15-20, and complete the table.

PROBLEMS

**5.** What is the input hp at 2 gpm and 400 psi? ______

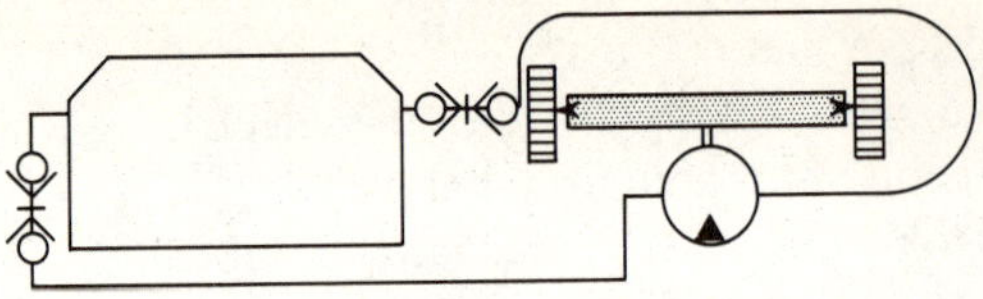

FIG. 15-37 *Circuit.*

**6.** What is the output hp at 2 gpm and 400 psi? ______

**7.** What is the motor mechanical efficiency at 2 gpm and 400 psi? ______

**8.** What is the input hp at 5 gpm and 400 psi? ______

**9.** What is the output hp at 5 gpm and 400 psi? ______

**10.** What is the motor mechanical efficiency at 5 gpm and 400 psi? ______

**11.** If this motor were driving a wheel, would the motor output hp be the input hp to the wheel? ______

VOLUMETRIC EFFICIENCY: From the information on Table 15-20, the volumetric efficiency for the motor can be determined.

TABLE 15-20 MOTOR PERFORMANCE

| *gpm* | *Supply pressure, psi* | *Torque, in.-lb* | *Torque, ft-lb* | *rpm* | *Input, hp* | *Output, hp* | *Mechanical efficiency, %* |
|---|---|---|---|---|---|---|---|
| 2 | 200 | | | | | | |
| 4 | 200 | | | | | | |
| 5 | 200 | | | | | | |
| 2 | 400 | | | | | | |
| 4 | 400 | | | | | | |
| 5 | 400 | | | | | | |
| 2 | 600 | | | | | | |
| 4 | 600 | | | | | | |
| 5 | 600 | | | | | | |
| 2 | 800 | | | | | | |
| 4 | 800 | | | | | | |
| 5 | 800 | | | | | | |

Torque will be determined by the Prony brake

$$\text{Torque (ft-lb)} = \frac{\text{torque (in.-lb)}}{12}$$

rpm is determined by a tachometer

$$\text{Motor input (hp)} = \frac{\text{pressure* (psi)} \times \text{gpm}}{1714}$$

* Pressure is the differential pressure between motor inlet and outlet as discussed in Sec. 8. Use pressure as indicated on pressure supply gauge less the return-line pressure.

 

NAME .................................................. DATE ............... CLASS ...........

The volumetric efficiency is the rpm delivered by a motor at a specific gpm and a low pressure* compared with the rpm delivered with the same gpm and a high pressure; rpm's are lost through slippage. Manufacturers will rate volumetric efficiencies from different standards. Uniform rating standards have not been established. For this exercise, find the volumetric efficiency based on rpm delivered at 0 psi at 4 gpm and 800 psi at 4 gpm.

**12.** What rpm did the motor deliver at 800 psi and 4 gpm? ______

**13.** What rpm did the motor deliver at 200 psi and 4 gpm? ______

**14.** What is the volumetric efficiency based on these values? ______ percent.

The overall efficiency $E$ of a motor or pump is the mechanical efficiency multiplied by the volumetric efficiency.

$$E_o = E_m \times E_v$$

and $\quad E_o \text{ (percent)} = E_m \times E_v \times 100$

where the subscripts $o$, $m$, and $v$ represent overall, mechanical, and volumetric respectively.

**15.** What is the percent overall efficiency of this motor based on 800 psi and 4 gpm?

______

**16.**

$$\text{Motor output (hp)} = \frac{\text{torque (ft-lb)} \times \text{rpm}}{5252}$$

$$\text{Mechanical efficiency} = \frac{\text{output (hp)}}{\text{input (hp)}}$$

* Actually it is the geometric displacement. The displacement, as determined by rpm at low pressure, will be essentially correct and can be used in lieu of the geometric. The formula is

$$\frac{\text{rpm @ 800 psi at 4 gpm}}{\text{rpm @ 0 psi at 4 gpm}} = \text{volumetric efficiency}$$

Volumetric efficiency $\times$ 100

= volumetric efficiency (percent)

**17.** Plot and label the input and output hp for 2 gpm on the graph (Fig. 15-38).

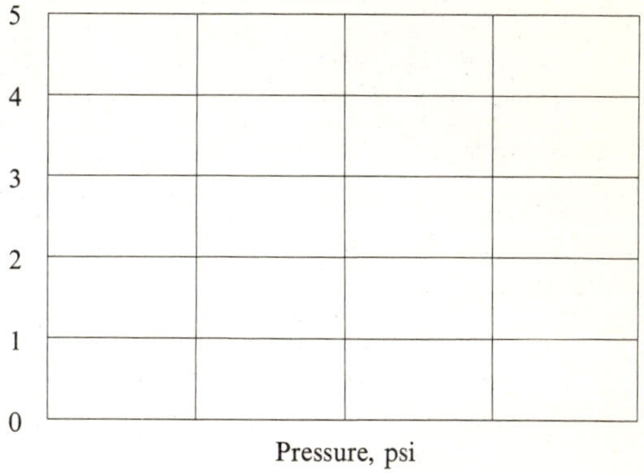


FIG. 15-38 *Demonstration graph.*

**18.** Plot and label the input and output hp for 5 gpm on the graph (Fig. 15-39).

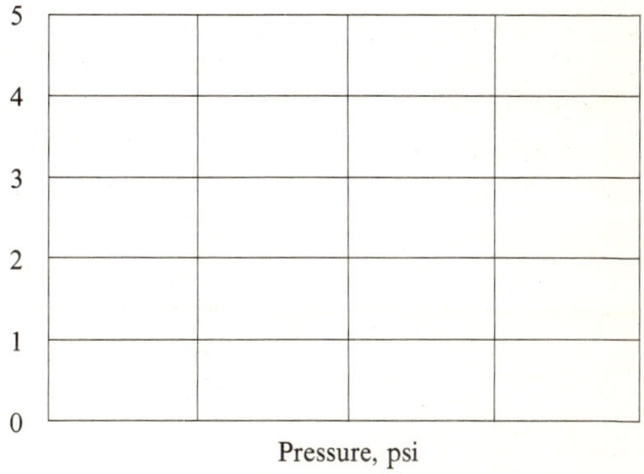


FIG. 15-39 *Demonstration graph.*

Identifying colors or solid lines and dotted lines may be used. These will help to distinguish the different lines.

**19.** Determine the following specifications:

(*a*) Manufacturer and type ______
(*b*) Model ______
(*c*) Maximum pressure ______
(*d*) Maximum gpm ______
(*e*) Maximum rpm ______
(*f*) Intake port size ______
(*g*) Tank port size ______
(*h*) Thread type ______

PURPOSE: To study and analyze a dynamometer circuit.

COMPONENTS: Power unit pump and electric motor, pump and motor installed on "table," tachometer, pressure gauge, and needle valve.

(*Note:* The pump installed in the unit will be referred to as the unit pump. The pump or motor installed on the table will be referred to as the table pump or motor. It will be necessary to alternate the table pump return line and the table motor return line in order to meter the flow from each.)

PROCEDURE

**1.** Set relief valve to maximum pressure.

**2.** Set panel flow control for full flow (position 1) to pressure supply 1.

**3.** Connect pressure supply 1 to the inlet port of the hydraulic motor, and connect the outlet port to the auxiliary return line.

**4.** Connect auxiliary supply to the inlet port on the hydraulic pump.

**5.** Install circuit tachometer to record the table pump and motor rpm.

**6.** Connect pump outlet to the metered return line with a needle valve installed.

**7.** Using pump flow control set delivery to pressure supply 1 at 4 gpm at 400 psi (use needle valve to induce the necessary load required to develop 400 psi).

**8.** Operate the unit as required to obtain the necessary information to complete Table 15-21. Use pump flow control to maintain a constant rate of flow at 4 gpm for this exercise.

**9.** Record the values for column 9 on Table 15-21 (output hp table motor). These cannot be determined from the available data; they must be found by using the procedure described in the section on the hydraulic motor. Thus the values for column 9 can be obtained from Table 15-4 (motor performance chart for 1.07 cu in. displacement).

**10.** Set panel flow control to "maximum flow" to pressure supply 1.

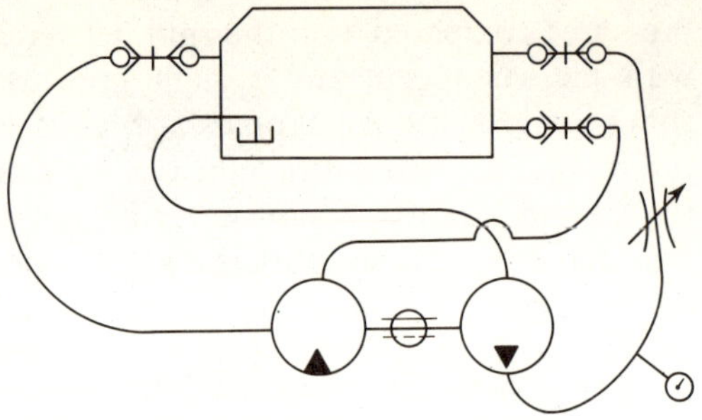

FIG. 15-40 *Dynamometer circuit.*

**11.** Plot the following on the graph (Fig. 15-41) and label.

A Input hp to electric motor
B Input hp to unit hydraulic pump
C Input hp to table hydraulic motor
D Input hp to table hydraulic pump

**12.** Plot the following on the graph (Fig. 15-42) and label.

A Output hp from electric motor
B Output hp from unit hydraulic pump
C Output hp from table hydraulic motor
D Output hp from table hydraulic pump

PROBLEMS

**13.** At 500 psi what is the input hp to the electric motor? ______

**14.** At 500 psi what is the output hp for the table pump? ______

**15.** What is the mechanical efficiency of this entire circuit at 500 psi? ______

Fluid power efficiencies when analyzed from the standpoint of input and output energies will not indicate a high percentage. This is only part of the problem in power transmission. Machine, or job, efficiency when using fluid power must be taken into account. Advantages of fluid power, such as low initial cost, weight, size, versatility, its ability to be modified or updated, and its appearance, must be considered. Often it does a job which would otherwise be impossible. When weighing these facts, it becomes apparent that fluid power is most often the best and the most efficient method of power transmission.

 

TABLE 15-21 DYNAMOMETER TEST DATA

| COLUMN 1 | 2 | 3 | 4 | 5 | 6 | 7 | 8 | 9 | 10 | 11 | 12 | 13 | 14 | 15 | 16 |
|---|---|---|---|---|---|---|---|---|---|---|---|---|---|---|---|
| *Pressure supply 1, psi* | *Input elec. motor, watts* | *Input elec. motor, hp* | *Output elec. motor, hp†* | *Pump elec. motor, rpm* | *Input unit pump, hp* | *Output unit pump, hp* | *Input table motor, hp* | *Output table motor, hp** | *Table pump and motor, rpm* | *Input table pump, hp* | *Table pump, delivered pressure* | *Table pump, delivered gpm* | *Table pump, output hp* | *Unit pump, delivered gpm* | *Return-line pressure, psi* |
| (Min.) | | | | | | | | | | | | | | 4 | |
| 200 | | | | | | | | | | | | | | 4 | |
| 300 | | | | | | | | | | | | | | 4 | |
| 400 | | | | | | | | | | | | | | 4 | |
| 500 | | | | | | | | | | | | | | 4 | |
| 600 | | | | | | | | | | | | | | 4 | |
| 700 | | | | | | | | | | | | | | 4 | |
| 800 | | | | | | | | | | | | | | 4 | |
| 900 | | | | | | | | | | | | | | 4 | |
| 1000 | | | | | | | | | | | | | | 4 | |

* See Circuit 3.
† Use motor efficiency value of 0.8 or 80 percent.

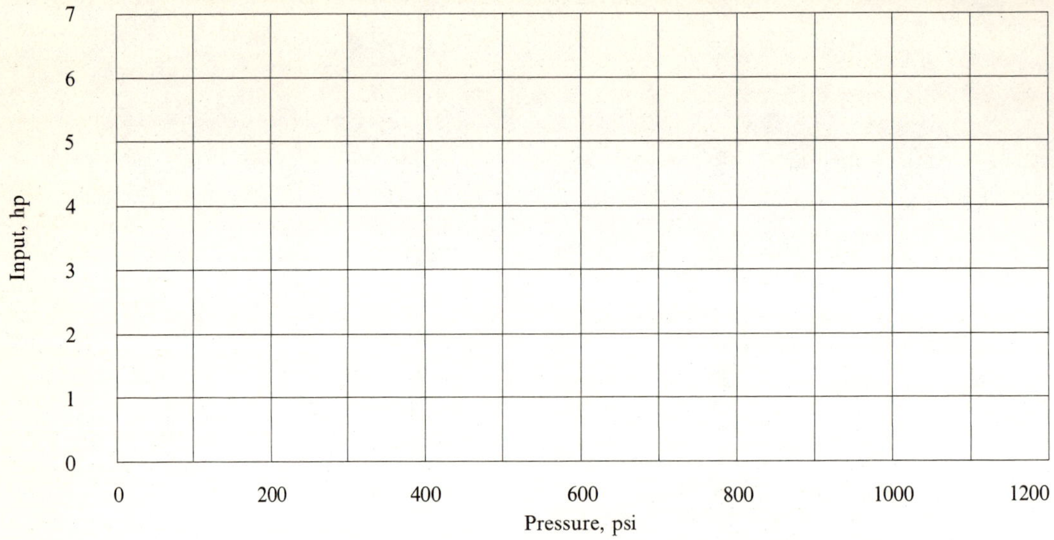


FIG. 15-41 *Demonstration graph.*

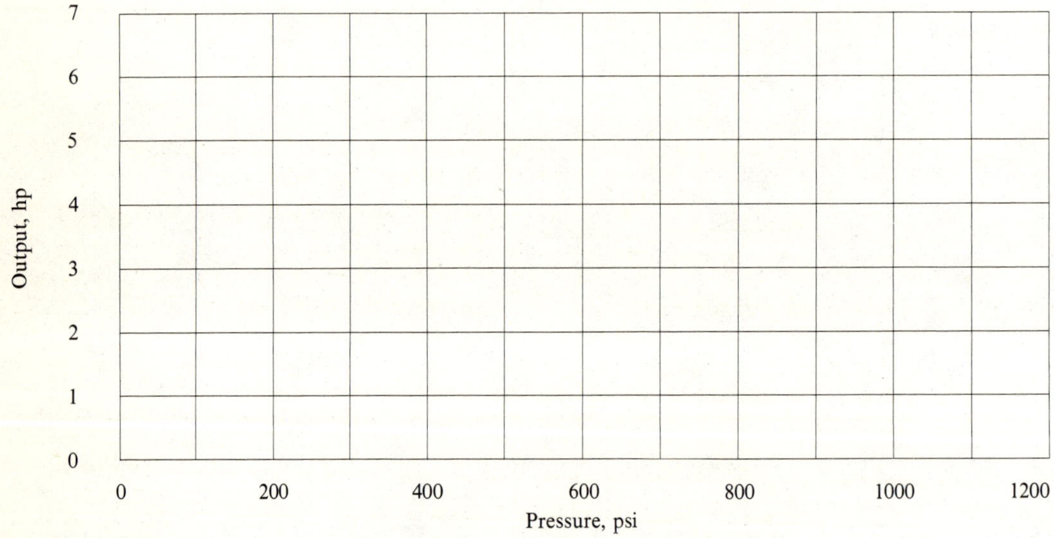


FIG. 15-42 *Demonstration graph.*

 

## 16 • DUAL-PRESSURE CIRCUIT

PURPOSE: Operation of a dual-pressure circuit with a pressure-reducing valve.

COMPONENTS: Two hydraulic motors, a pressure-reducing valve, two Prony brakes, two pressure gauges, and a tachometer.

PROCEDURE

**1.** Set panel flow to 4 gpm at pressure supply 1.

**2.** Using a tee, connect the pressure-reducing valve into the primary pressure line, which is connected to pressure supply 1 and motor 1 (which is installed in a Prony brake).

**3.** Connect the secondary (low) pressure line to motor 2, which is installed in a Prony brake. Install a pressure gauge in the line.

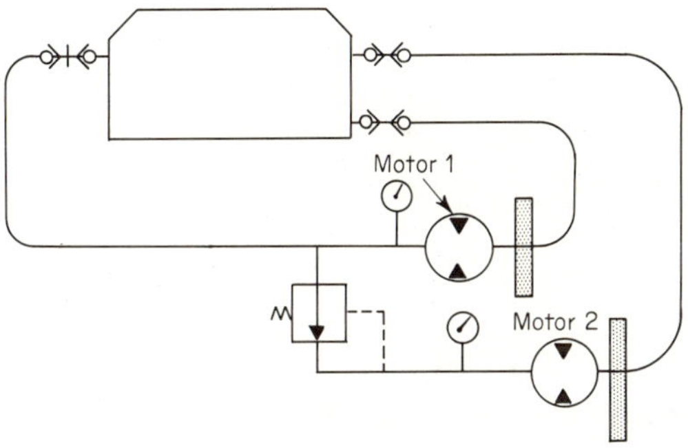


FIG. 15-43 *Dual pressure circuit.*

**4.** Preset the pressure-reducing valve to actuate at 500 psi.

**5.** Operate the motors as required to complete the data in Table 15-22.

**6.** Whenever possible, maintain equal rpm at the motors except in conditions 11 and 12. *Note:* Some of the conditions in the table cannot be met, since a pressure-reducing valve is not a flow control valve. Indicate on the table which conditions are not possible.

PROBLEMS

**7.** Which of the 10 conditions in Table 15-22 are not possible to obtain and why?

______________________________

______________________________

______________________________

______________________________

______________________________

______________________________

**8.** What was maximum hp delivered by motor 1 (maintaining equal flow and rpm at the motors)? ________

**9.** Under the above conditions, what was maximum hp delivered by motor 2? ________

TABLE 15-22 PRESSURE-REDUCING VALVE CIRCUIT

| CONDITION | FLOW, GPM | INLET PRESSURE, PSI | | TORQUE, FT-LB | | RPM | | OUTPUT, HP | |
|---|---|---|---|---|---|---|---|---|---|
| | | *Motor 1* | *Motor 2* | *Motor 1* | *Motor 2* | *Motor 1* | *Motor 2* | *Motor 1* | *Motor 2* |
| 1 | 4 | 200 | 0- | | | | | | |
| 2 | 4 | 200 | 200 | | | | | | |
| 3 | 4 | 0 | 200 | | | | | | |
| 4 | 4 | 400 | 0 | | | | | | |
| 5 | 4 | 400 | 400 | | | | | | |
| 6 | 4 | 0 | 400 | | | | | | |
| 7 | 4 | 500 | 500 | | | | | | |
| 8 | 4 | 600 | | | | | | | |
| 9 | 4 | 700 | | | | | | | |
| 10 | 4 | 800 | | | | | | | |
| 11 | 4 | 800 | | | | 0 | | | |
| 12 | 4 | 800 | | | | | 0 | | |

**10.** What was maximum hp delivered by motor 1 while receiving total flow under condition 12? ______

**11.** What was maximum hp delivered by motor 2 while receiving total flow under condition 11? ______

**12.** If this circuit used two identical cylinders (4-in.-dia. with 1-in. rod), what would be the maximum force possible with cylinders connected to the secondary (low) pressure line when (*a*) extending? ______; (*b*) retracting? ______

**13.** In Fig. 15-43 if a cylinder were installed where motor 1 is shown, would a pressure-reducing valve have any effect on the performance? ______ Why?

______

______

______

______

**14.** Name some industrial or mobile circuits where this valve might be used effectively.

______

______

______

## 17 • SEQUENCE VALVE

PURPOSE: To illustrate typical sequence-valve application.

COMPONENTS: Sequence valve, two cylinders, and a three-position four-way valve.

PROCEDURE

**1.** Connect a four-way valve to pressure supply 1 with the outlet to the metered return line.

**2.** Connect one control port to the inlet port on the sequence valve.

**3.** Connect the primary outlet port on the sequence valve to the blank end of cylinder 1 and the secondary outlet port on the sequence valve to the blank end of cylinder 2.

**4.** Connect the second control port on the four-way valve to the rod ends on cylinders 1 and 2, using needle valves and a tee to provide for a common return flow.

**5.** Install cylinder 1 in a loading device (if available) or install both cylinders in loading devices (if possible). Use a needle valve if loading device is not available.

**6.** Set sequence valve to actuate at 500 psi.

**7.** Operate circuit and complete data on Table 15-23.

PROBLEMS

**8.** At what pressure was the sequence valve set to actuate flow to its secondary line? ______

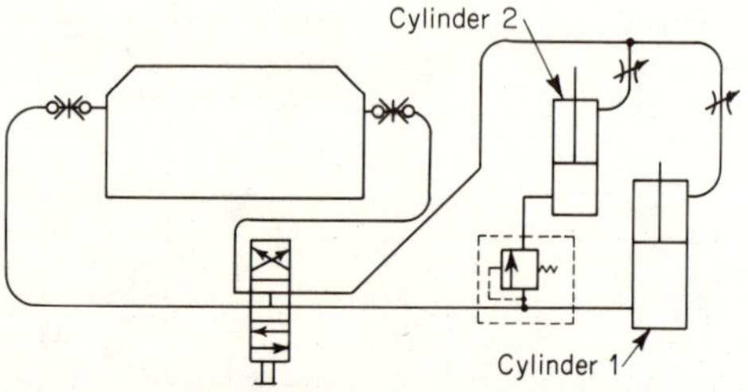

FIG. 15-44 *Sequence valve circuit.*

**9.** Does valve maintain pressure on primary line after being actuated? ______

**10.** How is the pressure loss through this valve held at a minimum?

______

______

______

______

**11.** Does the valve used in this circuit permit free reverse flow? ______ If it did not, how could this be accomplished externally?

______

______

______

______

 

TABLE 15-23 SEQUENCE CIRCUIT DATA*

| FLOW, GPM | PRESSURE SUPPLY 1, PSI | EXTENDING FORCE, LB | | RETRACTING FORCE, LB | | EXTENDING SPEED, FT/MIN | |
|---|---|---|---|---|---|---|---|
| | | *Cyl. 1* | *Cyl. 2* | *Cyl. 1* | *Cyl. 2* | *Cyl. 1* | *Cyl. 2* |
| 2 | 200 | | | | | | |
| 2 | 400 | | | | | | |
| 2 | 600 | | | | | | |
| 2 | 800 | | | | | | |
| 2 | 200 | | | | | | |
| 2 | 400 | | | | | | |
| 2 | 600 | | | | | | |
| 2 | 800 | | | | | | |

* Two 2-in.-dia. cylinders with 1⅛-in.-dia. rod.

**12.** Does the valve permit the same maximum pressure to be available on secondary cylinder 2 as is available on primary cylinder 1? ______

**13.** If the pressure required in the secondary line becomes greater than that set on the valve, will the pressure in the primary line increase correspondingly? ______

## 18 • UNLOADING CIRCUIT

PURPOSE: To analyze the use and purpose of an unloading circuit.

COMPONENTS: Unloading valve, four-way valve, hydraulic cylinder, needle valve, and pressure gauge.

PROCEDURE

**1.** Connect pressure supply 1 to inlet port of four-way valve.

**2.** Connect pressure supply 2 to inlet port on unloading valve with a pressure gauge installed.

**3.** Connect the pressure outlet on the unloading valve to the inlet line on the four-way valve using a tee and install a check valve to prevent reverse flow.

**4.** Connect a pilot line from the four-way valve inlet line to the pilot port on the unloading valve.

**5.** Connect the control ports on the four-way valve to the rod and blank end port on the hydraulic cylinder.

**6.** Mount cylinder in a loading device; if this is not available, install a needle valve to induce loads as required in Table 15-24.

**7.** Set unloading valve to actuate at 500 psi.

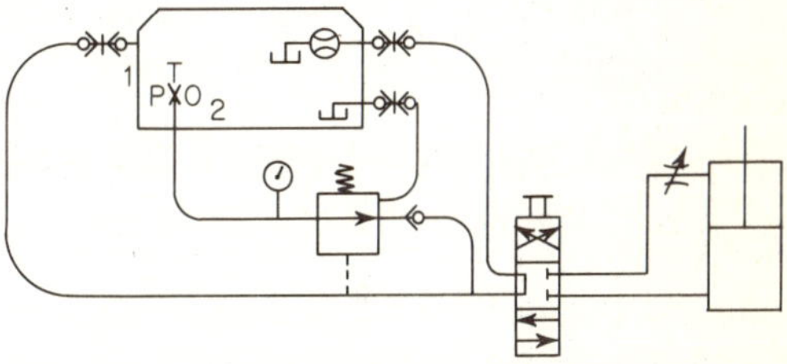


FIG. 15-45 *Unloading circuit.*

**8.** Set panel flow control to deliver 3 gpm to pressure supply 1 (balance of fluid will be available at pressure supply 2).

**9.** Operate circuit and complete data in Table 15-24.

PROBLEMS

**10.** At what pressure was the valve preset to unload? ______ psi.

**11.** Before the valve unloaded supply 2, was the pressure in supply line 1 and 2 equal while the cylinder was operating? ______

**12.** What was the pressure in the secondary line after unloading? ______

TABLE 15-24

| SUPPLY PRESSURE 1, PSI | SUPPLY PRESSURE 2, PSI | CYLINDER FORCE, LB | | FULL-FLOW EXTENDING | | UNLOADING EXTENDING | | FULL-FLOW RETRACTING | |
|---|---|---|---|---|---|---|---|---|---|
| | | *Extending* | *Retracting* | *ft/min* | *gpm* | *ft/min* | *gpm* | *ft/min* | *gpm* |
| 200 | | | | | | | | | |
| 400 | | | | | | | | | |
| 600 | | | | | | | | | |
| 800 | | | | | | | | | |

**13.** How is this valve able to unload with a minimum of pressure drop?

______

______

**14.** What happened to the cylinder speed immediately after the valve unloaded?

______

**15.** Was the pressure at the cylinder (pressure supply 2) affected immediately after the valve unloaded?

______

**16.** Was total pump flow used while cylinder was retracting? ______

**17.** If a dual pump were used for the two pressure sources, what kilowatts would be required just after the valve unloaded? (Assume that each pressure supply from the pump delivers 3 gpm and 4500 kw are used just prior to unloading.) ______

**18.** What would the electric motor input hp be (*a*) just prior to unloading? ______; (*b*) just after unloading? ______

**19.** Would the electric motor be capable of doubling the pressure where the low flow is 50 percent if it were operating at capacity just prior to unloading? ______

**20.** In what instance is this valve used in fluid power transmission?

______

______

______

______

## 19 • MOBILE CIRCUIT

PURPOSE: Analysis of mobile circuit.

COMPONENTS: Flow divider (set to divide flow on a 50–50 basis), two three-position four-way valves, two motors, and two Prony brakes.

PROCEDURE

**1.** Complete circuit as shown on Fig. 15-46. If flow divider is not available, panel flow control may be used by connecting to pressure supply 1 and 2.

*Note:* Panel flow control is not a true flow divider. It is a priority divider which provides a constant flow at supply 1 with the balance of the flow being delivered by the pump to supply 2. For example, if pump is delivering 6 gpm to the flow control valve

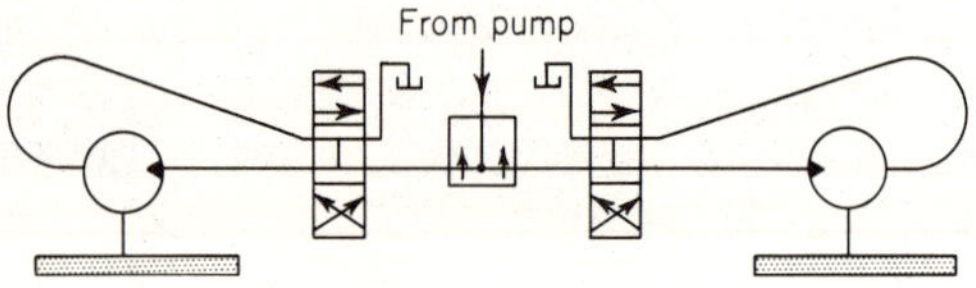


FIG. 15-46 *Mobil circuit.*

which is set to deliver 3 gpm to pressure supply 1, the balance (or 3 gpm) will be available at pressure supply 2. As pressure is increased and pump delivery is reduced to 5 gpm, supply 1 will still have 3 gpm available, whereas supply 2 will provide only 2 gpm. In a 50–50 flow divider no such priority exists. As the flow is reduced, each port would provide 50 percent of the total flow, which in the above case would be initially 3 gpm and finally 2½ gpm. A panel flow control controls the flow at the priority port, whereas the flow divider divides the total flow proportionately to its setting (percentages such as 50–50, 80–20, 60–40).

**2.** Operate as required to complete Table 15-25.

PROBLEMS

**3.** If a flow divider were used, would rpm remain constant at each motor regardless of pressure variation required at the motors? ______

**4.** Name other applications where a flow divider might effectively be used.

______

______

______

**5.** (*a*) If panel flow control were used, which motor would receive the primary flow? ______ (*b*) What is the disadvantage of this flow control in a circuit where equal speed is required of each actuator under varying pressure conditions?

______

______

______

______

______

**6.** Assume that a circuit, as shown in Fig. 15-46, is operated with components having the specifications listed below:

Motor: Gasoline, 20 hp, 2000 rpm max
Pump: Displacement 1 cu in., 2000 rpm max, 3000 psi max
Flow divider: Preset 50–50
Three-position four-way valve: Manually operated
Motors: Displacement 5 cu in./revolution

Also assume 90-percent volumetric efficiency.

(*a*) What would the pump deliver in gpm at 1000 rpm? ______

(*b*) At 1500 rpm? ______

(*c*) At 2000 rpm? ______

(*d*) What would motor rpm be while pump was being driven at 1000 rpm? ______

(*e*) At 1500 rpm? ______

(*f*) At 2000 rpm? ______

**7.** List some mobile applications where this type of circuit might be or is used:

______

______

______

______

______

TABLE 15-25

| TOTAL FLOW, GPM | PRESSURE, PSI | | GPM | | TORQUE, FT-LB | | HP | | RPM | |
|---|---|---|---|---|---|---|---|---|---|---|
| | *Motor 1* | *Motor 2* | *Motor 1* | *Motor 2* | *Motor 1* | *Motor 2* | *Motor 1* | *Motor 2* | *Motor 1* | *Motor 2* |
| 6 | 400 | 400 | 3 | 3 | | | | | | |
| 6 | 600 | 300 | 3 | | | | | | | |
| 6 | 600 | 200 | 3 | | | | | | | |
| 6 | 200 | 600 | | | | | | | | |
| 3 | | | | | | | | | | |

PURPOSE: To analyze pressure drop in components and fluid passages.

COMPONENTS: ¼-in., ½-in., and ¾-in. I.D. hose, four-way valve, sequence valve, and unloading valve.

PROCEDURE

**1.** Connect pressure differential gauge to each end of hose or on each side of the components listed above and plot a line showing the pressure drop on the graphs (Fig. 15-47).

**2.** The values for differential pressures are omitted on the graphs. Insert estimated values, which will serve the purpose, for the length of hose or individual component used.

PROBLEMS

**3.** Describe laminar flow

____________________

____________________

____________________

____________________

**4.** Describe turbulent flow

____________________

____________________

____________________

____________________

HYDRAULIC-HOSE PRESSURE DROP

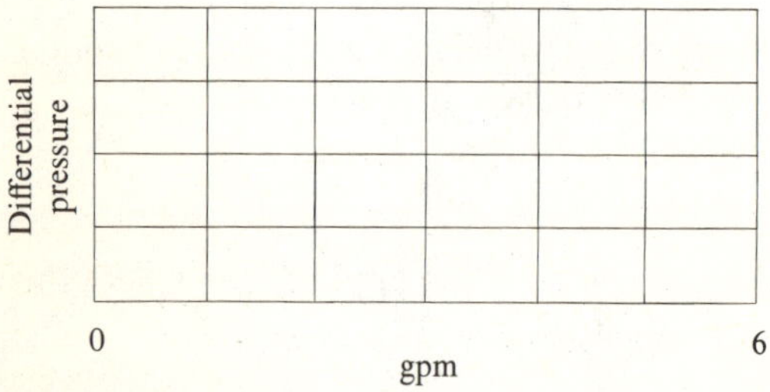


HOSE DATA

Hose length ____________________

Supply pressure: 200 psi, dashed line – – –.
Supply pressure: 600 psi, solid line ——.
Label hose diameter on graph.

SEQUENCE-VALVE PRESSURE DROP

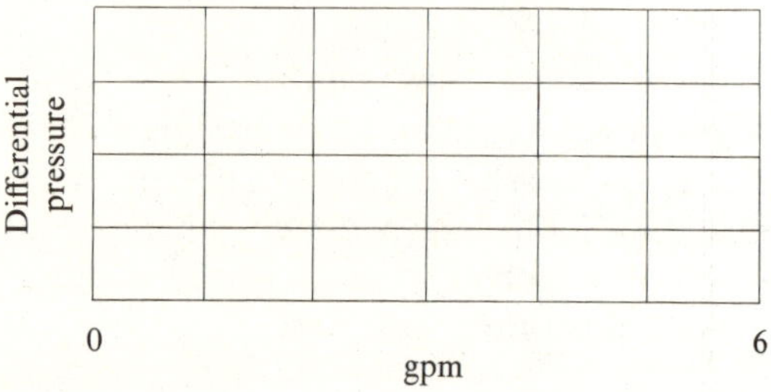


VALVE DATA

Mfr. ____________________

Model ____________________

Supply pressure: 200 psi, dashed line – – –.
Supply pressure: 600 psi, solid line ——.

FOUR-WAY-VALVE PRESSURE DROP

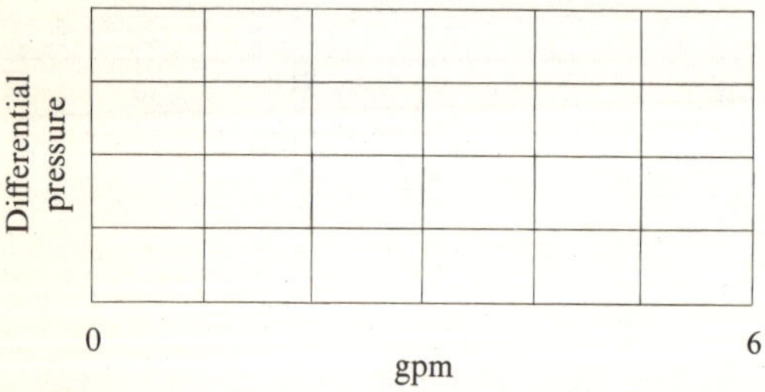


VALVE DATA

Mfr. ____________________

Model ____________________

Supply pressure: 200 psi, dashed line – – –.
Supply pressure: 600 psi, solid line ——.

UNLOADING-VALVE PRESSURE DROP

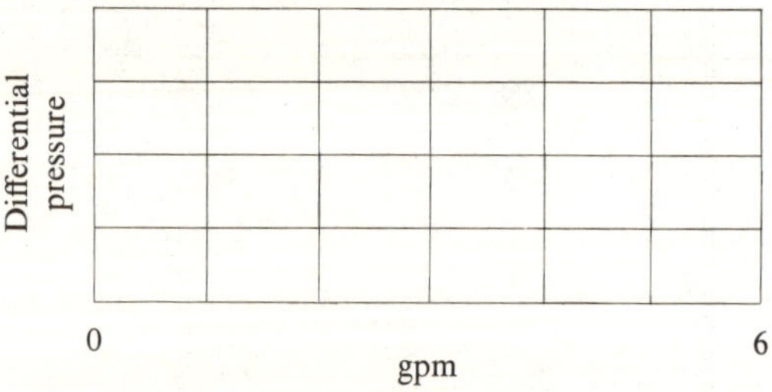


VALVE DATA

Mfr. ____________________

Model ____________________

Valve operating 400 psi, dashed line – – –.
Valve unloading, solid line ——.

FIG. 15-47

 

NAME .......................................... DATE .............. CLASS ...........

**5.** In laminar flow (2-gpm ½-in. hose), does pressure drop double when flow is doubled?

______________

**6.** In turbulent flow (3-gpm ¼-in. hose) what is the relationship between pressure drop and flow when flow is doubled?

______________________________________

______________________________________

**7.** What is the general relationship between "pressure drop" and "flow" through a valve at a constant pressure as the rate of flow is increased? ______________________________

______________________________________

______________________________________

______________________________________

**8.** What effect does increased pressure in a hose or hydraulic line have on the pressure drop (assuming constant volume)?

______________________________________

______________________________________

**9.** In a system why should pressure drop be kept at a minimum (except at the actuator)?

______________________________________

______________________________________

**10.** Since pressure drop is a loss of energy, what happens to the energy that is represented by the pressure drop (except actuators)?

______________________________________

______________________________________

# REVIEW QUESTIONS

# 16

**1.** Draw flow arrows for the basic circuit while the cylinder is retracting (Fig. 16-1).

**2.** Label the basic components on Fig. 16-1.

**3.** What is meant by pressure balance in (*a*) a hydraulic pump?

_______________

_______________

_______________

_______________

(*b*) a control valve?

_______________

_______________

_______________

**4.** A balanced vane pump is operating at 2000 psi and has 1 sq in. total (½ sq in. each side) effective vane area. What is the torque delivered in inch-pounds if the pump rotor has a 3-in. diameter? _______

**5.** How many gpm will a pump deliver which has a geometric displacement of 1.5 cu in. and whose volumetric efficiency is found to be 80 percent while the rpm are 2000? _______

**6.** What is the purpose of a sequence valve?

_______________

_______________

_______________

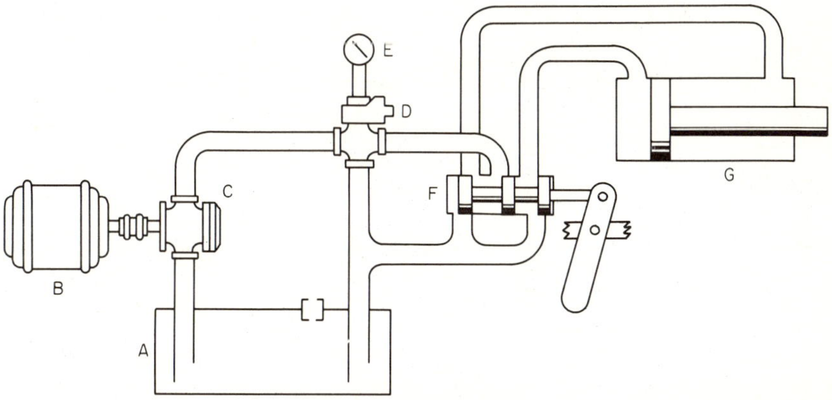


FIG. 16-1 *Basic circuit.*

**7.** Name uses for a counterbalance valve that is remotely operated and externally drained with a free reverse flow feature.

________________________________

________________________________

________________________________

________________________________

**8.** What are the advantages of a pilot-operated relief valve?

________________________________

________________________________

________________________________

________________________________

**9.** (*a*) How many inch-pounds of torque will be developed in Fig. 16-2? _______

(*b*) How many foot-pounds? _______

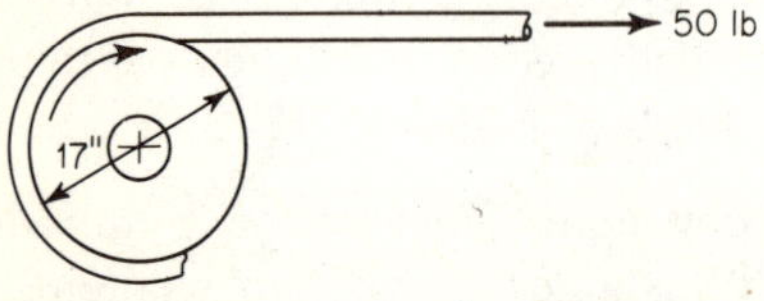


FIG. 16-2 *Torque development.*

**10.** What force is required at *X* (Fig. 16-3) to develop 30,000 ft-lb of torque? _______

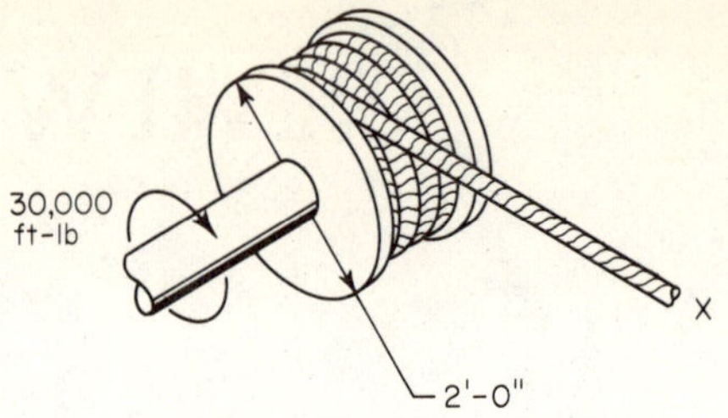


FIG. 16-3 *Force-torque relationship.*

**11.** (*a*) What will torque be at *X* (Fig. 16-4) under the given conditions? _______

(*b*) What will rpm be? _______

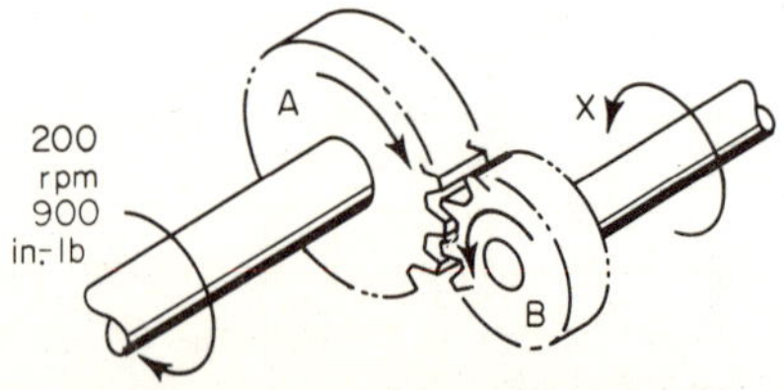


FIG. 16-4 *Torque-rpm relationship.*

**12.** Describe center conditions in Figs. 16-5 to 16-8. (Solid lines are pressure lines.)

________________________________

________________________________

________________________________

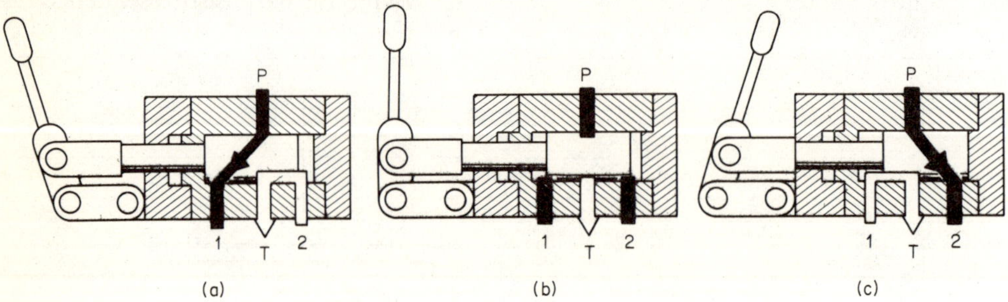


FIG. 16-5 *Three-position four-way valve.*

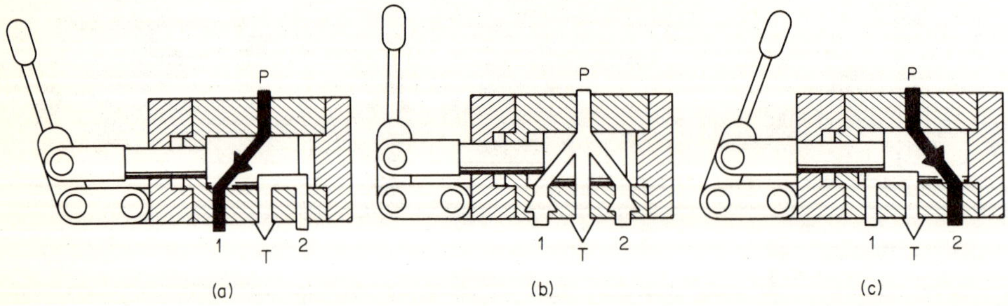


FIG. 16-6 *Three-position four-way valve.*

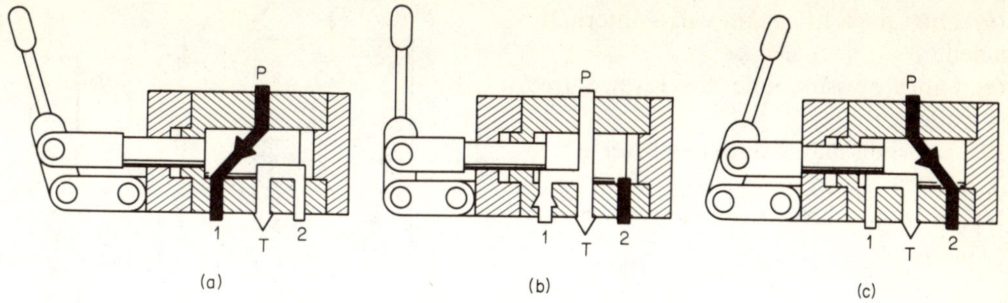


FIG. 16-7 *Three-position four-way valve.*

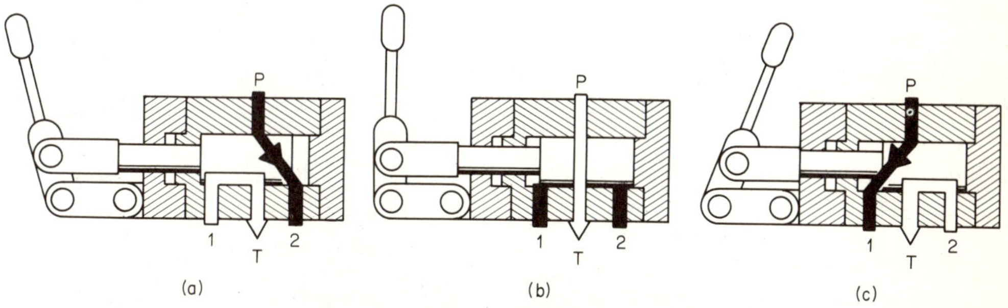


FIG. 16-8 *Three-position four-way valve.*

**13.** (*a*) Draw flow arrows for each sequence valve (Fig. 16-9) indicating the direction of flow in each of the two conditions.

(*b*) Is valve internally or pilot operated?

---

(*c*) Is valve internally or externally drained?

---

**14.** (*a*) Draw flow arrows (Fig. 16-10) for the two conditions of the valve operation.

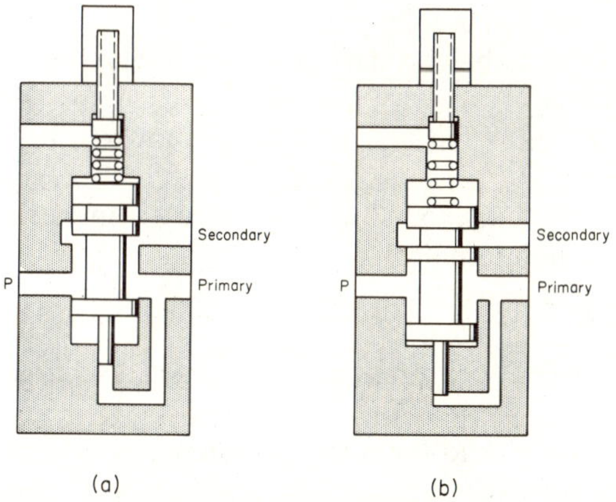


FIG. 16-9 *Sequence valve.*

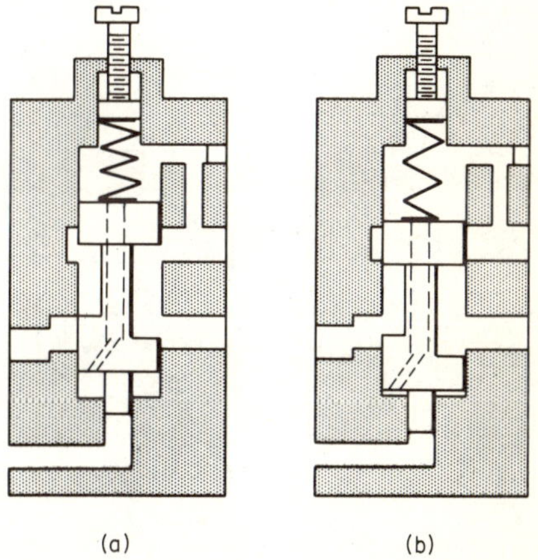


FIG. 16-10 *Sequence valve.*

(*b*) Label pilot line, pressure port, tank port, and work area port.

(*c*) Plug port to make valve internally drained.

(*d*) Why is internal spool drain necessary?

---

---

**15.** (*a*) Plug port to make valve directly operated (Fig. 16-11).

(*b*) Plug port to make valve internally drained.

(*c*) Label pressure inlet or reverse free flow outlet.

(*d*) Label discharge outlet or reverse free flow inlet.

(*e*) Draw flow arrows for the two operating conditions.

**16.** (*a*) Label external drain line (Fig. 16-12).

(*b*) Show flow arrows.

(*c*) Why is drain line necessary?

______________________

**17.** (*a*) What type of valve is shown in Fig. 16-13? ____________

(*b*) Is it externally or internally drained? ____________

(*c*) Is it pilot or directly operated? ____________

(*d*) Label secondary pressure port.

(*e*) Draw flow arrows where not shown.

**18.** (*a*) What type of valve is shown in Fig. 16-14? ____________

(*b*) Is it internally or externally drained? ____________

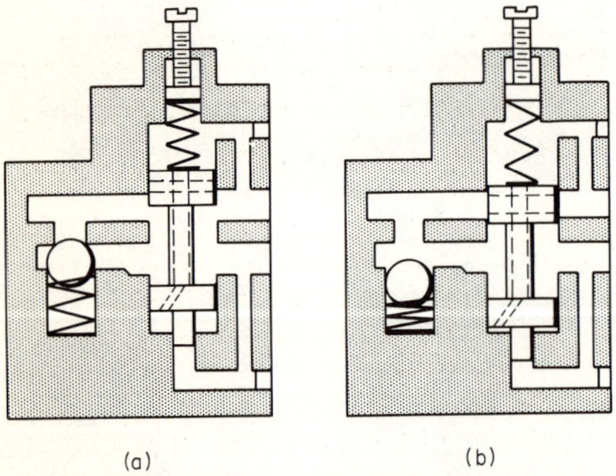

FIG. 16-11 *Free reverse-flow sequence valve.*

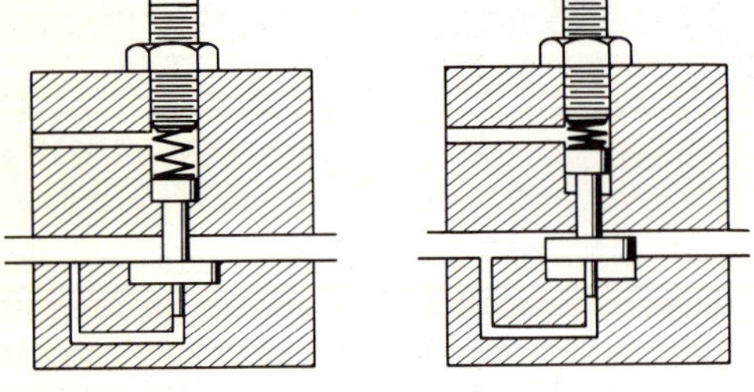

FIG. 16-12 *Pressure-reducing valve.*

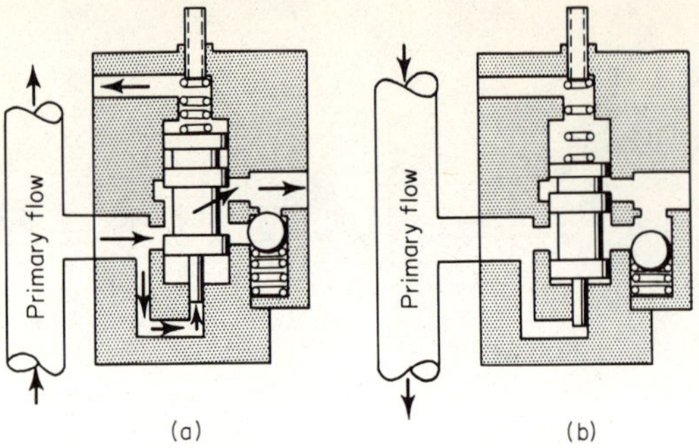

FIG. 16-13 *Free reverse-flow sequence valve.*

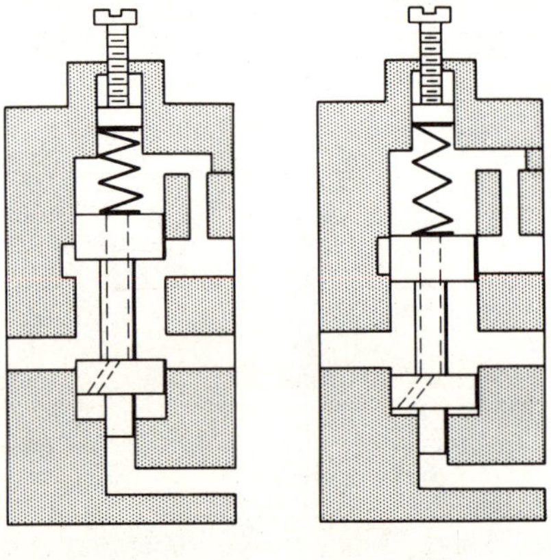

FIG. 16-14 *Counterbalance valve.*

(*c*) Label pressure inlet and discharge outlet ports.

(*d*) Label preset adjustment screw.

(*e*) Draw flow arrows for the two conditions shown.

**19.** (*a*) What type of valve is shown in Fig. 16-15? ____________

(*b*) Is it directly or remotely operated? ____________

(*c*) Label inlet pressure port.

(*d*) Label primary and secondary outlet ports.

**20.** (*a*) What type of valve is shown in Fig. 16-16? ____________

(*b*) Can you plug the valve to make it drain internally? ____________

(*c*) Must this valve be pilot operated? ____________

(*d*) Label pressure port and discharge port.

(*e*) Could there be two pressure ports on this valve? ________

(*f*) Draw flow arrows.

 

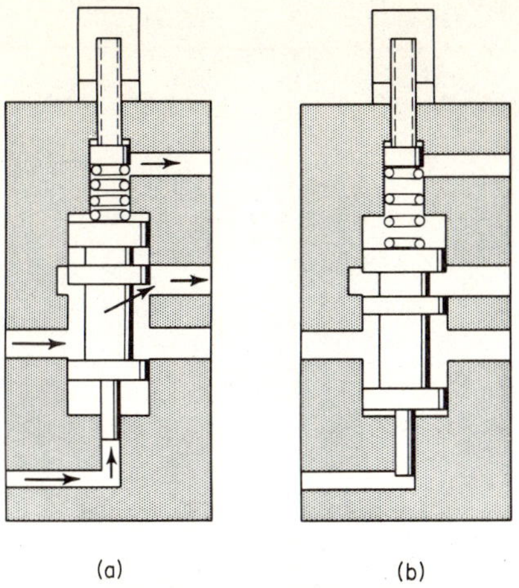


FIG. 16-15 *Sequence valve.*

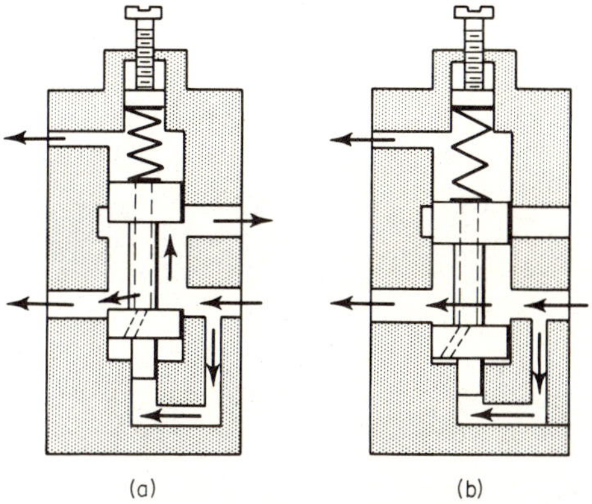


FIG. 16-16 *Sequence valve.*

**21.** List the components in Fig. 16-17.

A ______________________________

B ______________________________

C ______________________________

D ______________________________

E ______________________________

F ______________________________

G ______________________________

H ______________________________

I ______________________________

J ______________________________

K ______________________________

L ______________________________

Describe the circuit:

______________________________

______________________________

______________________________

______________________________

**22.** Give the cause, source, and corrections of the troubles listed in the accompanying table.

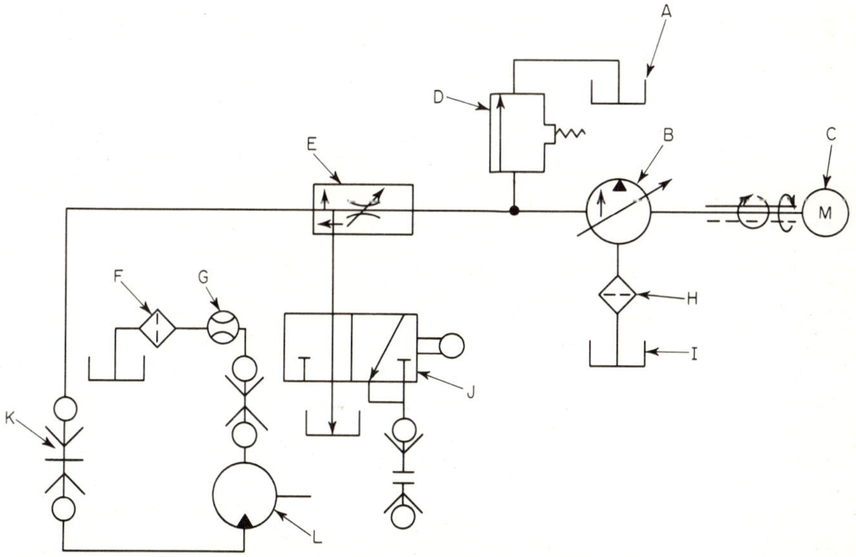


FIG. 16-17 *Basic motor circuit.*

| TROUBLE | CAUSE | SOURCE OF TROUBLE | CORRECTION |
|---|---|---|---|
| (*a*) Milk-colored hydraulic oil | | | |
| (*b*) Spongy actuator movements | | | |
| (*c*) Unusual pump noise | | | |
| (*d*) No flow from pump | | | |
| (*e*) Pump delivering flow but no pressure | | | |
| (*f*) Relief valve continuously open to tank | | | |

 

NAME ........................................ DATE .............. CLASS ...........

| TROUBLE | CAUSE | SOURCE OF TROUBLE | CORRECTION |
|---|---|---|---|
| (*g*) Sticky manual, four-way valve | | | |
| (*h*) Malfunctioning sequence valve | | | |
| (*i*) Malfunctioning pressure-reducing valve | | | |
| (*j*) Malfunctioning counterbalance valve | | | |
| (*k*) Cylinder will not hold load | | | |
| (*l*) Motor will not deliver torque | | | |
| (*m*) Motor will not deliver rpm | | | |

# POWER UNIT

# 17

## BASIC COMPONENTS

Figure 17-1 shows the panel layout of the instruments and the power-unit components. Following are a list of components and a brief description and general specifications.

### Reservoir

1. Tank size: 4.5 cu ft
2. Tank capacity: 33.66 gal
3. Oil capacity to the ¾ level of the sight gauge: 3.3 cu ft (29.0 gal)
4. Oil level gauge; sight type; high-low indication
5. Filler cap; combination strainer; breather type

### Hydraulic Suction Filter

1. Manufacturer: Gresen Model FB-215
2. Port sizes: 1¼ in. to 11½ in. NPTF
3. Type of filter element: reusable 100 wire mesh cloth
4. By-pass: internal spring loaded
5. Condition indicator: 0 to 30-in. Hg vacuum gauge on panel

This filter may be used as a return-line filter. If it is to be so used, a change must be made in the by-pass spring. Suction spring requires 5 lb to permit by-pass while return line spring requires 15 lb.

### Electric System

1. Motor type: induction
2. Manufacturer: Model 215; 5-hp, 3-phase, 220-volts; 1750 rpm (or as specified)
3. Starter: Allen-Bradley Model 709 BAA; maximum amps—27; maximum volts—230
4. On-off switch: Allen-Bradley Magnetic Model 2SA
5. Wattmeter: Simpson Model 1379; watts 0 to 9000
6. Thermostat: Minneapolis Honeywell remote bulb temperature controller with a range from 40 to 180°F

### Flexible Coupling

1. Type: rubber intermediate member
2. Make and designation: link-belt type B
3. Bore size: 1⅛ in. to ¾ in. with ¼ in. and 3/16 in. keyway and set screws

### Hydraulic Pump

1. Type: variable-volume pressure-compensated piston
2. Manufacturer: Denison Model PV004-11R-155
3. Capacity: 6 gpm maximum
4. Pressure: 500 to 1500 psi factory set

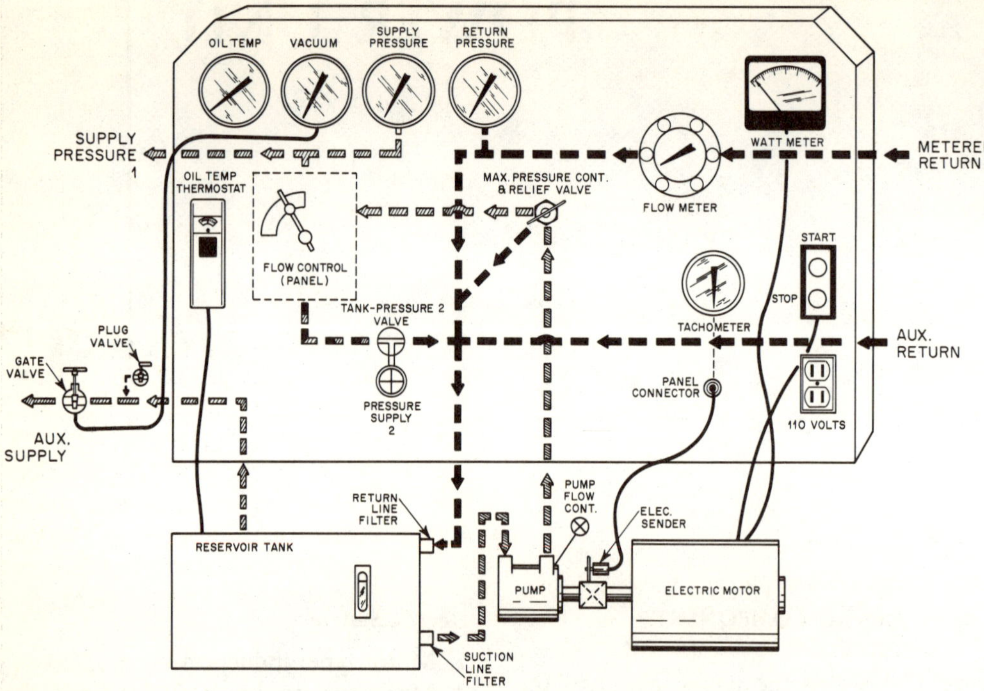


FIG. 17-1 *Panel layout.*

**Tachometer**

1. Type: direct reading frequency impulse with four-pin panel-connected remote sender
2. Manufacturer: Westach Model 4L12
3. Range: 0 to 2500 rpm

**Maximum Circuit Pressure Control and Relief Valve**

1. Manufacturer: Gresen in-line, Series JL adjustable
2. Capacity: up to 20 gpm
3. Port sizes: ¾-in. in-line; ½ in. by-pass NPTF

**Flowmeter**

1. Manufacturer: Erdco Engineering Corporation Model 2-3000BR
2. Range: 0 to 10 gpm

**Panel Flow Control Valve**

1. Type: manually operated with dial
2. Manufacturer: Brand Hydraulics, Series FC
3. Port sizes: ⅜ in.
4. Capacity: full adjustment 0 to 8 gpm

**Pressure Supply 1**

1. Connection: quick disconnect
2. Hose size and type: ½-in. single wire braid
3. Working pressure rating: 1250 psi
4. Burst pressure: 5000 psi

**Supply Pressure Gauge**

1. Type: dial
2. Manufacturer: USG, Marsh, or Marshalltown
3. Range: 0 to 2000 psi

**Auxiliary Supply**

1. Type: gate valve with cavitation valve
2. Size: 1-in. pipe

**Metered Return Line**

1. Connection: quick disconnect
2. Hose size and type: ½-in. I.D. single wire braid with ½-in. male pipe ends
3. Working pressure rating: 1250 psi
4. Burst pressure: 5000 psi

**Auxiliary Return**

1. Connection: quick disconnect

 

### Return Flow Pressure Gauge

1. Type: dial
2. Manufacturer: USG, Marsh, or Marshalltown
3. Range: 0 to 60 psi

### Return Filter

1. Manufacturer: Gresen Model FB-103
2. Port sizes: ¾ in.
3. Type of filter element: 33-$\mu$ nonreusable paper
4. By-pass: spring type set at 15 psi
5. Condition indicator: 0- to 100-psi pressure gauge

### Quick Disconnects

1. Manufacturer: Bruning Industrial SM-500
2. Recommended capacity: 15-gpm maximum
3. Recommended working pressure (maximum): 6000 psi

### Selector Valve

1. Manufacturer: Gresen Model SM-50
2. Type: two-position three-way
3. Capacity: up to 20 gpm
4. Port sizes: ½ in.
5. Pressure range: 0 to 2000 psi

## POWER-UNIT CIRCUIT

The circuit shown in Fig. 17-1 is designed with all the features which are normally encountered in a fluid power circuit. It consists of the prime mover, pump, reservoir, strainer, return filter, and control valve. Actuators with controls are needed to complete a circuit. With a variety of control components an infinite number of circuits can be assembled, analyzed, and studied.

### Reservoir

The reservoir provides for fluid storage and serves as a heat exchanger. To fill the reservoir, it is necessary to remove the table top. The filler can, which also acts as a breather, is equipped with a strainer to remove particles from fluid during filling. Once filled, the necessity for refilling is infrequent.

### Prime Mover

The prime mover is a 5-hp 220-volt three-phase electric motor. It drives a hydraulic pump at approximately 1800 rpm. A tachometer, mounted on the panel, is provided to determine the exact pump rpm.

### Pump

The hydraulic pump, driven by the electric motor, is a variable-displacement, pressure-compensated, axial-piston type. This pump has a theoretical displacement of 0.77 cu in. of fluid per revolution, and is capable of operation at 1500 psi continuously.

*Variable displacement* is accomplished by manually turning the adjusting screw. Zero discharge is obtained when the centerline of the cylinders and the plane of the cam plate are at right angles to each other.

*Pressure compensation* will accomplish automatically the results that were accomplished manually. If pressure builds up on the discharge side, it will force the cam plate to a no-flow condition. In this way, the flow will stop at whatever maximum pressure setting is set on the adjusting screw.

### Strainer

Between the reservoir and the pump is a strainer whose purpose is to remove contaminants from the fluid before they enter the pump. The strainer is equipped with a condition indicator, which is a vacuum gauge mounted on the panel. As the screen becomes clogged, the vacuum will increase, thereby indicating the condition of the strainer. Two to six in. Hg "vacuum" is normal operating condition. Oil will be by-passed if the vacuum should become greater than 10 in. Hg. A spring-loaded check valve will open at this vacuum and permit flow directly from the tank to the pump.

### Cavitation and Air Inlet Section

Next to the auxiliary supply outlet is a gate valve which demonstrates cavitation in the pump on the dynamometer unit. There is also an inlet valve to show effects of air in the system.

### Relief Valve (Maximum-pressure Gauge)

The pump provides flow into the system through a flexibile hose to the panel. The first component is a relief valve, which is also known as a maximum-pressure valve. These two terms are synonymous. This is a ball-and-spring-type valve with the maximum setting adjustable. One adjusts it by increasing or decreasing the spring tension. It is also equipped with a knurled lock bolt which can be hand tightened to prevent creep of the knob during the operation.

Upon obtaining pressures greater than that set on the spring, the pressure against the ball will overcome this spring tension, move the ball away from the seat, and permit the flow to return directly to the tank. When the pressure drops, the spring will reseat the ball and again provide flow to the system.

### Panel Flow Control

From the relief valve the flow is directed to a pressure-compensated flow control valve. (The valve is adjustable, so that regardless of the gpm that the pump delivers to the valve, a controlled amount can be delivered to the circuit at pressure supply 1.) It will deliver the set amount of flow regardless of the pressure required at the work end of the circuit. This can be from 0 to 100 percent or 0 to 6 gpm depending on the position of the indicator. The balance of flow is diverted back to the tank, or to pressure supply 2, depending upon the setting of the panel selector valve. The panel selector valve has two positions: tank (push in) and circuit (pull out). The excess flow can be diverted back to the tank, or to pressure supply 2. For multiple circuits this additional supply is used.

At position 8 full flow is directed to pressure supply 1, and at position 0 full flow is directed to pressure supply 2. Intermediate positions will direct flow to both stations according to dial setting. When flow is delivered to both stations, and either one requires a pressure greater than the setting on the relief valve, or that set on the pump, no flow will be available at either pressure supply station.

It should be noted that pressure supply 1 is the controlled flow station and 2 is the excess. For example, if the pump provides 6 gpm, and the valve is set for 3 gpm at 1, it will provide 3 gpm at each station. If the pump, through increased pressure requirement, drops to 5 gpm, the valve will still deliver 3 gpm to station 1, and station 2 will receive 2 gpm.

A pressure-supply gauge registers the pressure at pressure supply 1.

### Auxiliary Supply

An auxiliary supply is provided for pumps which are mounted on the table for dynamometer tests and analyzing.

### Return Line

Fluid in the metered return line flows through a meter which registers gpm.

### Panel Instruments

1. A pressure-supply gauge is provided to indicate supply line system pressures.
2. A return-line pressure gauge is provided on the panel and indicates the pressure in the system as it enters the metered return line. The auxiliary return flow and the metered return flow are common on the downstream side of the flowmeter.
3. An oil temperature gauge indicates the oil temperature which changes with relation to table circuits.
4. A thermostat is provided which can be set for the maximum oil temperatures under which the circuits will operate. Generally 140 to 160°F is used. When the oil is heated above this range, the electrical motor will automatically shut off.

 

5. A flowmeter is provided on the metered return line.
6. A tachometer is provided and connected to a four-pin panel connector. A remote electrical sender is provided with a 6-ft extension cord.

#### Return-line Filter

The fluid is directed, on its return to the tank, through a filter which removes contaminants from the fluid. This is a *filter,* as compared with a *strainer,* which is on the inlet line. A strainer is only capable of removing large particles (100-mesh screen or particles the size of coarse salt), where filters are capable of removing particles as fine as silt. Strainers must be coarse to prevent a vacuum and cavitation of the pump. A filter requires pressure to force the oil through it, but it presents no problem, since it is located on the return line. Normal pressure required is 4 to 8 psi. If the filter becomes contaminated, pressure will rise. If it exceeds 15 psi, a check valve will open and permit the flow directly to the tank. This results in no filtration of the oil and is indicated on the panel filter-condition indicator. The filter should be changed before it reaches this point.

#### Electrical Circuit

Electrical instrumentation consists of an on–off switch. A 100-volt outlet is provided for circuits which require its use. A wattmeter is installed to provide a determination of the power furnished by the electric motor to the prime pump.

$$\frac{\text{Watts}}{746} \times 0.80 = \text{eff. electric motor}$$

### POWER-UNIT INSTALLATION

1. Remove the table top panel and fill the reservoir with SSU 150 to 220 hydraulic oil. A sight gauge is provided. Fill to the upper one-third of this gauge. Twenty-six to twenty-eight gallons of oil will be required. A combination filler and breather cap is provided for this purpose. Its position is such that the table top must be removed in order to fill the tank. While this may seem awkward, it will minimize the collection of debris in the filler cap. Further, very infrequent refillings will be required.
2. The standard power unit is equipped with a 5-hp electric motor. This motor is an induction-type 220-volt 3-phase. A 5-hp single-phase 220-volt motor may be ordered as an alternate. An electric cord with a male plug is installed. This will permit moving the unit if required with respect to location within a room. The unit is equipped with casters to provide for this mobility.

In areas where permanent installations are required for 220-volt 3-phase motors the alternate 220-volt single-phase motor is available where mobility of the unit is desired or required.

*Be sure to check the switch for off position before installing, plugging into the receptacle, or wiring the unit.*

The rotation of the pump should immediately be checked. The installed pump is designed to provide flow with clockwise rotation on the pump (as one looks at the pump input shaft). If pump rotation is counterclockwise to the input shaft, change the pump rotation by changing the wiring to the electric motor.

A 110-volt outlet is provided on the unit. This is an independent circuit and will require an extension cord from the male plug provided on the machine to a receptacle in the building.

### POWER-UNIT OPERATING INSTRUCTIONS

When the unit is installed, it is ready to operate; it can be used for a wide variety of demonstrations and circuits. Also, it provides an adequate flow volume and sufficient pressure to operate an unlimited number of realistic circuits.

The entire unit, its components, and the

circuit in the unit are best analyzed by a series of demonstrations. However, several basic operating instructions should be noted at the outset. The unit is designed so that it can be started under any conditions, provided that installation instructions are completed. Several procedures are recommended as good practices in the laboratory or in the field.

1. Set the flow control valve on the panel to 0 flow or the 0 position before starting the unit. Zero flow can also be obtained by the pump flow control; however, it is more convenient to use the panel flow control valve.
2. Set the maximum flow control (relief valve) to any intermediate position and lock in this position by the knurled knob.
3. Check the oil level gauge.
4. Check the thermostat for the maximum oil temperature setting.
5. Check the quick disconnects for full lock position and the hose connection for tightness.
6. Start unit.
7. Induce flow to the circuit as required using the panel flow control.

If desired, or if the circuit requires, flow can be controlled by the pump flow control. Full flow from the pump is obtained when the control is in the out position, or rotating counterclockwise. By turning this control clockwise, flow will be reduced from maximum to zero flow. The pump is factory set for maximum pressure at 1500 psi.

 

# Appendix A

## GLOSSARY

These definitions are taken from the Glossary of Fluid Power Terms prepared by the National Fluid Power Association

AIR, STANDARD Air at a temperature of 68°F, a pressure of 14.70 psi absolute, and a relative humidity of 35 percent (0.0750 lb/cu ft). In gas industries the temperature of "standard air" is usually given as 60°F.

BERNOULLI'S LAW If no work is done on or by a flowing frictionless liquid, its energy, due to pressure and velocity, remains constant at all points along the stream line.

BOYLE'S LAW The absolute pressure of a confined body of gas varies inversely as the volume, provided that the temperature remains constant.

CHARLES' LAW The volume of a fixed mass of gas varies directly with absolute temperature, provided that the pressure remains constant.

FLUID POWER Power transmitted and controlled through use of a pressurized fluid.

HYDRAULICS Engineering science pertaining to liquid pressure and flow.

LAMINAR A flow situation in which motion occurs as a movement of one layer of fluid upon another.

METERED Flow at a controlled rate.

PASCAL'S LAW A pressure applied to a confined fluid at rest is transmitted with equal intensity throughout the fluid.

PNEUMATICS Engineering science pertaining to gaseous pressure and flow.

POUR POINT The lowest temperature at which a liquid will flow under specific conditions (ASTM Designation D91-47).

PRESSURE Force per unit area. It is usually expressed in pounds per square inch.

PRESSURE, ABSOLUTE The sum of atmospheric and gauge pressures.

PRESSURE, ATMOSPHERIC Pressure exerted by the atmosphere at any specific location. (Sea-level pressure is approximately 14.7 psi absolute.)

PRESSURE, BACK The pressure encountered on the return side of a system.

PRESSURE, GAUGE Pressure differential above or below atmospheric pressure.

TURBULENT A flow situation in which the fluid particles move in a random manner.

VISCOSITY INDEX A measure of the viscosity temperature characteristics of a fluid as referred to that of other fluids.

## GENERAL TERMS

ACCUMULATOR A storage chamber in which fluid energy may be accumulated and from which it may be withdrawn.

ACTUATOR A device for converting fluid energy into mechanical motion.

ADDITIVE A chemical compound or compounds added to a fluid to change its properties.

CUSHION A device which provides controlled resistance to flow.

CUSHION, CYLINDER A cushion built into a cylinder to restrict flow at the outlet port, thereby arresting the motion of the piston rod.

CYLINDER, PISTON A cylinder in which the movable element has a greater cross-sectional area than the piston rod.

CYLINDER, PLUNGER A cylinder in which the movable element has the same cross-sectional area as the piston rod.

CYLINDER, SINGLE ACTING A cylinder in which the fluid force can be applied to the movable element in only one direction.

CYLINDERS A device for converting fluid energy into linear motion. It usually consists of a movable element such as a piston and piston rod, plunger, or rod operating within a cylindrical bore.

CYLINDERS, DOUBLE ACTING A cylinder in which fluid force can be applied to the movable element in either direction.

CYLINDERS, DOUBLE ROD A cylinder with a single piston and a piston rod extending from each end.

EMULSION, WATER-OIL A stabilized mixture of two immiscible components, water and oil. It may contain additives. There are two types: oil-in-water and water-in-oil.

FILTER A device through which fluid is passed to separate matter held in suspension.

FLASH POINT The temperature at which a fluid first gives off sufficient flammable vapor to ignite when approached by a small flame or spark.

HOSE FITTINGS, REUSABLE A hose fitting that can be removed from a hose and be reused.

INHIBITOR Any substance which slows or prevents such chemical reactions as corrosion or oxidation.

INTENSIFIER A device which increases the pressure over that of the source pressure.

MANIFOLD A conductor which provides multiple connection ports.

MOTOR A device for converting fluid energy into rotary mechanical motion.

MOTOR, LIMITED ROTARY A rotary motor having limited motion.

MOTOR, ROTARY A motor capable of continuous rotary motion.

OIL-IN-WATER A dispersion of oil in a continuous phase of water.

PETROLEUM A fluid composed of petroleum oil. It may contain additives.

PHOSPHATE ESTER A fluid composed of phosphate esters. It may contain additives.

PHOSPHATE ESTER BASE A fluid which contains a phosphate ester as one of the major components.

PORT An internal or external terminus of a passage in a component.

PORT, AND 10050 An Air Force–Navy Aeronautical Design Standard in which a straight-thread port is used to attach tube fittings to various components. It employs an O-ring seal compressed in a special cavity.

PORT, PLAIN O RING A port which uses an O ring in a rectangular groove machined in the port face.

PORT, SAE A straight-thread machined port used to attach tube and hose couplings. It employs an O ring compressed in a wedge-shaped cavity. A standard of the Society of Automotive Engineers.

PRESSURE GAUGE An instrument which indicates the pressure in the system to which it is connected.

PRESSURE SWITCH A switch operated by a fluid, responsive to rise or drop in pressure.

PUMP A device for converting mechanical energy into fluid energy.

RESERVOIR A chamber used to store hydraulic fluid on which pumps, motors, and valves may be mounted. It is usually equipped with baffles, an air breather, and a liquid level gauge.

RESTRICTOR A device for producing a deliberate pressure drop or resistance in a line by reducing the cross-sectional flow area.

RESTRICTOR, CHOKE A restrictor, the length of which is relatively large with respect to its cross-sectional area.

RESTRICTOR, ORIFICE A restrictor, the length of which is relatively small with respect to its cross-sectional area. The orifice may be fixed or variable. Variable types are noncompensated, pressure-compensated, or pressure- and temperature-compensated.

SILICATE ESTER A fluid composed of organic silicates. It may contain additives.

SILICONE A fluid composed of silicones. It may contain additives.

SUBPLATE (BACK PLATE) An auxiliary ported plate for mounting components.

THREADS, DRYSEAL PIPE A pipe thread in which the truncations at the root and crest of the internal and external threads are controlled to give an interference fit by deformation in assembly.

THREADS, TAPER PIPE A tapered pipe thread machined to American (National) Pipe Standards in which the members perform as a seal and holding device when assembled.

VALVE A device which controls fluid flow conditions such as pressure, temperature, time, or rate.

VALVE, CHECK A valve which permits flow of fluid in only one direction.

VALVE, CLOSED CENTER A valve in which all ports are closed in the center position prohibiting any flow.

VALVE, COUNTERBALANCE A valve which maintains back pressure to prevent a load from falling.

VALVE, DECELERATION A valve which gradually reduces flow rate to provide deceleration.

VALVE, DIRECTIONAL CONTROL A valve whose primary function is to direct or prevent flow through selected passages.

VALVE, FLOW CONTROL A valve whose primary function is to control flow rate.

VALVE, NEEDLE A valve with an externally adjustable tapered closure which regulates the flow passage.

VALVE, OPEN CENTER A valve in which all ports are interconnected in the center position allowing flow in any direction.

VALVE, PILOT A valve applied to operate another valve or control.

VALVE, PILOT OPERATED A valve in which operating parts are actuated by pilot fluid.

VALVE, PRESSURE REDUCING A valve whose primary function is to limit outlet pressure.

VALVE, RELIEF A valve whose primary function is to limit system pressure.

VALVE, SEQUENCE A valve whose primary function is to direct flow in a predetermined sequence.

VALVE, SHUTOFF A valve which operates fully open or fully closed.

VALVE, SOLENOID-CONTROLLED–PILOT-OPERATED A valve which is operated by a solenoid-operated pilot valve.

VALVE, SOLENOID-OPERATED A valve which is positioned by one or more solenoids.

VALVE, TWO- AND THREE-POSITION A valve having two or three positions of flow control.

VALVE, TWO-, THREE-, OR FOUR-WAY A directional control valve having two, three, or four distinctive external working connections.

WATER-GLYCOL A fluid whose major constituents are water and one or more glycols or polyglycols.

# Appendix B

## GENERAL FORMULAS AND RULES

Area of a circle $= \dfrac{3.14 \times d^2}{4} = 0.7854 \times d^2$

Area of a pipe needed to provide for a given gpm to flow at a specific velocity $= \dfrac{\text{gpm} \times 0.3208}{\text{velocity (ft/sec)}}$

Cylinder displacement = area × stroke

Cylinder piston travel speed (in./min) $= \dfrac{\text{gpm} \times 231}{\text{area (sq in.)}}$

Cylinder piston travel speed (in./min) $= \dfrac{\text{cu in. of flow per min}}{\text{area (sq in.)}}$

Energy = work

Flow velocity through a pipe varies as the square of the inside diameter. Doubling the diameter increases the area four times; it decreases the flow velocity to one-fourth its original value.

Fluid hp supplied $= \dfrac{\text{pressure} \times \text{gpm}}{1714}$ = motor input hp or pump output

Friction of a liquid in a pipe varies as the square of the velocity; doubling the velocity will result in four times the friction.

Gallon = 231 cu in. and weighs 8.33 lb (water)

Horsepower = work at a rate of 33,000 ft-lb/min or 500 ft-lb/sec

1 hp = 746 watts = 0.746 kw

or 1 kw = 1.34 hp

Hydraulic cylinder stroke time (sec) $= \dfrac{\text{stroke (in.)}}{\text{speed (in./sec)}}$

Hydraulic motor hp $= \dfrac{\text{torque (in.-lb)} \times \text{rpm}}{63{,}025}$ = motor output hp

Hydraulic motor hp $= \dfrac{\text{torque (ft-lb)} \times \text{rpm}}{5252}$

Load loss = fluid hp supplied less fluid hp on return line

Motor mechanical efficiency $= \dfrac{\text{hp delivered (mechanical)}}{\text{fluid hp supplied}} \times 100$ = percent

Oil is forced or pushed into a pump by atmospheric pressure which is 14.7 psi at sea level; it decreases approximately 0.1 psi for every 90 feet of elevation.

Oil will compress at an approximate value of ½ percent per 1000 psi.

Piston speed (ft/min) = speed (in./sec) × 5

$$\text{Power} = \text{force} \times \frac{\text{distance}}{\text{time}}$$

Pressure = force per unit area (usually psi)

Pressure in a system is the result of resistance to flow; a pump does not pump pressure.

Psi (1) = 2.04 in. Hg or 27.7 in. water

$$\text{Pump mechanical efficiency} = \frac{\text{fluid hp supplied}}{\text{electric hp supplied (mechanical)}}$$

Torque (in.-lb) = psi × displacement (cu in. × ½)

Torque (in. lb) = hp × 63,025

$$\text{Velocity in a pipe (ft/sec)} = \frac{\text{gpm} \times 0.3208}{\text{area (sq in.)}}$$

Volume of a cylinder or cylinder capacity = area × length = displacement

Work = force × distance

# Appendix C

## GRAPHIC SYMBOLS

The symbols shown here have been submitted to the American Standards Association, Inc., by the following task group representing the listed organizations. The group was sponsored by the Institute of Electrical and Electronics Engineers and the American Society of Mechanical Engineers.

**Task Group**

| | |
|---|---|
| *J. L. Fisher, Jr., Chairman* | *Bellows-Valvair, Division of IBEC* |
| *H. C. Parsons, Vice-Chairman* | *Vickers Incorporated* |
| *M. E. Long, Secretary* | *The Penton Publishing Co.* |
| *O. J. Abbott* | *Colonial Broach & Machine Co.* |
| *W. R. Forster* | *Westinghouse Air Brake Co.* |
| *Clark Langworthy* | *L & L Manufacturing Co.* |
| *Z. J. Lansky* | *Parker-Hannifin Corp.* |
| *G. A. Larsen* | *Texaco, Inc.* |
| *F. L. Mackin* | *The Fluid Power Society* |
| *C. A. Nazian* | *U. S. Department of Defense* |
| *J. J. Pippenger* | *Racine Hydraulic Development Corp.* |
| *A. B. Riddiford, Jr.* | *John S. Barnes Corporation* |
| *Ransom Tyler* | *The Oilgear Company* |

**Organizations Represented**

*American Ordnance Association, by J. J. Pippenger, H. C. Parsons, and Ransom Tyler.*
*American Society for Metals, by H. C. Parsons.*
*American Society of Agricultural Engineers, by M. E. Long.*
*American Society of Mechanical Engineers, by H. C. Parsons and A. B. Riddiford, Jr.*
*American Society of Tool and Manufacturing Engineers, by J. J. Pippenger.*
*Engineering Society of Detroit, by O. J. Abbott.*
*Engineers and Scientists of Milwaukee, by Ransom Tyler.*
*Fluid Power Society, by J. L. Fisher, Jr., F. L. Mackin, J. J. Pippenger, A. B. Riddiford, Jr., and Ransom Tyler.*
*Instrument Society of America, by G. A. Larsen.*
*National Fluid Power Association, by J. L. Fisher, Jr., W. R. Forster, Clark Langworthy, Z. J. Lansky, M. E. Long, H. C. Parsons, and J. J. Pippinger.*
*Society of Automotive Engineers, by M. E. Long and H. C. Parsons.*
*U. S. Department of Defense, by C. A. Nazian.*

1. Line, working (main)

2. Line, pilot (for control)

3. Line, liquid drain

4. Centerline

5. Flow, direction of

6. Line with fixed restriction

7. Line, flexible

8. Plugged port

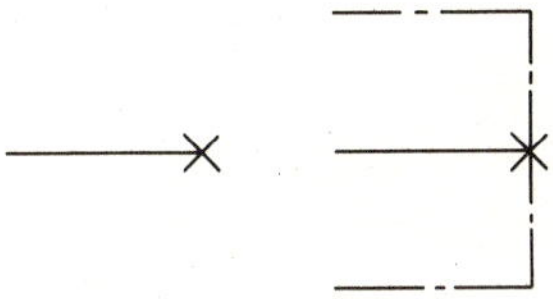

(a) (b)

9. Lines crossing

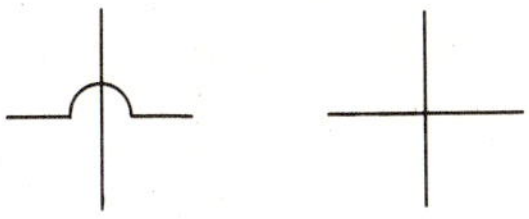

(a) (b)

10. Lines joining

(a) (b) (c)

11. An arrow through a symbol at approximately 45° indicates that the component can be adjusted or varied

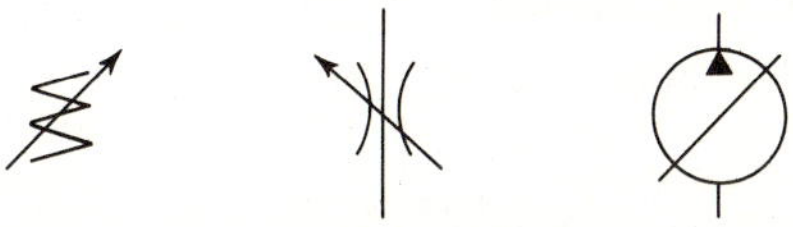

(a) (b) (c)

12. An arrow parallel to the short side of a symbol, within the symbol, indicates that the component is pressure compensated

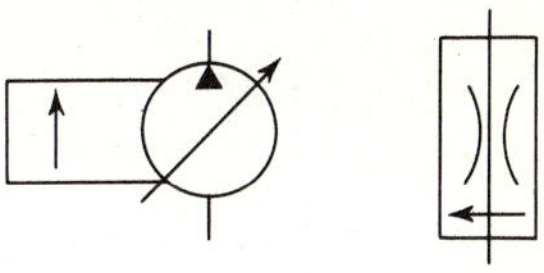

(a) (b)

13. Quick disconnect without checks

(a) (b)

14. Quick disconnect with two checks

(a) (b)

15. Quick disconnect with one check

(a) (b)

16. Rotating coupling

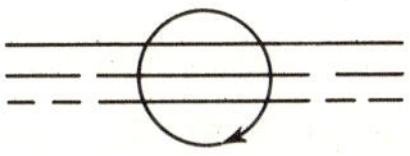

17. Energy storage and fluid storage; reservoir vented

18. Reservoir pressurized

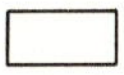

19. Reservoir with connecting lines; above fluid level

(a) (b)

20. Reservoir with connecting lines; below fluid level

(a) (b)

21. Accumulator

22. Accumulator, spring loaded

23. Accumulator, gas charged

24. Accumulator, weighted

25. Fluid conditioners (filter, strainer, heat exchanger)

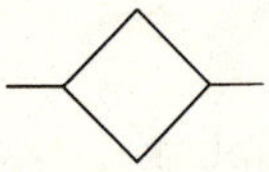

26. Inside triangles indicate heat dissipation

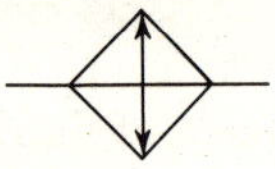

27. Filter, strainer

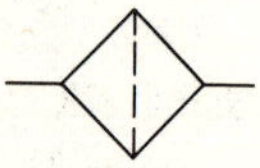

28. Cylinders, hydraulic and pneumatic (single acting)

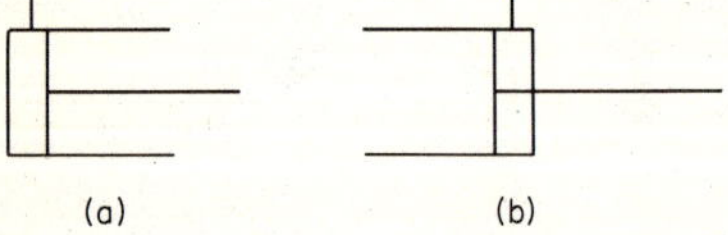

(a) (b)

29. Cylinders, double acting

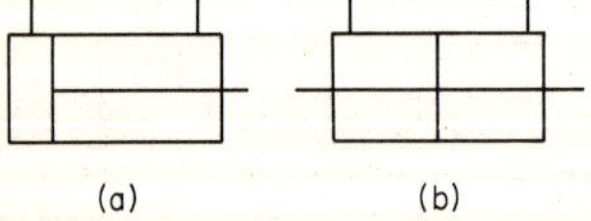

(a) (b)

30. Controls, spring

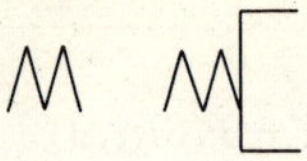

(a) (b)

31. Manual control (general symbol)

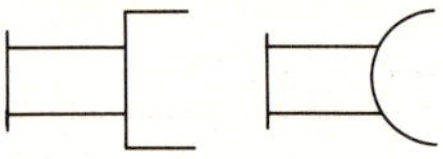

(a) (b)

32. Push button (manual)

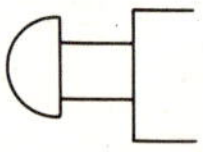

33. Push–pull lever (manual)

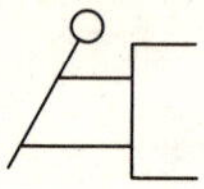

34. Mechanical

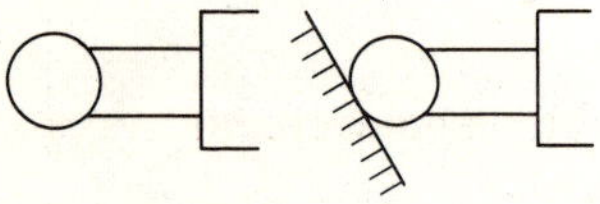

(a) (b)

35. Detent

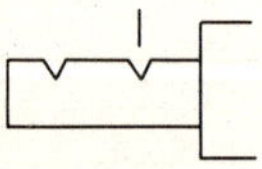

36. Pressure compensated

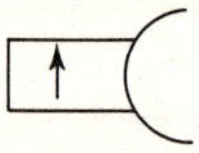

37. Electrical solenoid (single winding)

38. Reversing motor

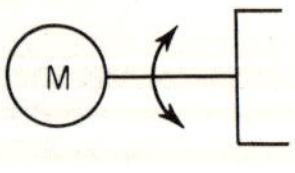

39. Pilot pressure

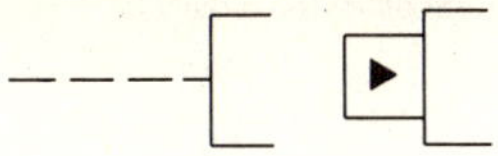

(a) (b)

40. Pilot controlled, spring centered

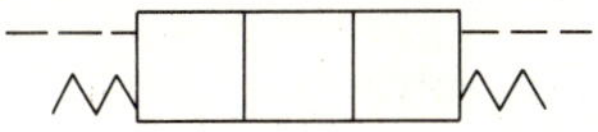

41. Hydraulic pump, fixed displacement

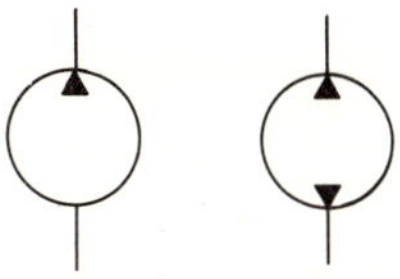

(a) (b)

42. Variable displacement, noncompensated

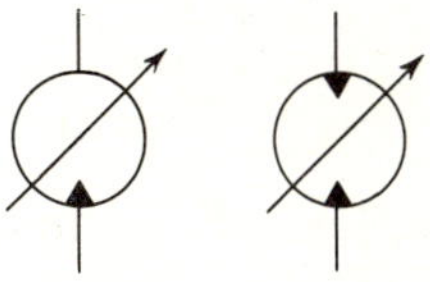

(a) (b)

43. Variable displacement, pressure compensated

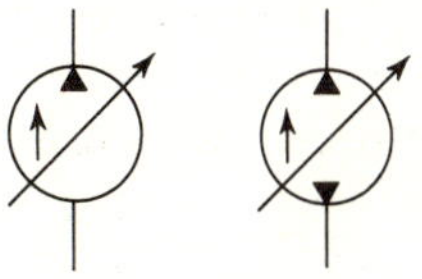

(a) (b)

44. Hydraulic motor

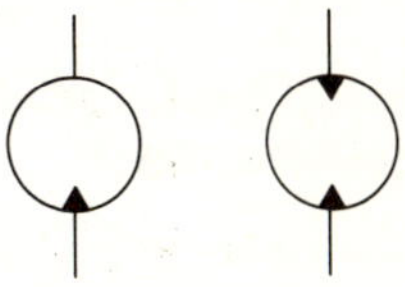

(a) (b)

45. Variable displacement

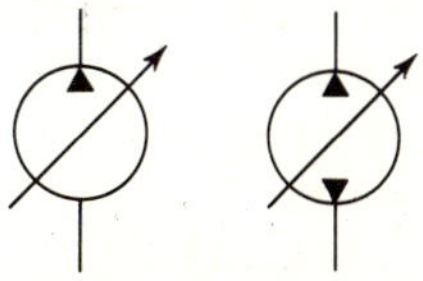

(a) (b)

46. Pump motor, hydraulic (dual purpose)

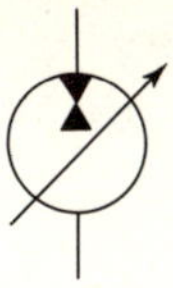

47. Electric motor

48. Pressure

49. Temperature

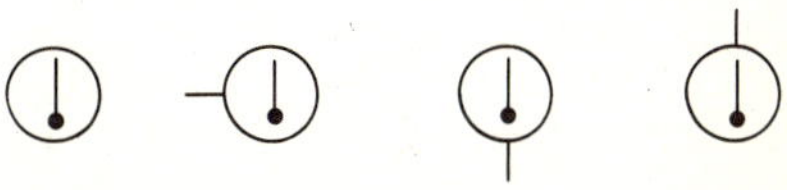

50. Flowmeter, flow rate

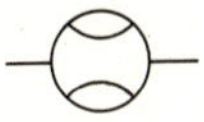

51. Pressure switch

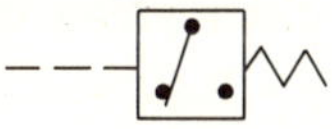

52. Float switch

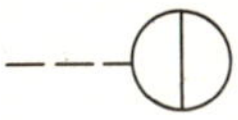

53. Envelopes

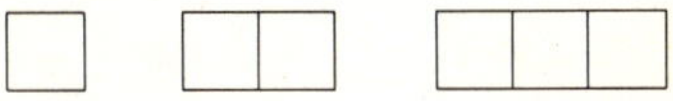

(a) (b) (c)

54. Ports

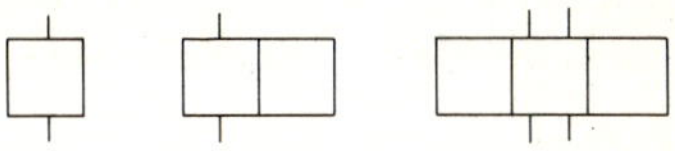

(a) (b) (c)

55. Ports, internally blocked

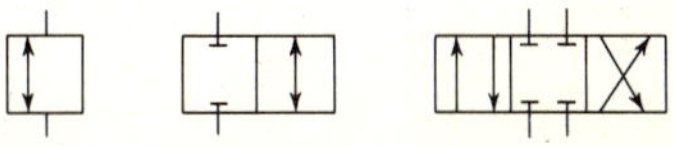

(a) (b) (c)

56. Flow paths, internally open

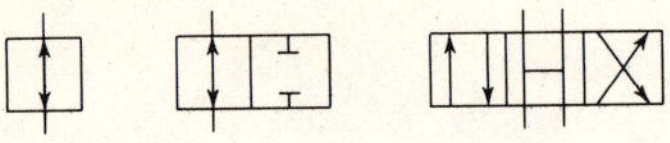

(a) (b) (c)

57. Two-way valves (manual shut-off), off

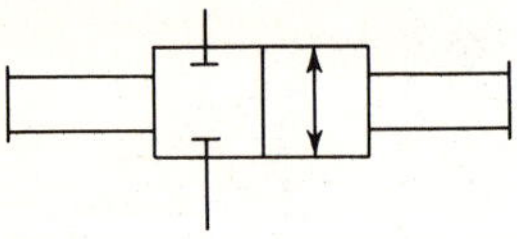

58. Two-way valves (manual shut-off), on

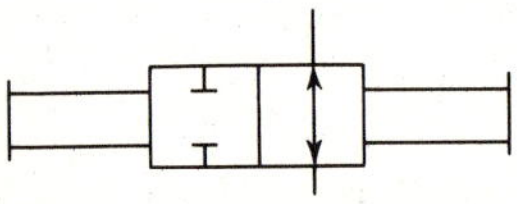

59. Check, simplified symbol

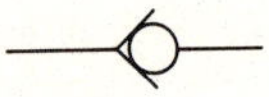

60. Check, pilot operated to open

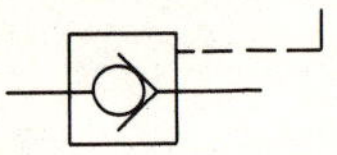

61. Check, pilot operated to close

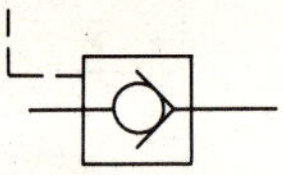

62. Two-way valves (infinite position)

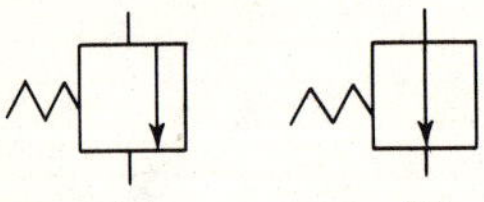

(a) (b)

63. Two-way valves (two-position)

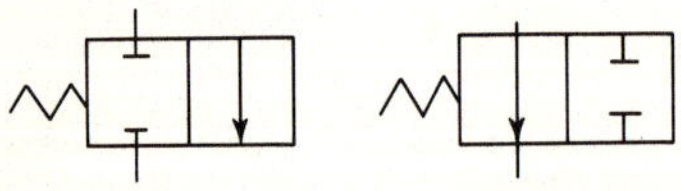

(a) (b)

64. Three-way valves normally open (two-position, manually operated)

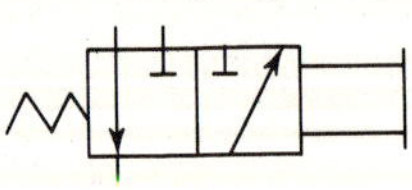

65. Three-way valves normally closed (two-position, solenoid operated)

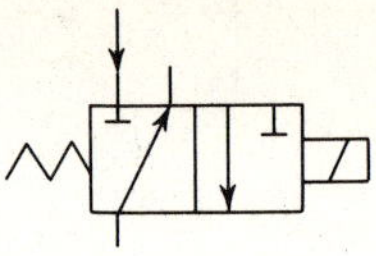

66. Four-way valves (two-position)

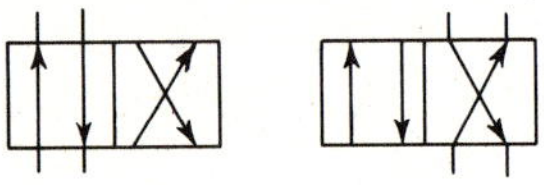

(a) (b)

67. Three-position valve, normal

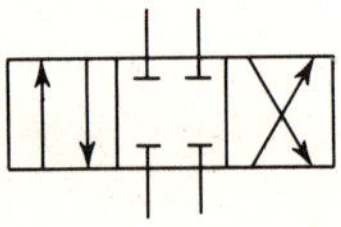

68. Three-position valve, actuated left

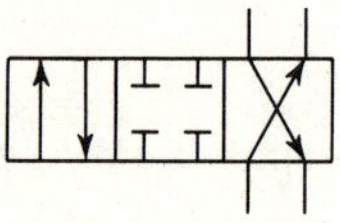

69. Three-position valve, actuated right

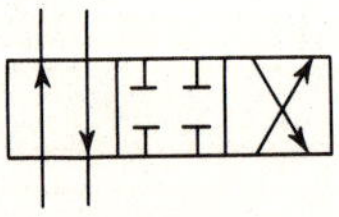

70. Typical flow paths for center condition of three-position valves

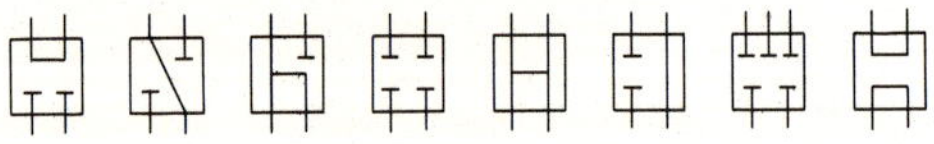

71. Pressure relief valve, simplified symbol

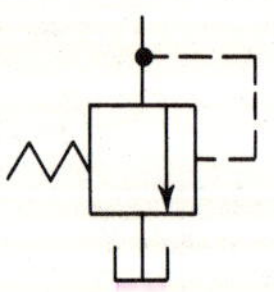

72. Pressure relief valve

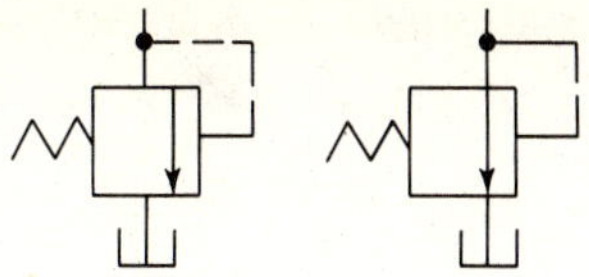

(a) (b)

73. Sequence valve

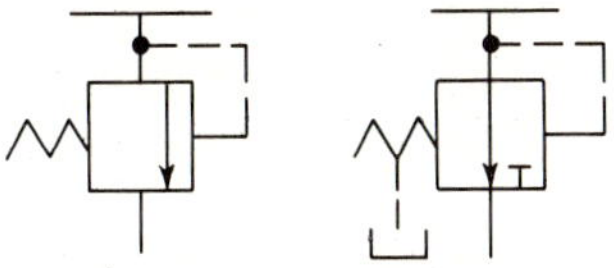

(a) (b)

74. Pressure-reducing valve

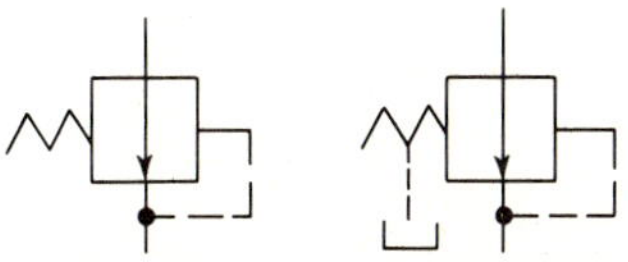

(a) (b)

75. Pressure-compensating flow control valve

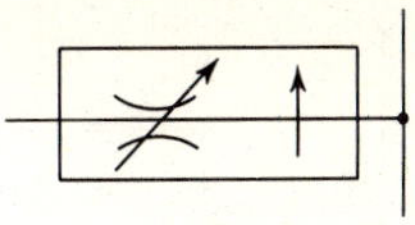

76. Flow divider, adjustable pressure compensated

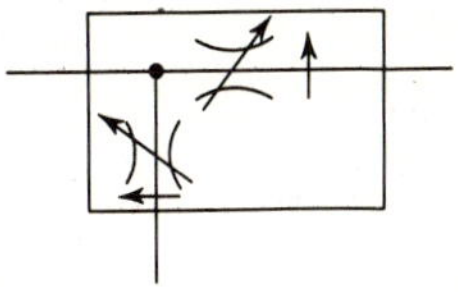

77. Counterbalance valve or load-lock valve; adjustable and pressure compensated with by-pass

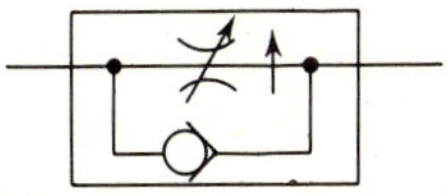

# Appendix D*

TABLE D-1 TECHNICAL DATA SHEET

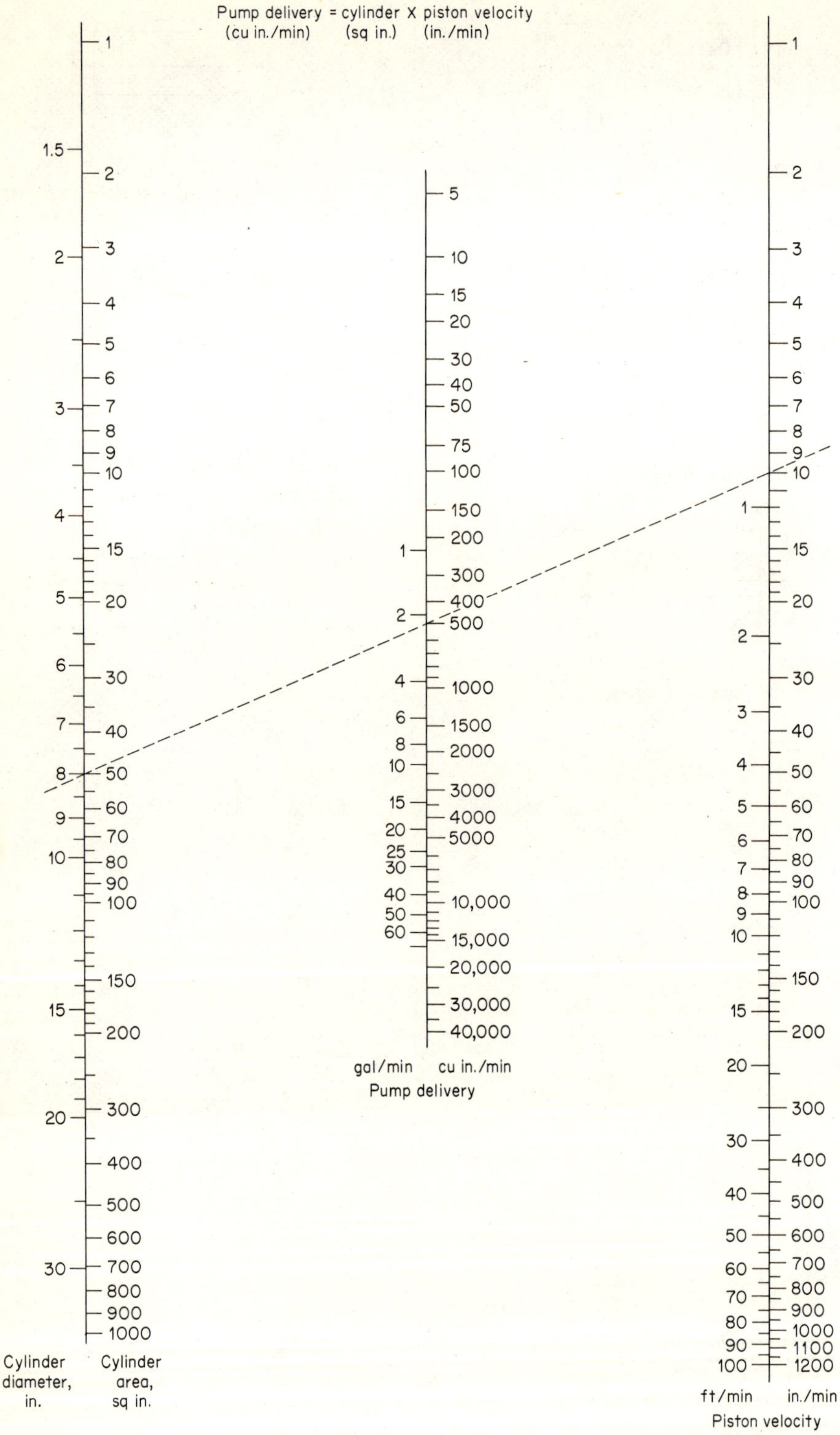


* All tables are courtesy of Cross Manufacturing Company, Inc.

# TABLE D-2 TECHNICAL DATA SHEET

$$\text{Time} = \frac{\text{cylinder displacement, cu in.}}{\text{pump delivery, gal/min}} \times 0.2597$$

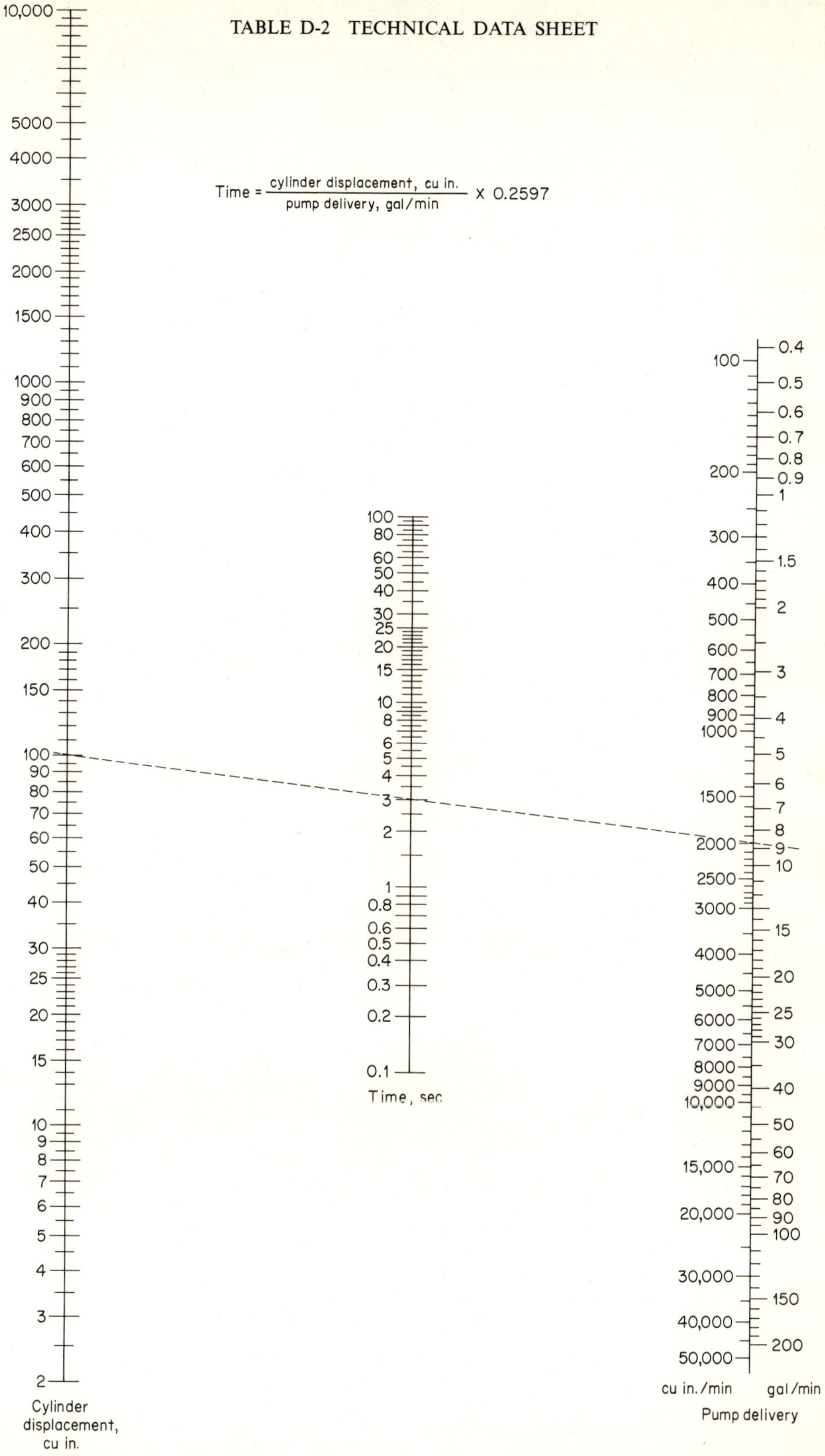

## TABLE D-3 TECHNICAL DATA SHEET

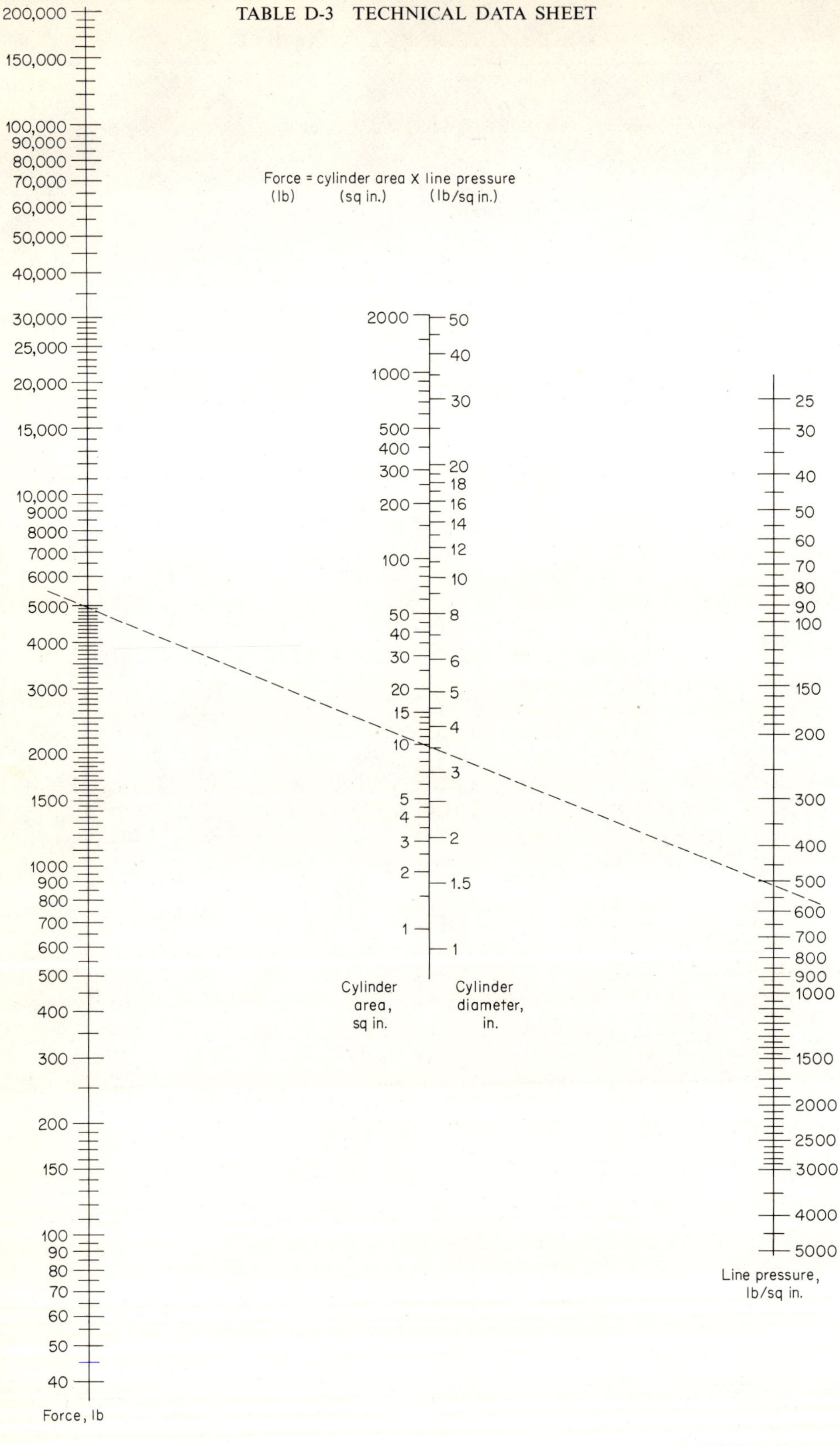

TABLE D-4 PRESSURE CHART (CYLINDER POWER, IN POUNDS, AT VARIOUS PRESSURES)

PUSH

| *Pump pressure* | *3,000* | *2,500* | *2,000* | *1,500* | *1,250* | *1,000* | *750* | *500* | *150* |
|---|---|---|---|---|---|---|---|---|---|
| 6″ Bore | 84,822 | 70,685 | 56,548 | 42,411 | 35,342 | 28,274 | 21,205 | 14,137 | 7,068 |
| 5″ Bore | 58,899 | 49,083 | 39,267 | 29,451 | 24,543 | 19,635 | 14,724 | 9,816 | 4,908 |
| 4″ Bore | 37,694 | 31,412 | 25,130 | 18,848 | 15,707 | 12,566 | 9,423 | 6,282 | 3,141 |
| 3½″ Bore | 28,861 | 24,051 | 19,241 | 14,431 | 12,026 | 9,621 | 7,215 | 4,810 | 2,405 |
| 3″ Bore | 21,204 | 17,670 | 14,136 | 10,602 | 8,835 | 7,068 | 5,301 | 3,534 | 1,767 |
| 2½″ Bore | 14,724 | 12,270 | 9,816 | 7,362 | 6,135 | 4,908 | 3,681 | 2,454 | 1,227 |
| 2″ Bore | 9,421 | 7,851 | 6,281 | 4,711 | 3,926 | 3,141 | 2,355 | 1,570 | 785 |
| 1½″ Bore | 5,295 | 4,413 | 3,531 | 2,649 | 2,208 | 1,767 | 1,323 | 882 | 441 |

PULL

| *Pump pressure* | *3,000* | *2,500* | *2,000* | *1,500* | *1,250* | *1,000* | *750* | *500* | *250* |
|---|---|---|---|---|---|---|---|---|---|
| 6″ Bore, 3″ Shaft | 63,618 | 53,015 | 42,412 | 31,809 | 26,507 | 21,206 | 15,904 | 10,603 | 5,301 |
| 6″ Bore, 2½″ Shaft | 70,098 | 58,415 | 46,732 | 35,049 | 29,207 | 23,366 | 17,524 | 11,683 | 5,841 |
| 6″ Bore, 2″ Shaft | 75,400 | 62,832 | 50,266 | 37,699 | 31,416 | 25,133 | 18,849 | 12,566 | 6,283 |
| 5″ Bore, 2″ Shaft | 49,478 | 41,232 | 32,986 | 24,740 | 20,617 | 16,494 | 12,369 | 8,246 | 4,123 |
| 5″ Bore, 1½″ Shaft | 53,604 | 44,670 | 35,736 | 26,802 | 22,335 | 17,868 | 13,401 | 8,934 | 4,467 |
| 4″ Bore, 2″ Shaft | 28,273 | 23,561 | 18,849 | 14,137 | 11,781 | 9,425 | 7,068 | 4,712 | 2,356 |
| 4″ Bore, 1½″ Shaft | 32,391 | 26,993 | 21,595 | 16,197 | 13,498 | 10,799 | 8,097 | 5,398 | 2,699 |
| 4″ Bore, 1¼″ Shaft | 34,011 | 28,343 | 22,675 | 17,007 | 14,173 | 11,339 | 8,502 | 5,668 | 2,834 |
| 3½″ Bore, 2″ Shaft | 19,440 | 16,200 | 12,960 | 9,720 | 8,100 | 6,480 | 4,860 | 3,240 | 1,620 |
| 3½″ Bore, 1½″ Shaft | 23,558 | 19,632 | 15,706 | 11,780 | 9,817 | 7,854 | 5,889 | 3,926 | 1,963 |
| 3½″ Bore, 1¼″ Shaft | 25,178 | 20,982 | 16,786 | 12,590 | 10,492 | 8,394 | 6,294 | 4,196 | 2,098 |
| 3″ Bore, 1½″ Shaft | 15,901 | 13,251 | 10,601 | 7,951 | 6,626 | 5,301 | 3,975 | 2,650 | 1,325 |
| 3″ Bore, 1¼″ Shaft | 17,521 | 14,601 | 11,681 | 8,761 | 7,301 | 5,841 | 4,380 | 2,920 | 1,460 |
| 3″ Bore, 1⅛″ Shaft | 18,218 | 15,182 | 12,146 | 9,110 | 7,592 | 6,074 | 4,554 | 3,036 | 1,518 |
| 2½″ Bore, 1¹⁄₁₆″ Shaft | 12,062 | 10,052 | 8,042 | 6,032 | 5,027 | 4,022 | 3,015 | 2,010 | 1,005 |
| 2″ Bore, 1¹⁄₁₆″ Shaft | 6,759 | 5,633 | 4,507 | 3,381 | 2,818 | 2,255 | 1,689 | 1,126 | 563 |
| 1½″ Bore, ¾″ Shaft | 3,974 | 3,312 | 2,650 | 1,988 | 1,657 | 1,326 | 993 | 662 | 331 |

*Note:* The pull at a given pressure will change according to shaft size.

*Check this list when ordering special cylinders.*

1. Bore of cylinders.
2. Stroke required.
3. Center distances when cylinders are closed or distance from base to clevis, or base to shaft end.
4. P.S.I. intended P.S.I. operation.
5. Action: single or double action. If single, is it push or pull application?
6. Application: how will cylinders be used?
7. If stroke is long it may be necessary to consider shaft size in connection with the application.
8. Usually cylinders are made to our specifications and design.
9. For special cylinders and applications, contact our Engineering Department.

## TABLE D-5 TECHNICAL DATA SHEET

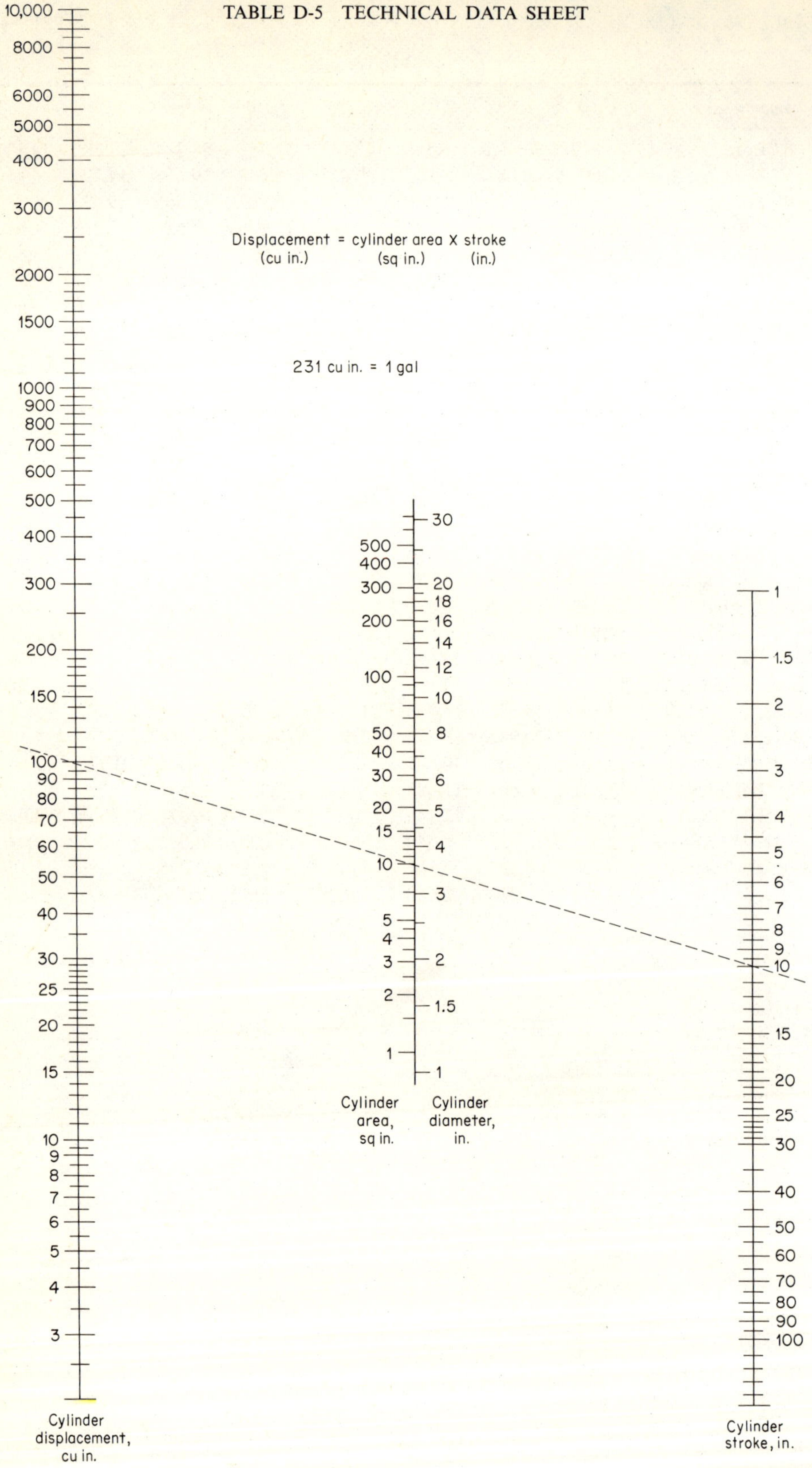

TABLE D-6 TECHNICAL DATA SHEET

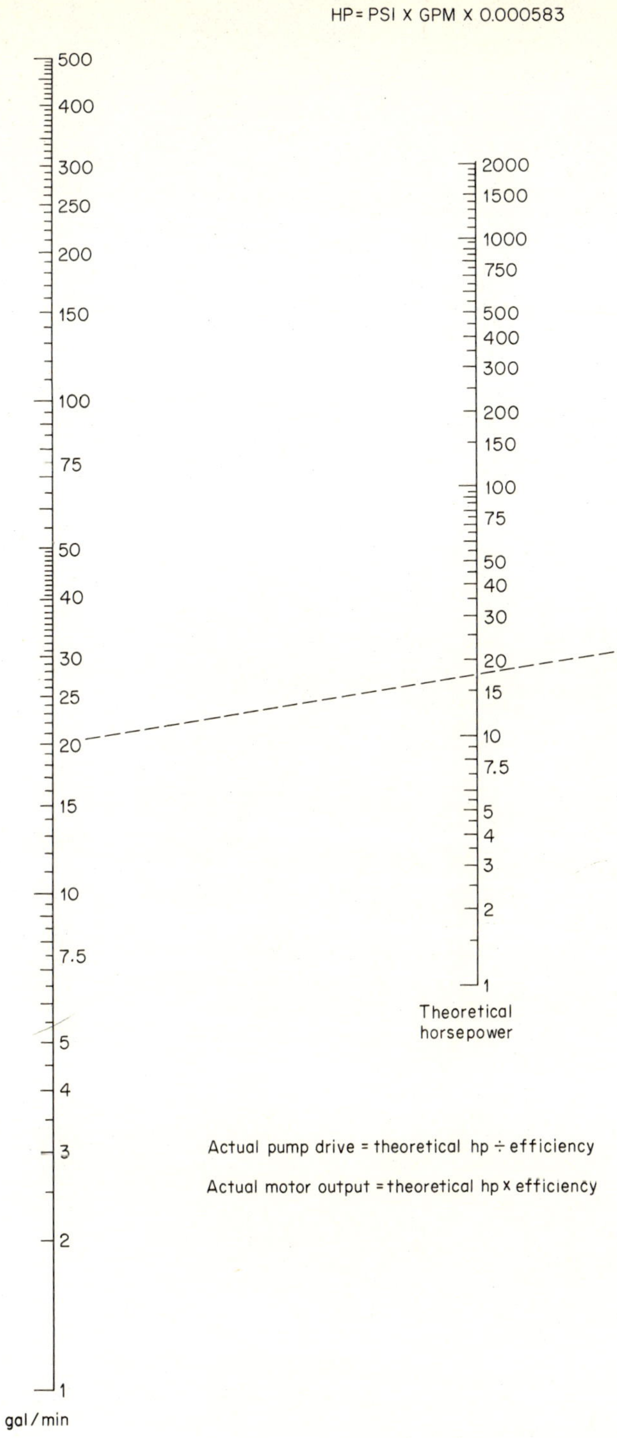

NOTES

NOTES

NOTES

# NOTES

NOTES

NOTES

NOTES

NOTES

NOTES

NOTES

NOTES

# NOTES

NOTES

NOTES

NOTES